Catalysts in Chemical Looping Technology for Energy Storage and Carbon Emission Reduction

Catalysts in Chemical Looping Technology for Energy Storage and Carbon Emission Reduction

Guest Editors

Chuande Huang

Bo Jiang

Xin Tian

Jiawei Hu

MDPI

Basel • Beijing • Wuhan • Barcelona • Belgrade • Novi Sad • Cluj • Manchester

Guest Editors

Chuande Huang
Dalian Institute
of Chemical Physics
Chinese Academy of Sciences
Dalian
China

Bo Jiang
School of Energy and
Power Engineering
Dalian University
of Technology
Dalian
China

Xin Tian
School of Power and
Mechanical Engineering
Wuhan University
Wuhan
China

Jiawei Hu
Shanghai Advanced
Research Institute
Chinese Academy of Sciences
Shanghai
China

Editorial Office

MDPI AG
Grosspeteranlage 5
4052 Basel, Switzerland

This is a reprint of the Special Issue, published open access by the journal *Catalysts* (ISSN 2073-4344), freely accessible at: http://www.mdpi.com.

For citation purposes, cite each article independently as indicated on the article page online and as indicated below:

Lastname, A.A.; Lastname, B.B. Article Title. *Journal Name* **Year**, *Volume Number*, Page Range.

ISBN 978-3-7258-6718-9 (Hbk)
ISBN 978-3-7258-6719-6 (PDF)
https://doi.org/10.3390/books978-3-7258-6719-6

Contents

About the Editors . . . . **vii**

Chuande Huang
Chemical Looping Technology for Energy Storage and Carbon Emissions Reductions
Reprinted from: *Catalysts* **2024**, *14*, 122, https://doi.org/10.3390/catal14020122 . . . . **1**

Shunshun Kang, Jinlin Deng, Xiaobo Wang, Kun Zhao, Min Zheng, Da Song, et al.
Plasma-Enhanced Chemical Looping Oxidative Coupling of Methane through Synergy between Metal-Loaded Dielectric Particles and Non-Thermal Plasma
Reprinted from: *Catalysts* **2023**, *13*, 557, https://doi.org/10.3390/catal13030557 . . . . **5**

Wenxi Chang, Yue Hu, Weibin Xu, Chuande Huang, Haonan Chen, Jiahui He, et al.
Recent Advances of Oxygen Carriers for Hydrogen Production via Chemical Looping Water-Splitting
Reprinted from: *Catalysts* **2023**, *13*, 279, https://doi.org/10.3390/catal13020279 . . . . **22**

Jingwei Wang, Xiaocen Liang, Zifan Xing, Haitao Chen, Yang Li, Da Song and Fang He
Ce-Doped $LaMnO_3$ Redox Catalysts for Chemical Looping Oxidative Dehydrogenation of Ethane
Reprinted from: *Catalysts* **2023**, *13*, 131, https://doi.org/10.3390/catal13010131 . . . . **51**

Dan Li, Yang Wang and Zhenshan Li
Limestone Calcination Kinetics in Microfluidized Bed Thermogravimetric Analysis (MFB-TGA) for Calcium Looping
Reprinted from: *Catalysts* **2022**, *12*, 1661, https://doi.org/10.3390/catal12121661 . . . . **65**

Shengpeng Xia, Chenyang Wang, Yu Chen, Shunshun Kang, Kun Zhao, Anqing Zheng, et al.
Sustainable Aromatic Production from Catalytic Fast Pyrolysis of 2-Methylfuran over Metal-Modified ZSM-5
Reprinted from: *Catalysts* **2022**, *12*, 1483, https://doi.org/10.3390/catal12111483 . . . . **84**

Da Song, Yan Lin, Kun Zhao, Zhen Huang, Fang He and Ya Xiong
Migration Mechanism of Lattice Oxygen: Conversion of CO_2 to CO Using $NiFe_2O_4$ Spinel Oxygen Carrier in Chemical Looping Reactions
Reprinted from: *Catalysts* **2022**, *12*, 1181, https://doi.org/10.3390/catal12101181 . . . . **94**

Zhen Huang, Yonghao Wang, Nanhang Dong, Da Song, Yan Lin, Lisheng Deng and Hongyu Huang
In Situ Removal of Benzene as a Biomass Tar Model Compound Employing Hematite Oxygen Carrier
Reprinted from: *Catalysts* **2022**, *12*, 1088, https://doi.org/10.3390/catal12101088 . . . . **114**

Minjun Wang, Shixiong Zhang, Ming Xia and Mengke Wang
A Theoretical Study of the Oxygen Release Mechanisms of a Cu-Based Oxygen Carrier during Chemical Looping with Oxygen Uncoupling
Reprinted from: *Catalysts* **2022**, *12*, 332, https://doi.org/10.3390/catal12030332 . . . . **127**

Ewelina Ksepko and Rafal Lysowski
Reactivity Study of Bimetallic Fe-Mn Oxides with Addition of TiO_2 for Chemical Looping Combustion Purposes
Reprinted from: *Catalysts* **2021**, *11*, 1437, https://doi.org/10.3390/catal11121437 . . . . **143**

About the Editors

Chuande Huang

Chuande Huang is an Associate Professor at the Dalian Institute of Chemical Physics (DICP) and member of the Youth Innovation Promotion Association, CAS. He received the Ph.D. degree in Engineering from DICP in 2017 and subsequently joined the Laboratory of Catalysts and New Materials at DICP. He is mainly engaged in applied fundamental research on developing high-temperature-resistant materials, single-atom catalysts, and the applications of these materials in the catalytic conversion of energy-related small molecules and non-toxic propellants. His academic contributions include over 40 publications in journals such as *Nature Communications*, the *Journal of the American Chemical Society*, and *Angewandte Chemie International Edition*, along with more than 30 applied invention patents.

Bo Jiang

Bo Jiang is an accomplished researcher in energy materials and hydrogen energy, with over 70 SCI papers published in prestigious journals including *Nature Communications*, *Advanced Functional Materials*, *Nano Energy*, and *Advanced Science*. His work has garnered more than 5000 citations (H-index 42), including 5 ESI highly cited papers, 2 ESI hot papers, and 4 cover papers. He holds 3 authorized invention patents and 2 utility model patents. Dr. Jiang serves as a Reviewer for leading journals such as *Advanced Materials* and *Advanced Functional Materials*, a Young Editorial Board Member for Wiley's *Carbon and Hydrogen and Carbon Neutralization*, and a Peer Reviewer for the National Natural Science Foundation of China. He has led multiple national and provincial projects, contributing to the Dalian Hydrogen Energy Demonstration Project. His accolades include the China Invention Association Award, Liaoning Provincial Outstanding Doctoral Thesis, and Dalian Municipal Outstanding Young Scientist Award.

Xin Tian

Xin Tian is currently an Associate Professor at the School of Power and Mechanical Engineering at Wuhan University. He received his PhD degree in thermal engineering from Huazhong University of Science and Technology in 2019. His main research interests focus on chemical looping-driven CO_2 capture and conversion. To date, he has published more than 50 academic papers, including 25 SCI-indexed papers as first or corresponding author in journals such as *ACS Catalysis*, *Applied Catalysis B: Environmental*, *Chemical Engineering Journal*, and *Combustion and Flame*. He is a recipient of several prestigious awards, including the Humboldt Research Fellowship for Postdoctoral Researchers and the "Chutian Scholar" of the Hubei Province.

Jiawei Hu

Jiawei Hu is now an Associate Professor at the Shanghai Advanced Research Institute (SARI), Chinese Academy of Sciences. He received his PhD in Chemical Engineering at Ghent University (Belgium) in 2018, and then worked as a postdoctoral fellow at the National University of Singapore for around 3 years. In 2021, he joined the Center for Low-Carbon Conversion Science and Engineering at SARI. He was awarded "Leading Talents of Shanghai (overseas)" in 2022. His research focuses on the development of new reaction configurations and advanced materials for CO_2 capture and catalytic conversion, on which he has published more than 20 research articles and invention patents.

catalysts

MDPI

Editorial

Chemical Looping Technology for Energy Storage and Carbon Emissions Reductions

Chuande Huang

CAS Key Laboratory of Science and Technology on Applied Catalysis, Dalian Institute of Chemical Physics, Chinese Academy of Sciences, Dalian 116023, China; huangchuande@dicp.ac.cn

Chemical looping (CL) technology, initially developed as an advanced combustion method, has been widely applied in various processes, including the selective oxidation of hydrocarbons (e.g., methane, ethane, and propane) and biomass, H_2O splitting, CO_2 splitting, air separation, and ammonia synthesis [1–6]. In most cases, metal oxides are introduced as oxygen carriers (OCs), which deliver the oxygen from oxidants (e.g., air, CO_2, and H_2O) to the reductants (e.g., hydrocarbons, biomass, H_2, and CO), avoiding the direct contact between these reactants [7–10]. Specially, metal nitrides or hydrides are applied as nitrogen carriers or hydrogen carriers, respectively, for the ammonia synthesis [11–13]. Such processes can decompose a reaction into multiple sub-reactions that are conducted in different reactors, which optimizes the reaction conditions independently and can inherently simplified subsequent separation processes [14,15]. In addition, the cyclic reaction mode is effective in restraining the catalyst deactivation induced by the coke deposition, which concurrently promotes the efficiency of the process.

Redox materials serve as both the catalyst and a medium for elemental (O, N, or H) transfer in the CL process [16–18]. Therefore, exploring suitable redox materials is one key issue for specific reactions. This Special Issue includes nine papers in total, of which eight are research articles and one is a review paper. The contributions of this collection are summarized below.

To address the problem of high reaction temperatures in methane conversion, Kang et al. (Contribution 1) proposed a plasma–catalyst hybrid system for the selective production of C_{2+} hydrocarbons, enabling C–H bond cleavage at ambient temperature. Detailed investigations show that the selection of dielectric materials is of crucial importance for the reaction and the performance of TiO_2 is superior to that of SiO_2 and Al_2O_3. When further decorated by MnO_x, the MnTi catalyst manifests the best C_{2+} yield of 27.29% and remains stable in 30 consecutive cyclic redox tests.

Wang et al. (Contribution 2) prepared a series of $La_{1-x}Ce_xMnO_{3+\delta}$ samples for the chemical looping oxidative dehydrogenation (CL-ODH) of ethane to ethylene. It was found that the substitution of La^{3+} by Ce^{4+} cations could significantly promote the Mn^{3+}/Mn^{4+} and $O_{lattice}/O_{adsorp}$ ratio at the catalyst surface. A possible mechanism is that the Mn^{3+}-$O_{lattice}$ redox couple is responsible for the selective production of ethylene from ethane. Therefore, $La_{0.7}Ce_{0.3}MnO_3$ shows the best performance of 22% ethane conversion, with 57% ethylene selectivity in the CL-ODH reaction.

Calcium looping is a promising technology for efficient CO_2 capture; however, measuring the kinetics in the fluidizing calcination of limestone remains challenging. Li et al. (Contribution 3) studied the effect of particle size (150–1250 μm), temperature (750–920 °C), and CO_2 concentration (0–30 vol.%) on the calcination characteristics of limestone. The results show that a higher temperature, lower CO_2 concentration, and smaller particles all contribute to the high calcination rate. The decomposition of limestone in particles is affected by both the chemical reaction and gas diffusion. It is inferred from the calculation results that the effects of chemical reaction and gas diffusion are overwhelming for particles with sizes below 80 μm and above 450 μm, respectively, while these two factors play a role in in the same order of magnitude for particles between 80 and 450 μm. In addition, improving the CO_2 concentration can greatly weaken the influence of particle size for calcination process.

The efficient utilization of biomass resources has been widely accepted as a useful strategy to address the concerns regarding the scarcity of fossil fuels and the related environmental problems. Xia et al. (Contribution 4) screened various metal-modified ZSM-5 (metal = Zn, Mo, Fe, and Ga) catalysts for the fast catalytic pyrolysis of bio-derived furans (2-methylfuran, MF). It is found that all these metal promoters could improve the conversion of MF and corresponding yield of aromatic, wherein the 2%Fe-ZSM-5 showed the highest aromatic yield of 40.03% at a WHSV of 2 h^{-1}. Mechanism studies reveal that such promotion arises from the synergy between Brønsted and Lewis acid sites. Therefore, generating new Lewis acid sites without a notable reduction in the Brønsted acid sites in ZSM-5 is key for designing a prospective catalyst for the aromatization of MF.

Taking benzene as a model compound of biomass tar, Huang et al. (Contribution 5) investigated the tar removal process using iron ore as an oxygen carrier. Compared with the blank experiment, the addition of hematite, functioning as both an oxygen donator and catalyst, can promote benzene conversion by 20%, which can be further enhanced to nearly 50% upon prolonging the residence time. In addition, the introduction of H_2O as a co-reactant is effective in reducing coke formation at the expense of lowering the benzene conversion, since the water vapor could reduce the residence time and weaken the catalytic activity by oxidizing the reduced iron species.

The migration of oxygen could exert great influence on the reactivity, selectivity, and stability of the oxygen carrier during redox cycles. In the study by Song et al. (Contribution 6), the oxygen migration mechanism of $NiFe_2O_4$ in chemical looping CO_2 splitting was investigated. It is suggested that the driving forces for diffusion of oxygen during the redox reaction are mainly attributed to the concentration gradient between the surface and the bulk. A highly reactive interface containing both lattice oxygen and metal atoms exists at the surface of OCs, which accounts for the outstanding redox performance of $NiFe_2O_4$. The movement of Ni particles into and out of a spinel matrix can improve the oxygen migration during the reaction.

Chemical looping combustion holds the virtue of inherent CO_2 separation by avoiding direct contact between fuel and air. Ksepko et al. (Contribution 7) prepared a series of oxygen carriers by mixing and calcination of the Fe_2O_3-MnO_2-TiO_2 powder. Among these OCs, the F50M35 sample with 50 wt.% of Fe_2O_3 and 35 wt.% of MnO_2 displays the highest oxygen capacity of ca. 16–17 wt.%. It is found that the addition of TiO_2 could inhibit the formation of the negative manganese-rich spinel phase by reacting with Fe_2O_3 and MnO_2 to generate Fe_2TiO_5 and $Fe_{0.6}Mn_{0.4}TiO_3$, which contributes to a better cycling stability. In addition, it is suggested that a milder reaction temperature between 850 °C and 900 °C is useful to take full advantage of the OCs' potential.

CuO is a promising OC for the chemical looping with oxygen uncoupling (CLOU) process, which could release gaseous O_2 for the total combustion of fuel. To determine the oxygen release mechanisms of CuO, Wang et al. (Contribution 8) conducted detailed DFT calculations over the (110) and (111) surfaces. The results indicated that the CuO (110) surface is much more active than (111) for the oxygen release reaction. Barriers of 1.89 eV and 3.22 eV need to be conquered for the O_2 formation and desorption, respectively, over the (110) surface. Instead, the diffusion of subsurface oxygen to the surface oxygen vacancies only exhibits a low energy barrier of 0.24 eV, which indicates that the surface reactions (O_2 formation and desorption) are the rate-limiting steps.

Chemical looping water-splitting (CLWS) is an attractive method for hydrogen production. In this process, reducing gaseous (methane, syngas, or diluted hydrogen), thermal reduction, or photo-assisted thermal reduction methods could be utilized to abstract the lattice oxygen from OC, which is then recovered by water oxidation, giving hydrogen with high purity by simple condensation of the water vapor. In the review by Chang et al. (Contribution 9), the recent advances on oxygen carriers are summarized in detail. It is suggested that the Fe- and Ce-based oxides are among the most studied redox materials. Composite oxides with perovskite structures (ABO_3) normally display outstanding performance by varying the A and B site cations.

Overall, chemical looping technology is effective in reducing the energy input for many reaction processes due to specific multi-step features. Although much progress has been achieved in these years, industrialized applications of this technology remain a great challenge. One main reason could be the lack of efficient redox materials. To tackle this issue, in situ characterizations, theoretical calculations, and thermodynamic analysis have been recommended for future studies; these are important for establishing clear structure–function relationships.

Funding: This work was supported by the National Natural Science Foundation of China (Grant No. 21706254), NSFC Center for Single-Atom Catalysis (Grant No. 22388102), and the Youth Innovation Promotion Association, CAS (Grant No. 2023189).

Acknowledgments: As a Guest Editor of the Special Issue "Chemical Looping Technology for Energy Storage and Carbon Emission Reduction", I would like to express my deep appreciation to all the authors and anonymous reviewers for their dedication.

Conflicts of Interest: The authors declare no conflicts of interest.

List of Contributions:

1. Kang, S.; Deng, J.; Wang, X.; Zhao, K.; Zheng, M.; Song, D.; Huang, Z.; Lin, Y.; Liu, A.; Zheng, A.; et al. Plasma-Enhanced Chemical Looping Oxidative Coupling of Methane through Synergy between Metal-Loaded Dielectric Particles and Non-Thermal Plasma. *Catalysts* **2023**, *13*, 557. https://doi.org/10.3390/catal13030557.
2. Wang, J.; Liang, X.; Xing, Z.; Chen, H.; Li, Y.; Song, D.; He, F. Ce-Doped $LaMnO_3$ Redox Catalysts for Chemical Looping Oxidative Dehydrogenation of Ethane. *Catalysts* **2023**, *13*, 131. https://doi.org/10.3390/catal13010131.
3. Li, D.; Wang, Y.; Li, Z. Limestone Calcination Kinetics in Microfluidized Bed Thermogravimetric Analysis (MFB-TGA) for Calcium Looping. *Catalysts* **2022**, *12*, 1661. https://doi.org/10.3390/catal12121661.
4. Xia, S.; Wang, C.; Chen, Y.; Kang, S.; Zhao, K.; Zheng, A.; Zhao, Z.; Li, H. Sustainable Aromatic Production from Catalytic Fast Pyrolysis of 2-Methylfuran over Metal-Modified ZSM-5. *Catalysts* **2022**, *12*, 1483. https://doi.org/10.3390/catal12111483.
5. Huang, Z.; Wang, Y.; Dong, N.; Song, D.; Lin, Y.; Deng, L.; Huang, H. In Situ Removal of Benzene as a Biomass Tar Model Compound Employing Hematite Oxygen Carrier. *Catalysts* **2022**, *12*, 1088. https://doi.org/10.3390/catal12101088.
6. Song, D.; Lin, Y.; Zhao, K.; Huang, Z.; He, F.; Xiong, Y. Migration Mechanism of Lattice Oxygen: Conversion of CO_2 to CO Using $NiFe_2O_4$ Spinel Oxygen Carrier in Chemical Looping Reactions. *Catalysts* **2022**, *12*, 1181. https://doi.org/10.3390/catal12101181.
7. Ksepko, E.; Lysowski, R. Reactivity Study of Bimetallic Fe-Mn Oxides with Addition of TiO_2 for Chemical Looping Combustion Purposes. *Catalysts* **2021**, *11*, 1437. https://doi.org/10.3390/catal11121437.
8. Wang, M.; Zhang, S.; Xia, M.; Wang, M. A Theoretical Study of the Oxygen Release Mechanisms of a Cu-Based Oxygen Carrier during Chemical Looping with Oxygen Uncoupling. *Catalysts* **2022**, *12*, 332. https://doi.org/10.3390/catal12030332.
9. Chang, W.; Hu, Y.; Xu, W.; Huang, C.; Chen, H.; He, J.; Han, Y.; Zhu, Y.; Ma, X.; Wang, X. Recent Advances of Oxygen Carriers for Hydrogen Production via Chemical Looping Water-Splitting. *Catalysts* **2023**, *13*, 279. https://doi.org/10.3390/catal13020279.

References

1. Zhang, L.; Hu, Y.; Xu, W.; Huang, C.; Su, Y.; Tian, M.; Zhu, Y.; Gong, H.; Wang, X. Anti-coke $BaFe_{1-x}Sn_xO_{3-\delta}$ Oxygen Carriers for Enhanced Syngas Production via Chemical Looping Partial Oxidation of Methane. *Energy Fuels* **2020**, *34*, 6991–6998. [CrossRef]
2. Zhu, X.; Imtiaz, Q.; Donat, F.; Müller, C.R.; Li, F. Chemical looping beyond combustion—A perspective. *Energy Environ. Sci.* **2020**, *13*, 772–804. [CrossRef]
3. Jerndal, E.; Mattisson, T.; Lyngfelt, A. Thermal Analysis of Chemical-Looping Combustion. *Chem. Eng. Res. Des.* **2006**, *84*, 795–806. [CrossRef]

4. Muhich, C.L.; Evanko, B.W.; Weston, K.C.; Lichty, P.; Liang, X.; Martinek, J.; Musgrave, C.B.; Weimer, A.W. Efficient generation of H_2 by splitting water with an isothermal redox cycle. *Science* **2013**, *341*, 540–542. [CrossRef]
5. Furler, P.; Scheffe, J.R.; Steinfeld, A. Syngas production by simultaneous splitting of H_2O and CO_2 via ceria redox reactions in a high-temperature solar reactor. *Energy Environ. Sci.* **2012**, *5*, 6098–6103. [CrossRef]
6. Li, F.; Zeng, L.; Fan, L.-S. Biomass direct chemical looping process: Process simulation. *Fuel* **2010**, *89*, 3773–3784. [CrossRef]
7. Zhang, L.; Xu, W.; Wu, J.; Hu, Y.; Huang, C.; Zhu, Y.; Tian, M.; Kang, Y.; Pan, X.; Su, Y.; et al. Identifying the Role of A-Site Cations in Modulating Oxygen Capacity of Iron-Based Perovskite for Enhanced Chemical Looping Methane-to-Syngas Conversion. *ACS Catal.* **2020**, *10*, 9420–9430. [CrossRef]
8. Zhao, X.; Zhou, H.; Sikarwar, V.S.; Zhao, M.; Park, A.-H.A.; Fennell, P.S.; Shen, L.; Fan, L.-S. Biomass-based chemical looping technologies: The good, the bad and the future. *Energy Environ. Sci.* **2017**, *10*, 1885–1910. [CrossRef]
9. Chen, S.; Zeng, L.; Mu, R.; Xiong, C.; Zhao, Z.-J.; Zhao, C.; Pei, C.; Peng, L.; Luo, J.; Fan, L.-S. Modulating lattice oxygen in dual-functional Mo–V–O mixed oxides for chemical looping oxidative dehydrogenation. *J. Am. Chem. Soc.* **2019**, *141*, 18653–18657. [CrossRef]
10. Gao, Y.; Wang, X.; Liu, J.; Huang, C.; Zhao, K.; Zhao, Z.; Wang, X.; Li, F. A molten carbonate shell modified perovskite redox catalyst for anaerobic oxidative dehydrogenation of ethane. *Sci. Adv.* **2020**, *6*, eaaz9339. [CrossRef] [PubMed]
11. Burrows, L.; Gao, P.-X.; Bollas, G.M. Thermodynamic feasibility analysis of distributed chemical looping ammonia synthesis. *Chem. Eng. J.* **2021**, *426*, 131421. [CrossRef]
12. Brown, S.; Hu, J. Review of chemical looping ammonia synthesis materials. *Chem. Eng. Sci.* **2023**, *280*, 119063. [CrossRef]
13. Gao, W.; Guo, J.; Wang, P.; Wang, Q.; Chang, F.; Pei, Q.; Zhang, W.; Liu, L.; Chen, P. Production of ammonia via a chemical looping process based on metal imides as nitrogen carriers. *Nat. Energy* **2018**, *3*, 1067–1075. [CrossRef]
14. Zeng, L.; Cheng, Z.; Fan, J.A.; Fan, L.-S.; Gong, J. Metal oxide redox chemistry for chemical looping processes. *Nat. Rev. Chem.* **2018**, *2*, 349–364. [CrossRef]
15. Luo, M.; Yi, Y.; Wang, S.; Wang, Z.; Du, M.; Pan, J.; Wang, Q. Review of hydrogen production using chemical-looping technology. *Renew. Sustain. Energy Rev.* **2018**, *81*, 3186–3214. [CrossRef]
16. Huang, C.; Wu, J.; Chen, Y.-T.; Tian, M.; Rykov, A.I.; Hou, B.; Lin, J.; Chang, C.-R.; Pan, X.; Wang, J.; et al. In situ encapsulation of iron(0) for solar thermochemical syngas production over iron-based perovskite material. *Commun. Chem.* **2018**, *1*, 55. [CrossRef]
17. Adanez, J.; Abad, A.; Garcia-Labiano, F.; Gayan, P.; de Diego, L.F. Progress in Chemical-Looping Combustion and Reforming technologies. *Prog. Energy Combust. Sci.* **2012**, *38*, 215–282. [CrossRef]
18. Tang, M.; Xu, L.; Fan, M. Progress in oxygen carrier development of methane-based chemical-looping reforming: A review. *Appl. Energy* **2015**, *151*, 143–156. [CrossRef]

catalysts

MDPI

Article

Plasma-Enhanced Chemical Looping Oxidative Coupling of Methane through Synergy between Metal-Loaded Dielectric Particles and Non-Thermal Plasma

Shunshun Kang [1,2,3,4], Jinlin Deng [1,2,3,4], Xiaobo Wang [1,2,3,4], Kun Zhao [1,2,3,4,*], Min Zheng [5], Da Song [1,2,3,4], Zhen Huang [1,2,3,4], Yan Lin [1,2,3,4], Anqi Liu [1,2,3,4], Anqing Zheng [1,2,3,4,*] and Zengli Zhao [1,2,3,4]

1 Guangzhou Institute of Energy Conversion, Chinese Academy of Sciences, Guangzhou 510640, China
2 CAS Key Laboratory of Renewable Energy, Guangzhou 510640, China
3 Guangdong Provincial Key Laboratory of New and Renewable Energy Research and Development, Guangzhou 510640, China
4 University of Chinese Academy of Sciences, Beijing 100049, China
5 State Key Laboratory of Complex Metal Resources Clean Utilization, Kunming University of Science and Technology, Kunming 650031, China
* Correspondence: zhaokun@ms.giec.ac.cn (K.Z.); zhengaq@ms.giec.ac.cn (A.Z.); Tel.: +86-0208-7057-716 (K.Z. & A.Z.); Fax: +86-0208-7057-737 (K.Z. & A.Z.)

Abstract: A plasma–catalyst hybrid system has been developed for the direct conversion of methane to C_{2+} hydrocarbons in dielectric barrier discharge (DBD) plasma. TiO_2 presented the highest C_{2+} yield of 11.63% among different dielectric materials when integrated with DBD plasma, which made us concentrate on the TiO_2-based catalyst. It was demonstrated that MnTi catalyst showed the best methane coupling performance of 27.29% C_{2+} yield with 150 V applied voltage, without additional thermal input. The catalytic performance of MnTi catalyst under various operation parameters was further carried out, and different techniques, such as X-ray diffraction, X-ray photoelectron spectroscopy, transmission electron microscopy, and H_2-temperature-programmed reduction were used to explore the effect of Mn loading on methane oxidative coupling (OCM) performance. The results showed that applied voltage and flow rate had a significant effect on methane activation. The dielectric particles of TiO_2 loaded with Mn not only synergistically affected the coupling reaction, but also facilitated charge deposition to generate a strong local electric field to activate methane. The synergy effects boosted the OCM performance and the C_{2+} yield became 1.25 times higher than that of the undoped TiO_2 under identical operating conditions in plasma, which was almost impossible to occur even at 850 °C on the MnTi catalyst in the absence of plasma. Moreover, the reaction activity of the catalyst was fully recovered by plasma regeneration at 300 °C and maintained its stability in for at least 30 consecutive cyclic redox tests. This work presents a new opportunity for efficient methane conversion to produce C_{2+} at low temperatures by plasma assistance.

Keywords: oxidative coupling of methane; chemical looping; synergy effects; plasma enhancement

1. Introduction

Light olefins, especially ethylene, are important organic raw materials, which are widely used in the production of plastics, drugs, fibers, etc. Ethylene is also regarded as a leading indicator in measuring the level of a country's petrochemical industry and plays a pivotal role in the national economy. Because of its abundant reserves and clean characteristics [1,2], it is considered that methane will replace oil as the primary energy in the future. Therefore, a significant amount of manpower and material resources have been invested in the comprehensive utilization of natural gas [3–5]. The production of low-carbon olefins from methane has attracted the attention of researchers [6]. The oxidative coupling of methane (OCM) offers a feasible approach for the direct conversion of methane to hydrocarbons [7–9]. However, because of the high reactivity of oxygen, which easily reacts with reactants and products, the generation of by-products of CO and CO_2 at high temperatures is unavoidable, and the C_{2+} yield of the OCM reaction is still limited to no more than 30%; thus, it is difficult to provide economic benefits in industry. On this basis, a replaceable technique known as chemical looping oxidative coupling of methane (CL-OCM) is investigated. As shown in Figure 1, the main advantage of CL-OCM is the replacement of gaseous oxygen with lattice oxygen in the catalyst (also known as the oxygen carrier), thus avoiding an excessive oxidation of methane. CL-OCM has become a hot research topic, since it can attain a higher C_2 yield (between 20 and 30%) compared to traditional methods, including the direct conversion of methane and the non-oxidative coupling of methane (NCM) [10]. However, CL-OCM typically requires high-temperature reaction conditions due to the high C–H bond activation barrier of methane, resulting in a low catalyst stability and a high energy consumption. Therefore, it is necessary to develop more modest technologies to assist the CL-OCM strategy.

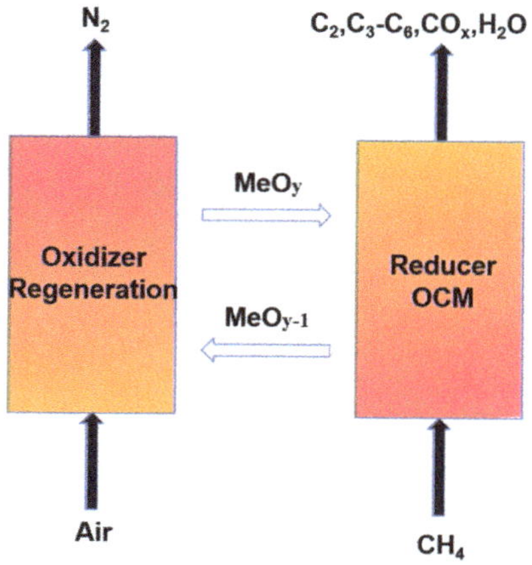

Figure 1. CL-OCM process.

As a system containing high-energy electrons and exhibiting low-temperature properties, plasma can effectively activate stable molecules at relatively low temperatures, thus providing a new approach toward CL-OCM. Plasma can be considered as a quasi-neutral system consisting of electrons, ions, radicals, and excitations of gas molecules formed by the ionization and dissociation of gas molecules under external energy, such as an electric field. Using plasma, energetic electrons collide with the reactants and activate them, producing large quantities of highly reactive substances to initiate the reaction [11]. Thus, some stable small molecules, such as CH_4 [12–14], CO_2 [15–18] and N_2 [19,20], can be effectively activated by plasma. Compared with the traditional thermal activation of methane, plasma shows a strong activation ability on methane [21]. Methane and products can avoid deep oxidation reactions at low temperatures (lower than 600 °C), which is beneficial to the selectivity of C_2 hydrocarbon products. Therefore, research on the plasma activation of methane has become increasingly active in recent years. However, the gas-phase non-catalytic reaction initiated by plasma is mainly carried out through the free radical mechanism, which has little controllability of the product distribution [22].

To control the selectivity of products and enhance the performance of methane activation, catalysts are usually introduced into the plasma system. The addition of catalysts in the plasma can combine the characteristics of low-temperature activation of plasma with the characteristics of product orientation of the catalysts, which provides a high potential to produce value-added products. At present, several dielectric materials, such as TiO_2 [23], Al_2O_3 [24,25], SiO_2 [13,24], and ZSM [26,27], have been used for methane activation in the presence of plasma. Microelectrodes formed under an external electric field inside dielectric particles will enhance the electric field, thus improving the methane conversion [24]. Moreover, the adsorption of porous material can also make the reactants relatively concentrated, which is expected to significantly improve the discharge energy efficiency [25,28]. Liu et al. [29] found that the presence of plasma vibrationally excited methane appeared to be primarily responsible for the improvement of methane conversion in the presence of catalysts. Vibrationally excited methane made a significant contribution to the formation of the products without catalysts, even though a large amount of electric field energy was consumed [30]. This was mainly due to the high internal energy; the vibrationally excited methane was more easily dissociated on the catalyst to produce methyl radicals, thus increasing the methane conversion. Therefore, the energy to overcome the methane activation barrier originates from the electric field rather than the external heat during the plasma reaction, thus avoiding the high-temperature condition.

In order to integrate the advantages of various catalysts, a more ideal approach is to form a composite of high-performance catalysts and dielectric materials. In this way, the local electric field generated by the high permittivity of catalysts can enhance the activation of methane, while simultaneously improving the selectivity of C_2 hydrocarbons [31,32]. However, so far, most of the studies on catalysts for methane oxidative coupling catalysts have been focused on thermal reactions at high temperatures, and few studies have reported their methane oxidation coupling activity under plasma conditions. Related studies have confirmed that catalysts with methane oxidation coupling performance at high temperatures also exhibit excellent OCM activity in the plasma [33]. Keller et al. [34] investigated the OCM activity of various metal oxides and found that manganese oxides presented excellent performance for C_2 formation with a low cost and environmental friendliness. Cheng et al. [35] employed Mg_6MnO_8 with the use of a low concentration of Li dopant as the catalyst for CL-OCM. The experiment showed that Mn-based oxides with Li dopant attained C_2 hydrocarbons with a selectivity of 50.4%. DFT calculations demonstrated that oxygen vacancies caused by Li-doping in Mg-Mn composite oxygen carriers were unfavorable for the adsorption of methyl radicals and, hence, increased the activation barrier of methyl radicals, thus significantly improving the selectivity of C_2 hydrocarbon. Jiang et al. [36] applied Na-doped $LaMnO_3$ as a catalyst in the CL-OCM process and found that Mn-based catalysts were well-suited catalysts, which could attain an ideal C_2 selectivity and yield. Moreover, Na_2WO_4 is considered as a suitable catalyst for the OCM reaction [37]. Previously, transition metal oxides impregnated with Na_2WO_4 were experimentally found to inhibit the formation of CO_x in the reaction, indicating that the introduction of Na_2WO_4 could result in the growth of C_2 selectivity [38]. Kiani et al. [39] prepared a Na_2WO_4/SiO_2 catalyst via a modified impregnation method and found that the OCM reaction occurred at the dispersed phase Na-WO_4 site. Due to the high thermal stability and superior C_2 yield, $Mn/Na_2WO_4/SiO_2$ is considered to be a promising catalyst to achieve commercial application [40,41]. Kim et al. [42] investigated the OCM activity of $Mn/Na_2WO_4/TiO_2$, and found that a superior C_2 yield was attained on $Mn/Na_2WO_4/TiO_2$ compared to on $Mn/Na_2WO_4/SiO_2$.

In this study, we investigated the effect of active metal loadings on different types of dielectric materials for methane conversion and C_{2+} production in dielectric barrier discharge (DBD) plasma at atmospheric pressure in chemical looping mode, with no oxygen co-feed. The regeneration process was carried out at 300 °C in an atmosphere of air under plasma. The impact of different input flows, temperatures, and applied voltages on the performance of the catalyst in the DBD reactor was also investigated. The catalysts were measured by X-ray diffraction (XRD) to confirm the crystal phases. H_2-temperature programmed reduction (H_2-TPR) and X-ray photoelectron spectroscopy (XPS) were performed to investigate the properties of oxygen species on catalysts. TEM–EDS was carried out to determine the element distributions. Finally, it was found that the MnTi catalyst resulted in a clear C_{2+} hydrocarbons enhancement in plasma when compared with a catalysis-only system. This discovery enables the room-temperature conversion of methane and provides an effective guideline for the study of methane activation in low-temperature plasma.

2. Results and Discussion

2.1. Effect of Dielectric Materials

All experiments were performed in chemical looping mode, meaning that no oxygen co-feed was present. The metal amount was not varied. According to the literature reported on plasma-catalytic systems, methane conversion and product selectivity would vary with different dielectric materials. It was found that the microelectrodes were generated inside the dielectric materials under the external electric field, which could improve the capability of charge transfer [43]. In terms of chemical reactions, the charge transfer capacity can be converted into the ability to activate methane, meaning that various dielectric materials can enhance the activity of methane conversion. Figure 2 shows the catalytic performance of SiO_2, Al_2O_3, and TiO_2 dielectric materials for CL-OCM in DBD plasma. SiO_2 exhibits the highest methane conversion of 37.75%, but contains CO and CO_2 as the main products (with a CO and CO_2 selectivity of 36.95% and 36.03%), resulting in a low C_{2+} yield of 9.12%, with a C_{2+} selectivity lower than 30%. While Al_2O_3 and TiO_2 exhibit an improvement for C_{2+} selectivity (mainly ethane and propane as the main products), C_{2+} yields of 9.85% and 11.63% are obtained, respectively. Compared to Al_2O_3, TiO_2 displays more remarkable oxidative coupling properties in plasma, which is illustrated by the further improvement in the selectivity of ethane and propane, and exhibits the highest C_{2+} yield among the used dielectric materials in this study. Thus, TiO_2 is considered as an ideal support for CL-OCM in plasma.

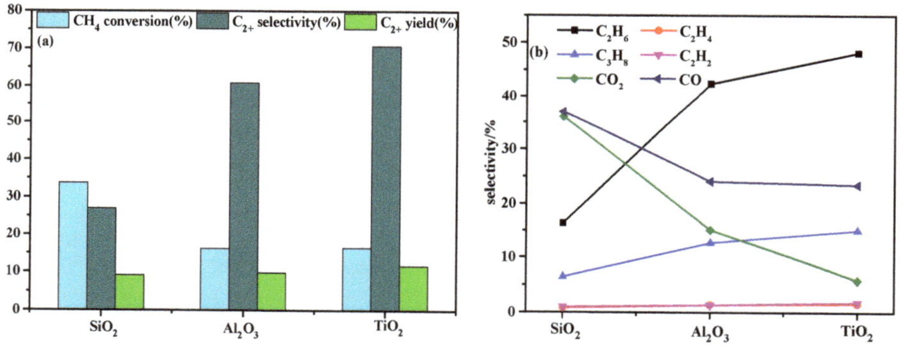

Figure 2. Performance of different dielectric materials in DBD plasma. (**a**) Oxidative coupling performance; (**b**) product distribution (temperature: ambient temperature, flow rate: 60 mL/min, applied voltage: 100 V).

In general, there is a positive correlation between the performance of catalysts and dielectric materials in the plasma–catalyst system. Since TiO_2 exhibited the best catalytic performance among the supports under plasma conditions, different metal loadings on TiO_2 support were further investigated. The catalytic performance of different metal loadings on TiO_2 support for CL-OCM in DBD plasma is shown in Figure 3. It can be seen that the loading of active metals on TiO_2 support contributes greatly to the increase in methane conversion and C_{2+} selectivity compared to pure TiO_2 in the plasma-catalytic system. The most striking improvement is nearly 100% of C_{2+} selectivity, with dominant products being ethane and propane, which account for more than 70% of the total products. As stated in previous research, manganese oxide and Na_2WO_4 would take part in the dehydrogenation reaction of methane and result in the formation of methyl radicals, which was the essential initial step in the OCM reaction [40,44,45]. However, a remarkable synergy between $Mn(NO_3)_2$ and Na_2WO_4 in the methane conversion could not be observed in the plasma system. The catalytic performance of different metal loadings on TiO_2 support in DBD plasma exhibits no obvious difference, and the order of activity of methane conversion and C_{2+} yield under different catalysts is MnTi > NaWTi > MNWTi. MnTi is chosen as the best catalyst judging from the C_{2+} yield.

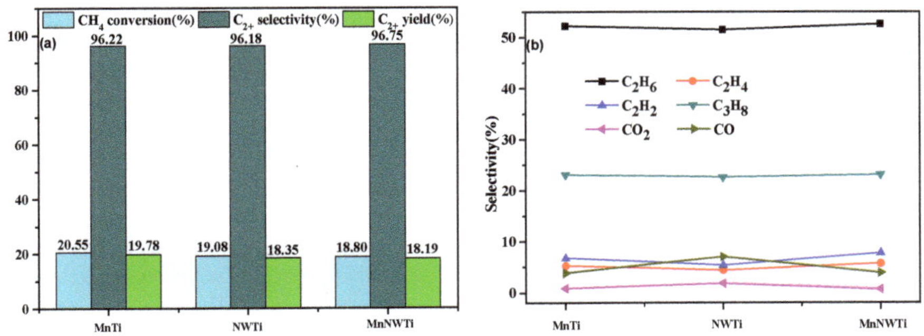

Figure 3. Performance of different metal loadings in DBD plasma. (**a**) Oxidative coupling performance; (**b**) product distribution (temperature: ambient temperature, flow rate: 60 mL/min, applied voltage: 100 V).

To investigate the role of plasma in methane conversion at low temperatures, comparison experiments were conducted, as shown in Figure 4. The C_{2+} yield of 19.78% can be obtained in the presence of plasma without additional thermal input, while the methane oxidative coupling performance of the MnTi catalyst is almost negligible, even at 850 °C on the MnTi catalyst without plasma. It generally known that C–H bond breakage of CH_4 cannot occur at ambient temperatures due to the high activation barrier; however, this can be enabled under plasma conditions. For the reaction under plasma conditions, electron collisions result in the formation of many excited state species with high internal energy. These excited state species have a lower C–H activation barrier, resulting in C–H breakage on the catalyst surface at a lower temperature. Thus, plasma plays a major role in activating the methane reaction, thus allowing the methane coupling reaction to occur at low temperatures.

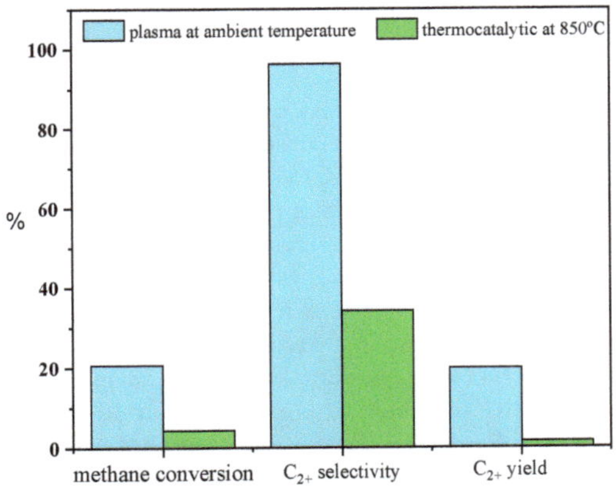

Figure 4. Catalytic performance of the MnTi catalyst with plasma and without plasma (flow rate: 60 mL/min, applied voltage: 100 V).

2.2. *Effect of Flow Rate*

Figure 5 shows the effect of different flow rates on the catalytic performance of the MnTi catalyst in the presence of DBD plasma. We can see that C_{2+} selectivity increases continuously with increasing input gas flow rate, but the selectivity of different products varies with the flow rate in a different trend. The selectivity of C_2H_4 and C_2H_2 increases with the increase in flow rate, while the selectivity of C_2H_6 and C_3H_8 remains stable. It is worth noting that the by-products of CO and CO_2 clearly decrease as the input flow increases, demonstrating the opposite effect of high input flow on C_{2+} selectivity. It could be inferred that under the condition of a high input flow, the free radicals formed by the ionization of CH_4 molecules under an electric field are easier to desorb from the surface of the catalyst into the gas phase, and avoid the subsequent excessive oxidation, resulting in the improvement of C_{2+} selectivity. However, from Figure 5a it can also be seen that methane conversion decreases with the increase in input flow, which is mainly because the high space velocity decreases the gas residence time and the conversion [31,46,47]. A higher residence time would increase the probability of collision between methane and high-energy electrons, thus improving the conversion of methane. Comprehensively considering the methane conversion and C_{2+} selectivity, the flow rate of 60 mL/min is found to be the optimum flow rate on the MnTi catalyst. The methane conversion reaches a maximum of 20.72% and C_{2+} products achieve the highest yield of 20.03%.

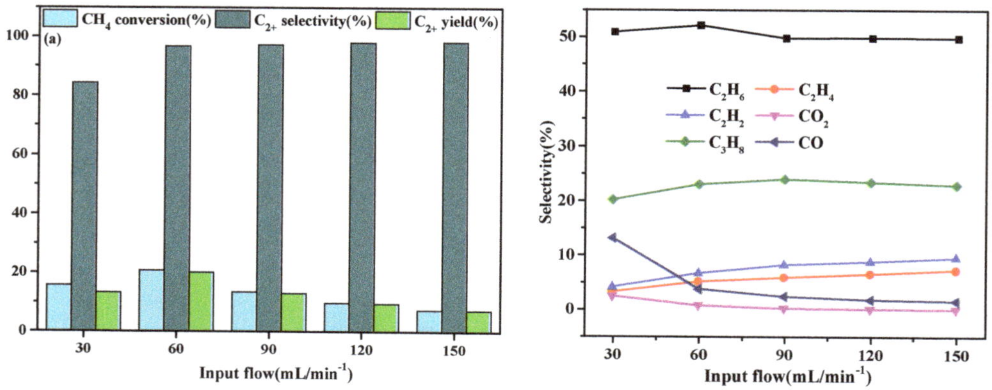

Figure 5. Effect of flow rate on the catalytic performance of the MnTi catalyst in DBD plasma. (**a**) Oxidative coupling performance; (**b**) product distribution (temperature: ambient temperature, applied voltage: 100 V).

2.3. *Effect of Temperature*

The variation of methane conversion and product selectivity with different temperatures is shown in Figure 6. A temperature of 200 °C is chosen as the upper limit to guarantee a reliable carbon balance. In light of the experimental results, the conversion of methane shows no significant difference. According to the previous literature, many efforts have been made to explore the effect of temperature on methane activation, but the relationship between temperature and catalytic performance is still controversial [33,34,48,49]. A similar result was observed by Liu et al., and the negative influence of temperature on methane conversion was attributed to the fact that a temperature rise favored the dehydration of the catalyst, which resulted in a decrease in OH^- concentration [33]. It was believed that hydroxyl species played a positive role in the activation of methane under an electric field. Similarly, from Figure 6b, it can also be found that the rise in temperature has little consequences for the distribution of the products. Therefore, the effect of temperature below 200 °C on methane conversion in DBD plasma can be considered negligible. This is because the dissociation energy of the C–H chemical bond is high up to 434 kJ/mol, which is almost impossible to be activated below 500 °C in traditional thermal reactions. Therefore, the breakage of the C–H bond in methane and subsequently generated methyl radicals is mainly initiated by high-energy electron collisions of plasma under 500 °C, rather than by additional thermal input.

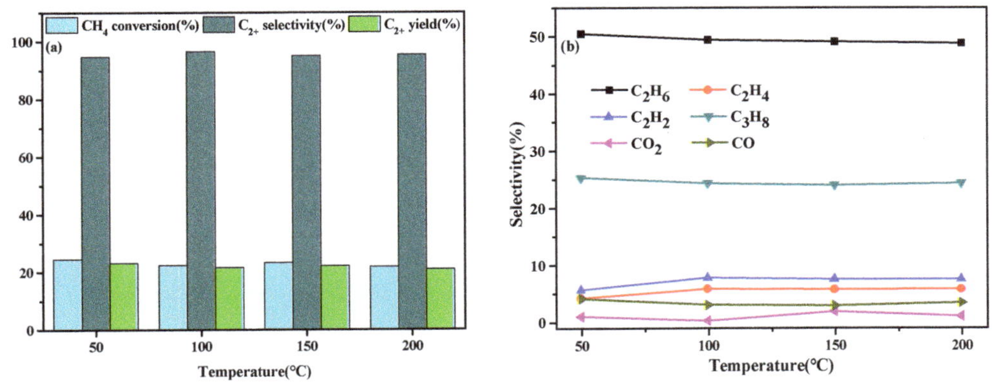

Figure 6. Effect of temperature on the catalytic performance of the MnTi catalyst in DBD plasma. (**a**) Oxidative coupling performance; (**b**) product distribution (flow rate: 60 mL/min, applied voltage: 100 V).

2.4. *Effect of Applied Voltage*

Figure 7 shows the catalytic performance of the MnTi catalyst in DBD plasma at different applied voltages, and 200 V is chosen as the upper limits to guarantee a reliable carbon balance. The conversion of methane increases with the increase in applied voltage. Lissajous plots of different applied voltages are shown in Figure 8, and the discharge power is calculated by the integral area of the Lissajous plots. According to the calculation, the increase in applied voltage will increase the discharge power, thus generating more high-energy electrons. The increased number of high-energy electrons may lead to a higher probability of bond breakage between carbon and hydrogen in the methane molecules, thus improving the degree of methane dissociation and conversion, which appears to be primarily responsible for a higher CH_4 activation [47]. However, the product distribution is independent of applied voltage and shows little difference with the variation of applied voltage, as shown in Figure 7b. Noticeably, the C_{2+} yield reaches a maximum of 30.04% at an applied voltage of 200 V, but the carbon balance of 85.28% is not perfect (the carbon balance of different applied voltages is displayed in Figure 9), indicating that the coke is formed due to methane cracking [31]. However, we still could obtain a 27.29% C_{2+} yield at a reliable carbon balance (91.4%) at 150 V applied voltage.

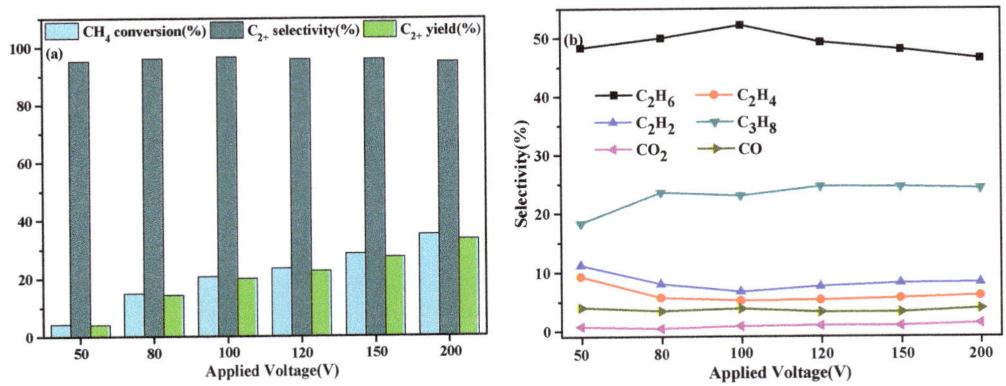

Figure 7. Effect of applied voltage on the catalytic performance of the MnTi catalyst in DBD plasma. (**a**) Oxidative coupling performance; (**b**) product distribution (temperature: ambient temperature, flow rate: 60 mL/min).

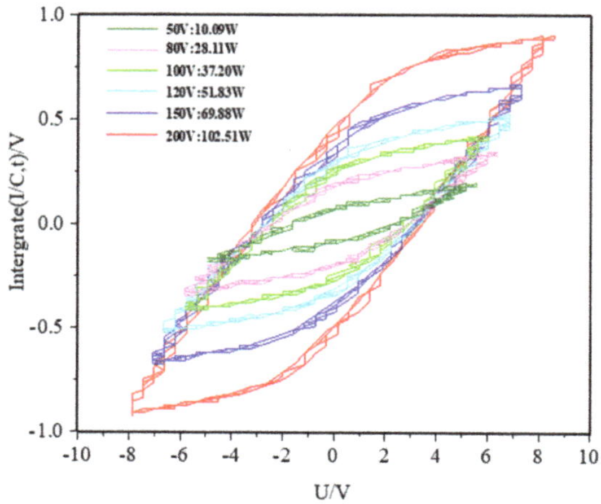

Figure 8. Lissajous plots of different applied voltages on the MnTi catalyst.

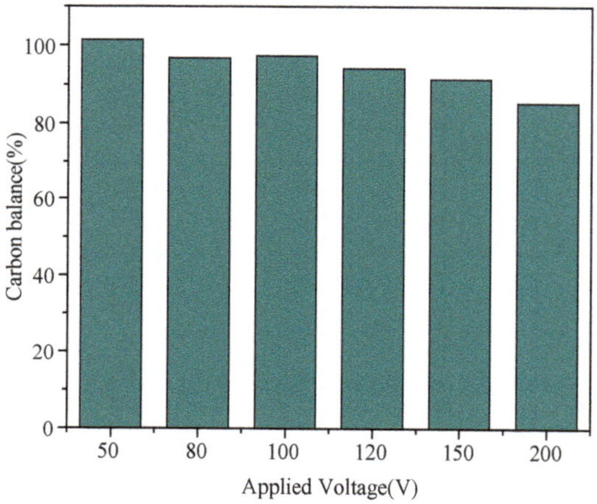

Figure 9. Carbon balance of different applied voltages on the MnTi catalyst.

3. Characterization

3.1. XRD

The XRD patterns of the as-prepared catalysts are shown in Figure 10. According to the XRD patterns, all the characteristic diffraction peaks of TiO_2 correspond well with the anatase phase in the JCPDS database (PDF-#84-1286). Moreover, new diffraction peaks also appear in the fresh MnTi catalyst with a structural change, indicating the emergence of a new phase after the loading of manganese nitrate. By comparing with the JCPDS database, these new diffraction peaks can be assigned to the rutile phase, which becomes the most dominant crystalline phase of the catalyst, judging from the intensity of the diffraction peaks. This change in crystalline phase structure is consistent with the thermal behavior of titanium dioxide [50]. The rutile phase is still the most dominant crystalline phase structure of the catalyst when the MnTi catalyst is reduced by CH_4, confirming the stability of the catalyst structure. It is worth noting that no characteristic diffraction peak for Mn in XRD patterns is observed, which might be due to low Mn loading or the high dispersion of Mn in the catalyst.

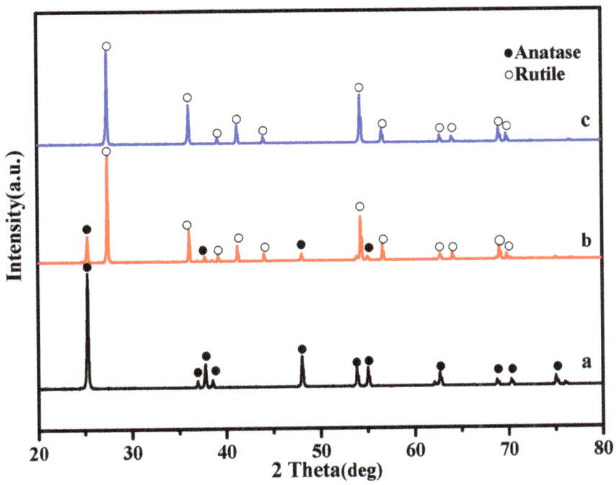

Figure 10. XRD patterns of (**a**) TiO_2, (**b**) fresh MnTi catalyst, and (**c**) reduced MnTi catalyst.

3.2. XPS

According to previous research, oxygen species play an important role in the process of oxidative coupling of methane [48]. Thus, it is necessary to explore the oxygen species on the catalyst surface, which can further explain the reasons for the improvement in methane activation and C_2 hydrocarbon selectivity through the synergetic effects between the catalyst and the plasma. The surface compositions and chemical states of the oxygen species were explored by XPS measurements, as displayed in Figure 11. For each TiO_2-based catalyst, four peaks are used for the fitting of the O 1s spectra. The peak at a lower binding energy between 528.69 eV and 529.61 eV is assigned to surface lattice oxygen (O^{2-}) [51]; the second peak existing between 531.65 eV and 529.68 eV is attributed to the surface adsorbed oxygen species (O_2^{2-}, O^-) [36]; the third peak with a binding energy of about 533.11–531.48 eV results from hydroxyls species OH^- [52]; and the forth peak with a binding energy higher than 534.0 eV is ascribed to adsorbed molecular water species H_2O [52]. The composition of the surface O species is displayed in Table 1. Among them, lattice oxygen species are considered as non-selective oxygen species, which contribute to the complete oxidation of methane and the generation of CO_x by-products, thus limiting C_{2+} selectivity [36]. Compared with TiO_2, it can be seen that Mn loading clearly increases the concentration of hydroxyl species, whilst simultaneously reducing the concentration of lattice oxygen. The decrease in lattice oxygen species indicates the increase in selective oxygen species, which favors the oxidative coupling of methane. Moreover, evidence from the literature also shows that hydroxyl species OH^- would react with different oxygen species to form catalytic active sites for OCM, which could significantly improve methane activation and C_2 selectivity. Moreover, the presence of hydroxyl species could create a micro-discharge on the catalyst surface, which further facilitates the activation of methane [30,33,45]. Compared to fresh MnTi catalyst, the concentration of the surface adsorbed oxygen species (O_2^{2-}, O^-) decreases after reduction with CH_4, confirming that the surface adsorbed oxygen species (O_2^{2-}, O^-) takes part in the oxidative coupling of methane. Moreover, it can be observed that there is a peak shift of the oxygen species towards a low binding energy when compared to pure TiO_2 catalyst, indicating the improvement in oxygen mobility and the formation of oxygen vacancies on the MnTi catalyst. The existence of oxygen vacancies can reduce the adsorption energy of methyl radicals, thereby increasing the selectivity for C_{2+} hydrocarbons.

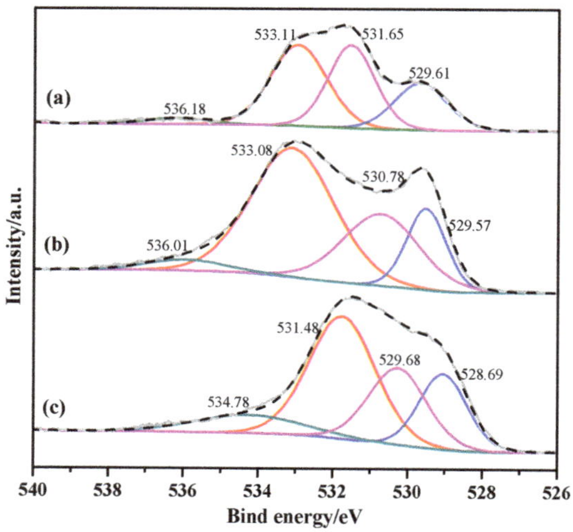

Figure 11. XPS scans on O 1s for (**a**) TiO_2, (**b**) fresh MnTi catalyst, and (**c**) reduced MnTi catalyst.

Table 1. Composition of the surface O species for fresh and spent MnTi catalyst and TiO_2.

Sample	Oxygen Distribution/%			
	O^{2-}	O_2^{2-}/O^-	OH^-	H_2O
TiO_2	21.94	37.96	36.05	4.06
Fresh 5%$Mn(NO_3)_2$-TiO_2	17.02	28.41	45.33	8.63
Reduced 5%$Mn(NO_3)_2$-TiO_2	17.82	22.73	49.26	10.18

3.3. TEM–EDS

TEM-EDS was carried out to explore the dispersion of Mn on the fresh MnTi catalyst, as displayed in Figure 12. The TEM images of the fresh MnTi catalyst reflect that Mn nanoparticles are dispersed uniformly on rod TiO_2 and no obvious Mn clusters can be found, confirming the high dispersion of Mn element on the catalyst surface. EDS mapping also demonstrates that Mn is distributed uniformly on the fresh MnTi catalyst. This explains the absence of Mn element in the XRD pattern. As stated in previous research, the stability and reaction activity of catalysts are closely related to the dispersion of the loaded metal. A higher dispersion of the metal would expose more active site edges or corners, which are more reactive on the metal surface, and hence be more conducive to improving the reactivity of the catalyst in different reactions [53]. Therefore, a higher Mn dispersion will be more favorable for the formation of C_{2+} hydrocarbons due to the large number of unsaturated active sites that are more active for the breakage of C-H bonds.

3.4. H_2-TPR

The reduction properties correlated with oxygen release characteristics of catalysts were tested by H_2-TPR technology, as shown in Figure 13. It can be observed that TiO_2 presents the low reductive property between 50 to 300 °C, and an obvious reduction peak of TiO_2 emerges from 300 °C and exhibits a peak at 625 °C, owing to the reduction of O_2^{2-} species at 387 °C and coordinatively unsaturated Ti^{3+} at 625 °C, respectively [51]. However, in the MnTi catalyst profile, a new reduction peak region emerges at ~551 °C, which contributes to the reduction process of Mn_2O_3 [36,54]. It was reported that Mn_2O_3 played a crucial role in enhancing the OCM activity of catalysts, which could take part in the hydrogen abstraction from CH_4 and trigger CH_4 activation. Moreover, the total reduction peak area increases as Mn is induced, confirming that the catalytic performance of the catalyst is enhanced. It is important to note that the reduction peaks of Ti^{3+} shift from

625 °C to 551 °C after Mn is loaded, demonstrating the improvement in oxygen release. This also proves, to some extent, that the formation of oxygen vacancies facilitates the formation of C_2 hydrocarbon products.

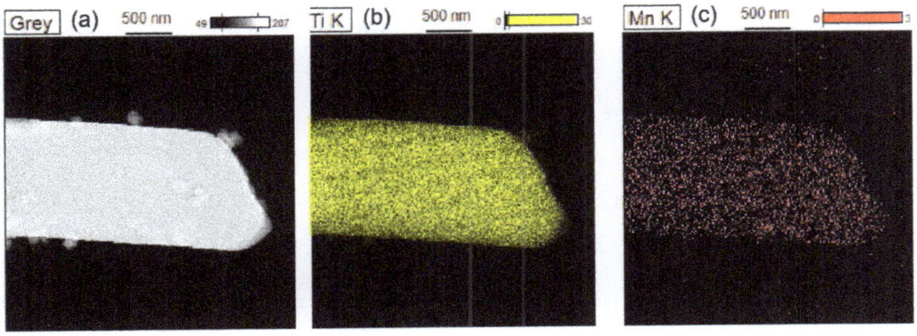

Figure 12. EDS mapping of the fresh MnTi catalyst (**a**) TEM images of MnTi catalyst, (**b**) Ti element distribution, and (**c**) Mn element distribution.

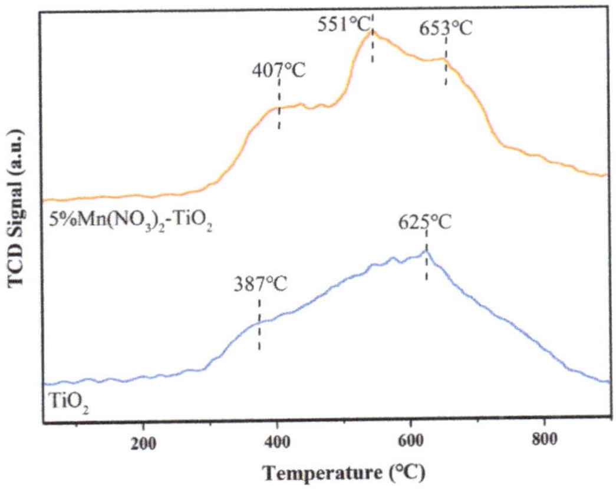

Figure 13. H_2-TPR profiles.

A possible reaction mechanism for CL-OCM on the MnTi catalyst is displayed in Figure 14. Because of the high relative permittivity, TiO_2 particles are polarized in the plasma discharge and a local electric field is formed around the particle contact point, causing an increase in methane conversion. Additionally, manganese oxide will take part in the hydrogen abstraction of methane, which is the first step of CL-OCM. In the traditional thermo-catalytic reaction, methane activation requires a high temperature due to the high activation barrier of the C-H bond. For the reaction in plasma, CH_4 activation is initiated by electronic collision to form substantial species, including radicals, ions, and vibrationally or electronically excited species. The existence of vibrationally excited methane appears to be primarily responsible for the improvement in methane conversion and C_{2+} selectivity in the presence of catalysts. Compared to the thermo-catalytic reaction, vibrationally excited methane molecules with elevated internal energy have a lower C-H activation barrier, and can, thereby, be activated on the Mn site at a much lower temperature, demonstrating that the combination of TiO_2 with the Mn species has a synergistic effect for CL-OCM reaction. Moreover, Mn loading can increase the amounts of hydroxyl species and oxygen vacancies, which can create a micro-discharge on the catalyst surface, thus facilitating the activation of methane. The increase in oxygen vacancies can reduce the adsorption energy of methyl radicals, thereby increasing the selectivity for C_{2+} hydrocarbons. It is noteworthy that coking becomes severe with increasing reduction time, which indicates that the redox-active lattice is consumed relatively quickly.

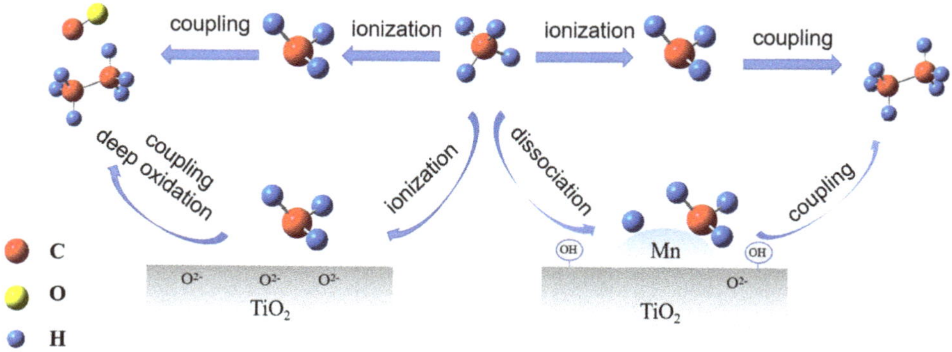

Figure 14. Possible reaction mechanism for CL-OCM on the MnTi catalyst.

3.5. Chemical Looping Redox Experiments

Since the stability and recyclability of catalysts are of utmost importance for catalysts in industrial chemical looping applications, 30 consecutive cyclic redox tests were explored on the MnTi catalyst. The catalytic performance of the catalyst on methane conversion, C_{2+} selectivity, and C_{2+} yield in the plasma–catalyst hybrid system is displayed in Figure 15. It can be seen that methane conversion and C_{2+} yield remain stable at average values of ~21% and ~20%, respectively, while the C_{2+} products selectivity simultaneously stays above 95% during 30 consecutive cyclic redox tests, demonstrating that the reactivity of the MnTi catalyst is fully recovered after the regeneration by plasma, and it presents good regeneration ability in the plasma–catalyst hybrid system. Moreover, compared to thermal regeneration, plasma regeneration could realize catalyst regeneration at a lower temperature (300 °C). According to relevant literature, the good regeneration of catalysts in plasma could be ascribed to the fact that reactive oxygen species were formed [48]. Reactive oxygen species, on the one hand, reacted more easily with carbon deposition to generate CO and CO_2, avoiding catalyst deactivation; on the other hand, they oxidized the reduced catalyst to recover OCM performance.

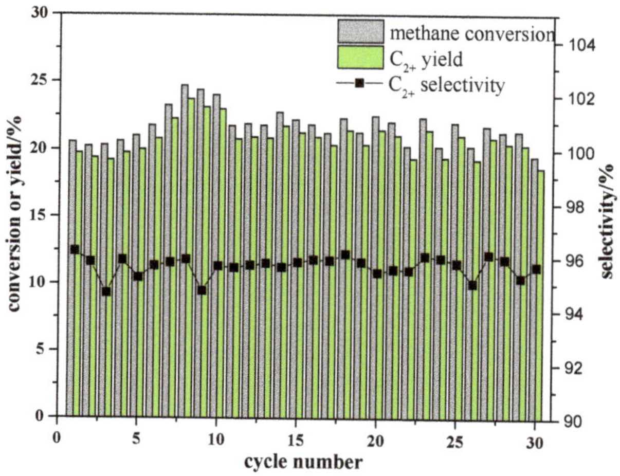

Figure 15. CLOCM of the MnTi catalyst (temperature: ambient temperature, flow rate: 60 mL/min, applied voltage: 100 V).

4. Experimental

4.1. Preparation of the Catalyst

5%$Mn(NO_3)_2$-TiO_2 (MnTi), 10%Na_2WO_4-TiO_2 (NWTi), and 5%$Mn(NO_3)_2$ + 10%Na_2WO_4-TiO_2 (MnNWTi) used in this study were prepared by the impregnation method, using aluminum oxide (Al_2O_3, Maclin), silica sand (SiO_2, 6–10 mesh, Maclin), anatase titanium oxide (TiO_2, 99.0% pure, Maclin), manganese nitrate hydrate ($Mn(NO_3)_2$, 50 wt% in water, Maclin), and sodium tungstate dihydrate ($Na_2WO_4{\cdot}2H_2O$, 99.5% pure, Maclin). To obtain the MnTi catalyst, TiO_2 was impregnated with aqueous solutions of $Mn(NO_3)_2$ as precursors, yielding 5 wt% $Mn(NO_3)_2$, and then dried at 105 °C until the water evaporated completely. Finally, the catalyst was calcined at 850 °C for 6 h and the MnTi catalyst was obtained. The production process of MnNWTi and NWTi followed the same procedure as described above.

4.2. Material Characterization

The prepared catalysts were measured by X-ray diffraction (XRD) using a Japan Science D/max-R Diffractometer (Tokyo, Japan) with Cu Kα radiation (40 kV, 40 mA, λ = 0.15406 nm) to confirm the structural features and crystal phases. The catalysts were scanned with a 2θ range of 10~80° at a rate of 10°/min.

The valence state of the oxygen species of fresh and reduced catalysts was investigated by X-ray photoelectron spectroscopy (ESCALAB 250Xi, Thermo Fisher Scientifific Inc, Waltham, MA, USA) with an Al Ka X-ray source (hv = 1486.6 eV) operating under the conditions of 20 kV and 10 mA.

The microstructure and lattice parameters of the catalyst were determined by transmission electron microscopy (JEO-2100F, JEOL, Akishima, Japan), and elemental distribution was confirmed by energy dispersive spectroscopy. Before analysis, the catalyst was treated in ethanol by ultrasonic dispersion for 20 min to obtain a uniform suspension.

H_2-temperature programmed reduction (H_2-TPR) technology was employed to determine the reducibility of catalysts with a Quantachrome ChemStar™ (Graz, Austria) instrument rigged with a thermal conductivity detector (TCD). Catalyst powder with a weight of 50 mg was loaded in a U-shape quartz tube. In order to remove water molecules and air molecules adsorbed on the catalyst surface, the catalyst was kept at 300 °C for 1 h under a helium atmosphere as a pretreatment. The heating rate of the catalysts was 10 °C/min and the flow rate of He carrier gas was 35 mL/min. After pretreatment, the catalyst was heated from 50 °C to 900 °C at a heating rate of 10 °C/min, in 10% H_2 and 90% Ar at a flow rate of 30 mL/min.

4.3. Catalytic Test

The CL-OCM experiment was tested with a plasma-catalytic DBD reactor system at atmospheric pressure, as displayed in Figure 16. The DBD reactor was an alundum tube (O.D: 25 mm, I.D: 20 mm) consisting of a stainless-steel rod as the inner electrode centered in the reactor. The high-frequency discharge probe of the plasma generator (CTP-2000K, Nanjing, China), whose panel could read the input voltage, and voltage regulator were directly connected to the stainless-steel rod, and the outer electrode was connected to the outer capacitor (or resistance). The discharge parameters were recorded using a digital oscilloscope (DPO 2024B, Tektronix) to draw Lissajous plots for the calculation of the output power of plasma. The calculation formula is as follows:

$$P(out) = f \times C \times k \times kx \times ky \times S$$

In this experiment, f = 10.04 kHz, C = 0.47 μF, k = 1000, kx, ky = 1, and S is the integral area of the Lissajous plots.

The catalyst pellets with a particle size of 20~50 mesh (2.0 g) were charged into the stainless mesh (>300 mesh) at the bottom of the stainless-steel rod. The stainless-steel mesh was placed in the grid area outside the alumina tube to ensure that the catalyst was within the discharge area. All experiments were conducted at atmospheric pressure. During the reaction step, the DBD reactor was set at the experimental temperature and pure N_2 was used for purging. The reaction gas (40% volume fraction CH_4, balance Ar) flowed through the reactor using a mass flow controller. After the introduction of CH_4, the plasma was turned on for 10 min and the gaseous products were collected. The regeneration process was carried out at 300 °C for 10 min in conditions of 21% O_2 balanced with N_2 under plasma with an applied voltage of 100 V. The gaseous products were analyzed by a gas chromatograph (GC-2014, Shimadzu in Japan).

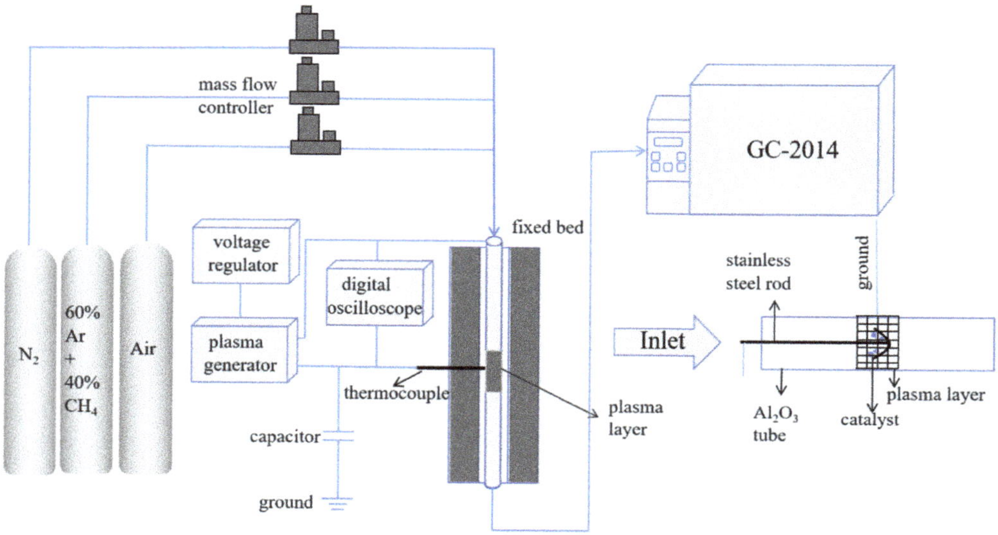

Figure 16. Reaction equipment diagram (the part on the bottom right is a magnification of the catalyst placement).

The conversion of methane, selectivity, and the yield of the C_2 product were determined by the following Equations (1)–(3):

$$X(CH4)(\%) = \frac{moles\ of\ CH4\ converted}{moles\ of\ CH4\ feed} \times 100 \tag{1}$$

$$\mathrm{S}(\mathrm{Cx})(\%) = \frac{x \times moles\ of\ Cx\ produced}{moles\ of\ CH4\ converted} \times 100 \tag{2}$$

$$Y(Cx)(\%) = \frac{x \times moles\ of\ Cx\ produced}{moles\ of\ CH4\ feed} \times 100 \tag{3}$$

5. Conclusions

In this study, the catalytic performance of different dielectric materials on methane oxidative coupling reaction was investigated in a DBD reactor. TiO_2 presented the best methane coupling performance among the various dielectric materials studies, such as SiO_2, Al_2O_3, and TiO_2. Hence, different active metal loadings on TiO_2 were tested. In light of the experimental results, metal loadings were capable of increasing methane conversion while simultaneously improving the C_{2+} selectivity. Among all tested catalysts, the MnTi catalyst was considered as the desirable catalyst judging from methane conversion and C_{2+} production. The high Mn dispersion, which was more conducive to improving the reactivity of the catalyst, was confirmed by TEM–EDS and XRD analysis. Moreover, the XPS spectra for O 1s revealed that Mn loading increased the concentration of hydroxyl species and reduced the concentration of non-selective oxygen species, which resulted in the growth of methane conversion and C_{2+} selectivity. Finally, H_2-TPR spectra showed that Mn loading supplied more oxygen species, which was also beneficial for the formation of methyl radicals and C_{2+} hydrocarbons.

We further optimized various operation parameters to explore the OCM activity of the MnTi catalyst in plasma under different flow rates, applied voltages, and temperatures. A higher methane conversion was realized by increasing applied voltage and lowering flow rates. Moreover, the variation of operation parameters showed little effect on the product distribution, indicating that the selectivity of products was mainly affected by the properties of the catalysts rather than the operation parameters. Finally, excellent regenerative properties of the MnTi catalyst during chemical looping were confirmed through 30 consecutive cyclic redox tests, showing a methane conversion of about 21%, a 96% C_{2+} hydrocarbons selectivity, and a 20% C_{2+} hydrocarbon yield. Overall, the combination of TiO_2 with a high Mn metal dispersion in plasma reactors shows excellent OCM activity, which is a novel technology that achieves high catalytic properties and recyclability in methane oxidation coupling reactions.

Author Contributions: S.K.: Data curation, Formal analysis, Writing—original draft. J.D.: Data curation. X.W.: Writing—Review and editing. K.Z.: Conceptualization, Methodology, Funding acquisition, Writing—review and editing. M.Z.: Data curation. D.S.: Formal analysis. Z.H.: Writing—Review and editing. Y.L.: Formal analysis. A.L.: Formal analysis. A.Z.: Writing—Review and editing. Z.Z.: Writing—Review and editing. All authors have read and agreed to the published version of the manuscript.

Funding: This research was funded by the National Natural Science Foundation of China (51876205, 22279144, 51866003); the Youth Innovation Promotion Association, CAS (2019341); and the Yunnan Basic Research Program Project (2019FB071).

Data Availability Statement: Not applicable.

Conflicts of Interest: There is no conflict of interest.

References

1. Boswell, R. Is gas hydrate energy within reach? *Science* **2009**, *325*, 957–958. [CrossRef] [PubMed]
2. McFarland, E. Unconventional chemistry for unconventional natural gas. *Science* **2012**, *338*, 340–342. [CrossRef] [PubMed]
3. Gu, H.; Gao, Y.; Iftikhar, S.; Li, F. Ce stabilized Ni–SrO as a catalytic phase transition sorbent for integrated CO_2 capture and CH_4 reforming. *J. Mater. Chem. A* **2022**, *10*, 3077–3085. [CrossRef]
4. Yuan, K.; Wang, Y.; Li, K.; Zhu, X.; Wang, H.; Jiang, L.; Wei, Y.; Shan, S.; Zheng, Y. $LaFe_{0.8}Co_{0.15}Cu_{0.05}O_3$ Supported on Silicalite-1 as a Durable Oxygen Carrier for Chemical Looping Reforming of CH_4 Coupled with CO_2 Reduction. *ACS Appl. Mater. Interfaces* **2022**, *14*, 39004–39013. [CrossRef]
5. Li, Z.; Feng, X.; Gu, Z.; Lu, C.; Li, D.; Zhu, X.; Jiang, L.; Deng, G.; Li, K. Enhanced performance of the CeO_2MgO oxygen carrier by NiO for chemical looping CO_2 splitting. *Fuel Process. Technol.* **2022**, *225*, 107045. [CrossRef]
6. Holmen, A. Direct conversion of methane to fuels and chemicals. *Catal. Today* **2009**, *142*, 2–8. [CrossRef]
7. Hou, Y.; Han, W.; Xia, W.; Wan, H. Structure Sensitivity of $La_2O_2CO_3$ Catalysts in the Oxidative Coupling of Methane. *ACS Catal.* **2015**, *5*, 1663–1674. [CrossRef]
8. Wang, H.; Schmack, R.; Paul, B.; Albrecht, M.; Sokolov, S.; Rümmler, S.; Kondratenko, E.V.; Kraehnert, R. Porous silicon carbide as a support for Mn/Na/W/SiC catalyst in the oxidative coupling of methane. *Appl. Catal. A-Gen.* **2017**, *537*, 33–39. [CrossRef]
9. Zhang, Z.; Ji, S.F. Numerical simulation of particle/monolithic two-stage catalyst bed reactor with beds-interspace distributed dioxygen feeding for oxidative coupling of methane. *Comput. Chem. Eng.* **2016**, *90*, 247–259. [CrossRef]
10. Guo, X.; Fang, G.; Li, G.; Ma, H. Direct, nonoxidative conversion of methane to ethylene, aromatics, and hydrogen. *Science* **2014**, *344*, 616–619. [CrossRef]
11. Goujard, V.; Tatibouët, J.-M.; Batiot-Dupeyrat, C. Carbon Dioxide Reforming of Methane Using a Dielectric Barrier Discharge Reactor: Effect of Helium Dilution and Kinetic Model. *Plasma Chem. Plasma Process.* **2011**, *31*, 315–325. [CrossRef]
12. Hu, X.; Liu, Y.; Dou, L.; Zhang, C.; Zhang, S.; Gao, Y.; Tu, X.; Shao, T. Plasma enhanced anti-coking performance of Pd/CeO_2 catalysts for the conversion of methane. *Sustain. Energy Fuels* **2022**, *6*, 98–109. [CrossRef]
13. Kim, J.; Go, D.B.; Hicks, J.C. Synergistic effects of plasma-catalyst interactions for CH_4 activation. *Phys. Chem. Chem. Phys.* **2017**, *19*, 13010–13021. [CrossRef] [PubMed]
14. Guo, Z.; Yi, Y.; Wang, L.; Yan, J.; Guo, H. Pt/TS-1 Catalyst Promoted C–N Coupling Reaction in CH_4–NH_3 Plasma for HCN Synthesis at Low Temperature. *ACS Catal.* **2018**, *8*, 10219–10224. [CrossRef]

15. Bogaerts, A.; Berthelot, A.; Heijkers, S.; Kolev, S.; Snoeckx, R.; Sun, S.; Trenchev, G.; Van Laer, K.; Wang, W. CO_2 conversion by plasma technology: Insights from modeling the plasma chemistry and plasma reactor design. *Plasma Sources Sci. Technol.* **2017**, *26*, 063001. [CrossRef]
16. Debek, R.; Azzolina-Jury, F.; Travert, A.; Mauge, F. A review on plasma-catalytic methanation of carbon dioxide-Looking for an efficient catalyst. *Renew. Sustain. Energy Rev.* **2019**, *116*, 109427. [CrossRef]
17. Liu, J.L.; Wang, X.; Li, X.S.; Likozar, B.; Zhu, A.M. CO_2 conversion, utilisation and valorisation in gliding arc plasma reactors. *J. Phys. D Appl. Phys.* **2020**, *53*, 253001. [CrossRef]
18. Qin, Y.; Niu, G.H.; Wang, X.; Luo, D.B.; Duan, Y.X. Status of CO_2 conversion using microwave plasma. *J. CO2 Util.* **2018**, *28*, 283–291. [CrossRef]
19. Barboun, M.; Mehta, P.; Herrera, F.; Go, D.; Schneider, W. Distinguishing Plasma Contributions to Catalyst Performance in Plasma-Assisted Ammonia Synthesis. *ACS Sustain. Chem. Eng.* **2019**, *7*, 32. [CrossRef]
20. Zhu, X.B.; Liu, J.; Hu, X.L.; Zhou, Z.J.; Li, X.B.; Wang, W.T.; Wu, R.B.; Tu, X. Plasma-catalytic synthesis of ammonia over Ru-based catalysts: Insights into the support effect. *J. Energy Inst.* **2022**, *102*, 240–246. [CrossRef]
21. Xu, C.; Tu, X. Plasma-assisted methane conversion in an atmospheric pressure dielectric barrier discharge reactor. *J. Energy Chem.* **2013**, *22*, 420–425. [CrossRef]
22. Thanyachotpaiboon, K.; Chavadej, S.; Caldwell, T.A. Conversion of Methane to Higher Hydrocarbons in AC Nonequilibrium Plasmas. *AIChE J.* **2010**, *44*, 2252–2257. [CrossRef]
23. Bouchoul, N.; Fourré, E.; Duarte, A.; Tanchoux, N.; Louste, C.; Batiot-Dupeyrat, C. Plasma-metal oxides coupling for CH_4-CO_2 transformation into syngas and/or hydrocarbons, oxygenates. *Catal. Today* **2021**, *369*, 62–68. [CrossRef]
24. Kim, J.; Jeoung, J.; Jeon, J.; Kim, J.; Mok, Y.S.; Ha, K.S. Effects of dielectric particles on non-oxidative coupling of methane in a dielectric barrier discharge plasma reactor. *Chem. Eng. J.* **2019**, *377*, 119896. [CrossRef]
25. Bouchoul, N.; Touati, H.; Fourré, E.; Clacens, J.-M.; Batonneau-Gener, I.; Batiot-Dupeyrat, C. Plasma-catalysis coupling for CH_4 and CO_2 conversion over mesoporous macroporous Al_2O_3: Influence of the physico-chemical properties. *Appl. Catal. B-Environ.* **2021**, *295*, 120262. [CrossRef]
26. Han, Q.; Tanaka, A.; Matsumoto, M.; Endo, A.; Kubota, Y.; Inagaki, S. Conversion of methane to C2 and C3 hydrocarbons over TiO_2/ZSM-5 core-shell particles in an electric field. *RSC Adv.* **2019**, *9*, 34793–34803. [CrossRef] [PubMed]
27. Krawczyk, K.; Młotek, M.; Ulejczyk, B.; Schmidt-Szałowski, K. Methane conversion with carbon dioxide in plasma-catalytic system. *Fuel* **2014**, *117*, 608–617. [CrossRef]
28. Zhang, X.; Liu, Y.; Zhang, M.; Yu, T.; Chen, B.; Xu, Y.; Crocker, M.; Zhu, X.; Zhu, Y.; Wang, R.; et al. Synergy between β-Mo2C Nanorods and Non-thermal Plasma for Selective CO_2 Reduction to CO. *Chem* **2020**, *6*, 3312–3328. [CrossRef]
29. Liu, L.; Das, S.; Zhang, Z.; Kawi, S. Nonoxidative Coupling of Methane over Ceria-Supported Single-Atom Pt Catalysts in DBD Plasma. *ACS Appl. Mater. Interfaces* **2022**, *14*, 5363–5375. [CrossRef]
30. Takanabe, K.; Iglesia, E. Rate and selectivity enhancements mediated by OH radicals in the oxidative coupling of methane catalyzed by $Mn/Na_2WO_4/SiO_2$. *Angew. Chem. Int. Ed. Engl.* **2008**, *47*, 7689–7693. [CrossRef]
31. Kasinathan, P.; Park, S.; Choi, W.; Hwang, Y.K.; Chang, J.S.; Park, Y.K. Plasma-Enhanced Methane Direct Conversion over Particle-Size Adjusted MOx/Al_2O_3 (M = Ti and Mg) Catalysts. *Plasma Chem. Plasma Process.* **2014**, *34*, 1317–1330. [CrossRef]
32. Kim, S.S.; Kwon, B.; Kim, J. Plasma catalytic methane conversion over sol–gel derived Ru/TiO_2 catalyst in a dielectric-barrier discharge reactor. *Catal. Commun.* **2007**, *8*, 2204–2207. [CrossRef]
33. Liu, C.; Marafee, A.; Mallinson, R.; Lobban, L. Methane conversion to higher hydrocarbons in a corona discharge over metal oxide catalysts with OH groups. *Appl. Catal. A-Gen.* **1997**, *164*, 21–33. [CrossRef]
34. Keller, G.E.; Bhasin, M.M. Synthesis of ethylene via oxidative coupling of methane: I. Determination of active catalysts. *J. Catal.* **1982**, *73*, 9–19. [CrossRef]
35. Cheng, Z.; Baser, D.S.; Nadgouda, S.G.; Qin, L.; Fan, J.A.; Fan, L.-S. C2 Selectivity Enhancement in Chemical Looping Oxidative Coupling of Methane over a Mg–Mn Composite Oxygen Carrier by Li-Doping-Induced Oxygen Vacancies. *ACS Energy Lett.* **2018**, *3*, 1730–1736. [CrossRef]
36. Jiang, S.C.; Ding, W.X.; Zhao, K.; Huang, Z.; Wei, G.Q.; Feng, Y.Y.; Lv, Y.J.; He, F. Enhanced Chemical looping oxidative coupling of methane by Na-doped $LaMnO_3$ redox catalysts. *Fuel* **2021**, *299*, 120932. [CrossRef]
37. Zeng, D.; Wang, C.; Liu, T.; Ou, W.; Xiao, R. Enhanced performance of chemical looping methane oxidative coupling by the synergistic effect of TiO_2 doped $Na_2WO_4/Mn_2O_3/SiO_2$ oxygen carriers. *Fuel Process. Technol.* **2022**, *234*, 107320. [CrossRef]
38. Lunsford, J.H. The catalytic oxidative coupling of methane. *Angew. Chem. Int. Ed. Engl.* **1995**, *34*, 970–980. [CrossRef]
39. Kiani, D.; Sourav, S.; Wachs, I.E.; Baltrusaitis, J. Synthesis and molecular structure of model silica-supported tungsten oxide catalysts for oxidative coupling of methane (OCM). *Catal. Sci. Technol.* **2020**, *10*, 3334–3345. [CrossRef]
40. Elkins, T.; Hagelin-Weaver, H. Characterization of $Mn–Na_2WO_4/SiO_2$ and $Mn–Na_2WO_4/MgO$ catalysts for the oxidative coupling of methane. *Appl. Catal. A-Gen.* **2015**, *497*, 96–106. [CrossRef]
41. Arndt, S.; Otremba, T.; Simon, U.; Yildiz, M.; Schubert, H.; Schomäcker, R. $Mn–Na_2WO_4/SiO_2$ as catalyst for the oxidative coupling of methane. What is really known? *Appl. Catal. A-Gen.* **2012**, *425*, 53–61. [CrossRef]
42. Kim, G.J.; Ausenbaugh, J.T.; Hwang, H.T. Effect of TiO_2 on the Performance of Mn/Na_2WO_4 Catalysts in Oxidative Coupling of Methane. *Ind. Eng. Chem. Res.* **2021**, *60*, 3914–3921. [CrossRef]

43. Chen, H.; Lee, H.; Chen, S.; Chao, Y.; Chang, M. Review of plasma catalysis on hydrocarbon reforming for hydrogen production—Interaction, integration, and prospects. *Appl. Catal. B-Environ.* **2008**, *85*, 1–9. [CrossRef]
44. Hou, S.; Cao, Y.; Xiong, W.; Liu, H. Site requirements for the oxidative coupling of methane on SiO_2-supported Mn catalysts. *Ind. Eng. Chem. Res.* **2006**, *45*, 7077–7083. [CrossRef]
45. Chawdhury, P.; Bhargavi, K.; Subrahmanyam, C. A single-stage partial oxidation of methane to methanol: A step forward in the synthesis of oxygenates. *Sustain. Energy Fuels* **2021**, *5*, 3351–3362. [CrossRef]
46. Seyed, M.N.; Savadkoohi, H.A.; Feizabadi, S.Y. Methane Conversion to C2 Hydrocarbons Using Dielectric-barrier Discharge Reactor: Effects of System Variables. *Plasma Chem. Plasma Process.* **2008**, *28*, 189–202. [CrossRef]
47. Miao, Y.; Kreider, P.; Reddick, I.; Pommerenck, J.; Collin, R.; AuYeung, N.; von Jouanne, A.; Jovanovic, G.; Yokochi, A. Methane Coupling to Ethylene and Longer-Chain Hydrocarbons by Low-Energy Electrical Discharge in Microstructured Reactors. *Ind. Eng. Chem. Res.* **2021**, *60*, 6950–6958. [CrossRef]
48. Lee, H.; Lee, D.H.; Ha, J.M.; Kim, D.H. Plasma assisted oxidative coupling of methane (OCM) over Ag/SiO_2 and subsequent regeneration at low temperature. *Appl. Catal. A-Gen.* **2018**, *557*, 39–45. [CrossRef]
49. Huu, P.; Gil, S.; Patrick, D.; Anne, G.-F.; Khacef, A. Plasma-catalytic hybrid reactor: Application to methane removal. *Catal. Today* **2015**, *257*, 86–92.
50. Gunawidjaja, R.; Anderson, B.; Eilers, H. Structural and spectroscopic characterization of irreversible phase changes in rapidly heated precursors of europium-doped titania nanoparticles. *J. Solid State Chem.* **2018**, *258*, 15–23. [CrossRef]
51. Zhang, P.; Xiong, J.; Wei, Y.C.; Li, Y.F.; Zhang, Y.L.; Tang, J.J.; Song, W.Y.; Zhao, Z.; Liu, J. Exposed {0 0 1} facet of anatase TiO_2 nanocrystals in Ag/TiO_2 catalysts for boosting catalytic soot combustion: The facet-dependent activity. *J. Catal.* **2021**, *398*, 109–122. [CrossRef]
52. Zhao, K.; Li, L.W.; Zheng, A.Q.; Huang, Z.; He, F.; Shen, Y.; Wei, G.Q.; Li, H.B.; Zhao, Z.L. Synergistic improvements in stability and performance of the double perovskite-type oxides $La_2{-}xSrxFeCoO_6$ for chemical looping steam methane reforming. *Appl. Energy* **2017**, *197*, 393–404. [CrossRef]
53. Xie, P.; Pu, T.; Nie, A.; Hwang, S.; Purdy, S.C.; Yu, W.; Su, D.; Miller, J.T.; Wang, C. Nanoceria-Supported Single-Atom Platinum Catalysts for Direct Methane Conversion. *ACS Catal.* **2018**, *8*, 4044–4048. [CrossRef]
54. Yang, N.Z.; Guo, R.T.; Pan, W.G.; Chen, Q.L.; Wang, Q.S.; Lu, C.Z. The promotion effect of Sb on the Na resistance of Mn/TiO_2 catalyst for selective catalytic reduction of NO with NH_3. *Fuel* **2016**, *169*, 87–92. [CrossRef]

catalysts

MDPI

Review

Recent Advances of Oxygen Carriers for Hydrogen Production via Chemical Looping Water-Splitting

Wenxi Chang [1,2], Yue Hu [2,3], Weibin Xu [2,3], Chuande Huang [2,*], Haonan Chen [1], Jiahui He [1], Yujia Han [2,3], Yanyan Zhu [1,*], Xiaoxun Ma [1] and Xiaodong Wang [2,*]

1 School of Chemical Engineering, Northwest University, International Scientific and Technological Cooperation Base of MOST for Clean Utilization of Hydrocarbon Resources, Chemical Engineering Research Center for the Ministry of Education for Advance Use Technology of Shanbei Energy, Xi'an 710069, China
2 CAS Key Laboratory of Science and Technology on Applied Catalysis, Dalian Institute of Chemical Physics, Chinese Academy of Sciences, Dalian 116023, China
3 School of Chemical Engineering, University of Chinese Academy of Sciences, Beijing 100049, China
* Correspondence: huangchuande@dicp.ac.cn (C.H.); zhuyanyan@nwu.edu.cn (Y.Z.); xdwang@dicp.ac.cn (X.W.)

Abstract: Hydrogen is an important green energy source and chemical raw material for various industrial processes. At present, the major technique of hydrogen production is steam methane reforming (SMR), which suffers from high energy penalties and enormous CO_2 emissions. As an alternative, chemical looping water-splitting (CLWS) technology represents an energy-efficient and environmentally friendly method for hydrogen production. The key to CLWS lies in the selection of suitable oxygen carriers (OCs) that hold outstanding sintering resistance, structural reversibility, and capability to release lattice oxygen and deoxygenate the steam for hydrogen generation. Described herein are the recent advances in designing OCs, including simple metal oxides (e.g., Fe, Zn, Ce, and Ti-based metal oxides) and composite metal oxides (e.g., perovskite, spinel, and garnets), for different CLWS processes with emphasis on the crucial parameters that determine their redox performance and future challenges.

Keywords: chemical looping; water-splitting; hydrogen; oxygen carrier; redox performance

1. Introduction

An urgent desire to reduce reliance on fossil fuels has been spurred in recent years due to the steeply increased concentration of greenhouse gases in the atmosphere [1]. Hydrogen, as an alternative, has been regarded as a promising fuel that possesses a high calorific value (121.00 MJ/kg) that is 2.8 times and four times greater than petroleum and coal, respectively, while giving clean water as the only combustion product [2–4]. More importantly, hydrogen is an important chemical raw material for various industrial processes, such as crude oil refining, synthesis of ammonia and methanol, etc. [5]. Currently, the production of hydrogen is mainly based on fossil fuels, such as steam methane reforming (SMR) and coal gasification (CG) [2,6,7]. However, both commercial routes are energy-intensive, rendering massive CO_2 emissions reaching as high as 10–16 and 22–35 $kgCO_2e/kgH_2$ for the SMR and CG processes, respectively [8–11]. Although carbon capture and storage technologies can potentially reduce the CO_2 emissions, this would inevitably induce a significant increase in energy penalty and manufacturing costs [12,13].

As a reverse reaction of hydrogen combustion, hydrogen production directly from water-splitting is considered a sustainable approach. Intuitively, direct thermal splitting of water ($H_2O(l) \rightarrow H_2(g) + 1/2O_2(g)$, ΔH^{θ}_{298} = 286 kJ/mol) is the simplest method to implement this goal. However, thermodynamic analysis shows that a positive Gibbs free energy is maintained for this reaction until a temperature higher than 4370 K (Figure 1), revealing an extremely tough process to break the O–H bond in H_2O molecules [14].

Kogan et al. [15] proposed that water conversion only reached ca. 25% at a temperature of 2500 K and pressure of 0.05 bar, and rapid cooling treatment was needed to suppress the reverse reaction between generated H_2 and O_2, causing an increased energy burden. A study showed that hydrogen produced by pyrolyzing the mixture of water and argon was lower than 3% percentage of all off-gases at 2000~2500 °C [16]. Recent reports suggested that water-splitting driven by light and electricity enabled production of hydrogen at ambient temperature [17,18]. Compared with the photocatalytic process that is still in its infancy, electrochemical water-splitting could provide roughly 5% of the world's hydrogen energy [19]. The purity of hydrogen generated in the electrolysis process is high, but the electrolysis process consumes a large amount of power, and the electricity cost accounts for around 80% of the overall water electrolysis process, rendering the scale-up of this technology strongly depending on the availability of low-cost and green electricity [20–22]. Therefore, research is ongoing to explore more efficient methods for hydrogen production with low carbon emissions and energy penalty.

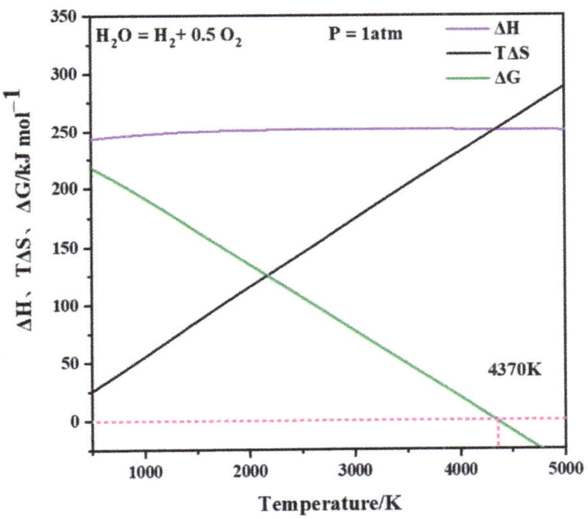

Figure 1. Variation of ΔH, TΔS, and ΔG with the temperature for direct water-splitting at 1 atm. Adapted with permission from Ref. [14], copyright 2007, American Chemical Society.

Chemical looping technology that was initially designed as an advanced combustion method has shown great potential for energy storage and production of chemicals due to the virtue of process intensification, leading to improved process efficiency and notably reduced CO_2 emission and investment expenses [23–25]. In recent years, pioneering reports have verified that applying the chemical looping concept to water-splitting, known as chemical looping water-splitting (CLWS), enables efficient hydrogen production at temperatures below 1300 °C. Compared with continuous reaction mode, CLWS reaction, utilizing metal oxides as oxygen carriers (OCs), is fulfilled by two or three spatiotemporal separated steps, including abstracting lattice oxygen from OCs via reduction treatment, regeneration by water with production of hydrogen, and in some cases, further oxidation by air to fully recover the OCs. Such a multistep reaction not only affords great convenience in separating hydrogen from products, but also requires less useful work and reaction free energy changes for water decomposition [26,27]. Furthermore, the introduction of OCs as redox catalysts significantly lowers the threshold to break the O–H bond of H_2O and improves the reaction kinetics [28–30].

In general, the yield of hydrogen in a single redox cycle is positively correlated with the oxygen capacity of OCs, which in turn is closely related to the reduction method to remove the lattice oxygen. According to the distinct strategies in abstracting lattice oxygen from OCs, CLWS can be classified into processes including a two-step thermochemical water-splitting cycle (thermal reduction, Figure 2A) [31], a methane chemical looping process (CH_4 reduction, Figure 2B), a chemical looping water gas shift reaction (CO reduction, Figure 2C) [32], a syngas (mixture of H_2 and CO) chemical looping process (syngas reduction, Figure 2D) [33], and a photo-thermochemical cycle (photochemical reduction, Figure 2E) [34]. As for CLWS reaction, the economic efficiency of hydrogen production is substantially determined by redox properties of OCs, including mechanical strength, oxygen capacity, reactivity, selectivity, and structural stability, as well as low toxicity and low cost [35,36]. A series of reviews have been proposed in recent years that comprehensively summarized the potential applications of chemical looping technology [37], reaction mechanism of OCs [38], and the recent advantages in OCs for chemical looping conversion of alkanes [39–41]. As for hydrogen production via the chemical looping method, Voitic et al. [42] have reviewed the progress with emphasis on the iron-based OCs, while Luo et al. [43] detailed the system integration, economic analysis, and OCs for the chemical looping reforming process (cofeeding the fuel and water to reactor) and fuel-driven chemical looping water-splitting. Herein, the review gives a detailed survey on the current state of OC development for hydrogen production via different CLWS processes. The effect of the inherent properties of OCs on feed-gas conversion or product selectivity and future challenges in designing the OCs are summarized.

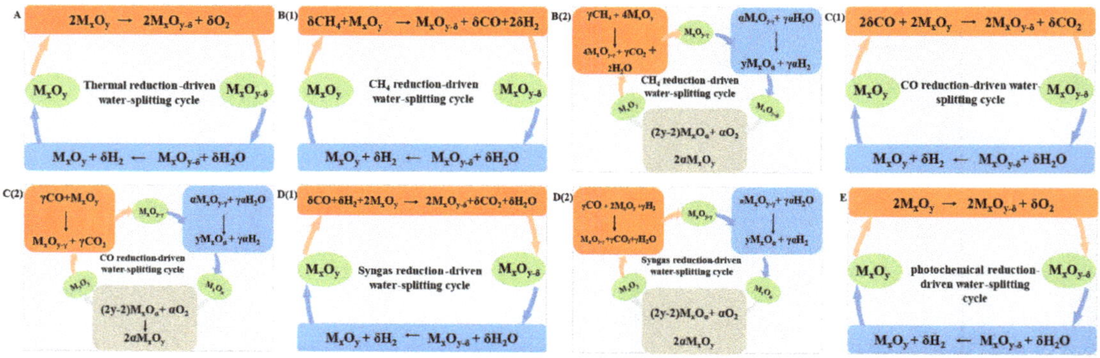

Figure 2. Chemical looping water-splitting process with various reducing methods: (**A**) two-step water-splitting thermochemical cycle, (**B**) methane chemical looping process, (**C**) chemical looping water gas shift process, (**D**) syngas chemical looping process, and (**E**) photo-thermochemical cycle.

2. Processes for Chemical Looping Water-Splitting

2.1. Two-Step Thermochemical Water-Splitting

To address the problems of high reaction temperature and low conversion for direct water-splitting, in 1966 Funk et al. [26] proposed the strategy of two-step thermochemical water-splitting (TCWS) for hydrogen production. The key to this process lies in selecting suitable metal oxides as OCs, which are thermally reduced at high temperature (>1300 °C) to release oxygen and subsequently recovered by H_2O at a temperature generally lower than 1100 °C with production of hydrogen. As a result, direct contact between hydrogen and oxygen is avoided, which bypasses the issue of hydrogen separation and enables higher water conversion at lower temperatures. In particular, when concentrated solar energy is utilized to heat the OCs (two-step solar thermochemical cycles), the solar energy is transformed to chemical energy with a theoretical solar-to-fuel efficiency reaching higher than 70% [14]. However, the high reaction temperature and thermal shocks induced by temperature swing pose a huge challenge in designing the OCs [44,45].

The Fe_3O_4/FeO redox pair was one of the earliest OCs used in a two-step TCWS cycle. Based on the thermodynamic analyses, Nakamura et al. [46] and Steinfeld et al. [47] reported that it was feasible to apply Fe_3O_4/FeO for hydrogen generation via thermochemical cycle at temperatures of ca. 2000 °C (Figure 3). However, these oxides were subjected to severe sintering under high operating temperatures. To address this issue, some active metals or inert supports were introduced to the OCs in an attempt to lower the thermal reduction temperature or improve their sintering resistance. Han et al. [48] found that removal of lattice oxygen from ferrites was realized at lower temperature of 1200 °C after doping of Ni, Mn, and Cu while water-splitting was implemented at 800 °C. Among the screened materials, $NiFe_2O_4$ displayed the highest rate for hydrogen production with an average H_2 yield of 0.442 mL/g (the amount of produced H_2 per gram of oxygen carrier in each cycle). Structural analysis revealed that no obvious changes occurred after these redox reactions, which demonstrated a high stability of this material. Kodama et al. [49,50] also suggested that Ni-modified ferrites in $A_xFe_{3-x}O_4$ (A = Mn, Co, Ni) exhibited better reactivity than Co or Mn promoters, and a scaling relationship between Ni content (from 0 to 1) with corresponding performance was noticed. When Ni loading increased to x = 1, a maximum hydrogen productivity of ca. 10 mL/g is reached. Their follow-up studies showed that loading of ferrites on ZrO_2 support significantly slowed down the sintering process, leading to stable hydrogen production. Ishihara et al. [51] reported that combination of $NiFe_2O_4$ with YSZ (yttrium-stabilized zirconia) support was capable of promoting the redox stability and oxygen mobility, resulting in a negligible decline in hydrogen production in 10 continuous cycles with reduction temperature of 1773 K and water-splitting temperature of 1473 K. Miller et al. [52] proposed that a mass ratio of 1/3 for $Co_{0.67}Fe_{2.33}O_4$/YSZ could effectively inhibit the agglomeration of OCs and enable stable hydrogen production during eight cycles. In contrast, the bare $Co_{0.67}Fe_{2.33}O_4$ oxides was inactivated in the second cycle.

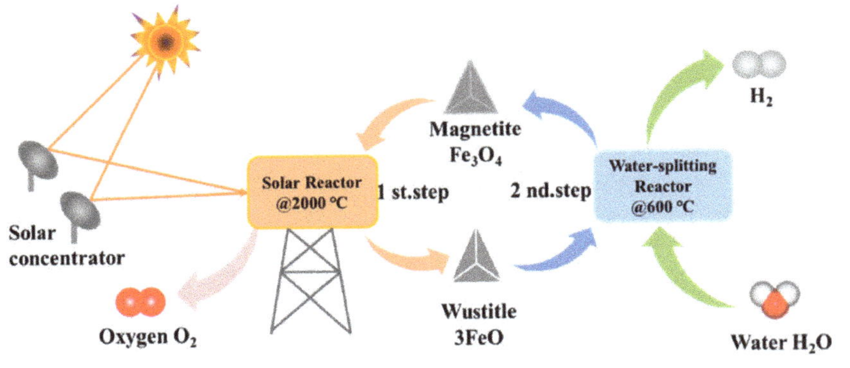

Figure 3. Schematic for two-step solar thermochemical water-splitting using Fe_3O_4/FeO redox system. Adapted with permission from Ref. [47], Copyright 1999, Elsevier.

In 1977, Bilgen's team [16] applied a ZnO/Zn redox pair to the two-step TCWS cycle. Experimental results proved that thermal decomposition of ZnO into gaseous Zn (boiling point of 907 °C) and O_2 was achieved at 2338 K using a solar furnace. Thermodynamic calculations showed that the theoretical solar-to-fuel efficiency (without heat recovery) of the water-splitting cycle could reach 51% when the redox cycle was conducted with thermal reduction temperature of 2338 K (ZnO decomposition) and water-splitting at 1500 K. However, practical application of ZnO in TCWS remains greatly challenging due to the facile recombination of Zn and O_2 before introduction of water; thus, a quench treatment is necessary to separate Zn from O_2 by rapid condensation of Zn gas into solid form. The numerical simulation results by Palumbo et al. [53] showed that a conversion of 80% was achieved for ZnO decomposition when a high solar flux density of 2 MW/m^2 and quenching rate of 2.0×10^7 K/s were applied. Another challenge for application of ZnO/Zn redox pairs lies in the formation of a ZnO shell during the water-splitting reaction ($Zn + H_2O \rightarrow ZnO + H_2$), which significantly slows down the reaction kinetics [54].

Therefore, recent works have been devoted to kinetic study on this process with the aim to improve the hydrogen production efficiency. Ernst et al. [55] investigated the reaction between submicron zinc and ~50 mol% H_2O/Ar at 603–633 K. It was found that the initial reaction was a surface-controlled process with a fast rate and apparent activation energy of 42.8 ± 7.4 kJ/mol, while the subsequent reaction was a slow diffusion control process due to the restriction of surface ZnO layer on the migration of Zn^{2+}. Lv et al. [56] investigated the effect of particle size on the water-splitting process and found that zinc powder with different sizes exhibited similar hydrolysis mechanisms, but that reducing the particle size could significantly enhance the reaction rate.

CeO_2, as a benchmark OC for two-step TCWS, has attracted particular attention since 2006 due to its outstanding redox properties. Abanades et al. [57] found that Ce_2O_3 obtained by thermal reduction of CeO_2 can be facilely oxidized by water with a conversion reaching 100%. However, a temperature of above 2000 °C is needed for complete reduction of CeO_2 to Ce_2O_3, which inevitably resulted in serious sintering of OCs. To bypass this issue, in most cases for TCWS applications, the reduction temperature was lowered to below 2000 °C, leading to formation of non-stoichiometric $CeO_{2-\delta}$. Steinfeld et al. [58] reported that 2.5~2.9 mL of O_2 can be released per gram of ceria in a single redox cycle (2.5~2.9 mL/g), corresponding to δ value of 0.039~0.044, when thermal reduction was conducted at ca. 1600 °C. This rendered a hydrogen evolution rate reaching a peak value of 7.6×10^2 mL/min during the following water-splitting step (<900 °C), which resulted in an overall solar-to-hydrogen efficiency of 0.7%. Costa Oliveira et al. [59] investigated the effect of CeO_2 morphology on water-splitting reaction and discovered that H_2 yield over three-dimensionally ordered macroporous (3DOM) CeO_2 was approximately 32% greater than that of conventional CeO_2 foam, suggesting that engineering the morphology was effective in adjusting the reaction kinetics and oxygen mobility. Although much progress has been achieved these years, the low oxygen capacity and high reduction temperature of pure ceria greatly limited the hydrogen productivity and process efficiency, which impeded the feasibility of ceria in practical STWS applications. To address this issue, significant research has been performed to tailor the redox properties of Ce-based OCs by tuning the oxygen vacancies and lattice strain via doping foreign cations [60–63]. Gokon et al. [64] compared the performance of 30 mol% M-doped CeO_2 (M = Fe, Co, Ni, or Mn) in two-step TCWS cycles and found that all dopants except for Ni greatly increased the oxygen evolution during thermal reduction at 1500 °C. However, the yield of H_2 is lower than that of bare CeO_2, which may be ascribed to severe sintering of these OCs at high temperature. Tamaura et al. [65] investigated the performance of CeO_2-MO_x solid solution (M = Mn, Fe, Ni, Cu) with Ce/M molar ratio of 9 for two-step TCWS. As shown in Figure 4A, introducing of di- or tri-valent transition metal cations into CeO_2 notably promoted H_2 generation by enhancing releasing of lattice oxygen. The same group further screened a series of $Ce_{0.9}M_{0.1}O_{2-\delta}$ ceramics (M = Mg, Ca, Sr, Sc, Y, Dy, Zr, and Hf), and found that the evolution of O_2 (reduction of Ce^{4+} to Ce^{3+}) in the thermal reduction step was improved by doping cations with higher valence state and smaller effective ionic radius (Figure 4B) [66,67]. Furthermore, the amount of H_2 produced in the water-splitting step was dominated by the amount of O_2 released in the reduction step. Instead, Abanades and coworkers [68] showed that addition of tri-valent dopants, including Y, La, Pr, and Gd, mainly contributed to improve the sintering resistance rather than the reducibility of ceria, but the ceria/zirconia mixed oxide was capable of greatly enhancing the reduction of bulk Ce^{4+}. After detailed study on the correlation between zirconium content and redox properties of $Ce_{1-\delta}Zr_{\delta}O_2$ ($0 \leq \delta \leq 0.5$), it was found that the ratio of Ce^{3+}/Ce^{4+} in thermally reduced oxides gradually increased with Zr content while an optimal composition of $Ce_{0.75}Zr_{0.25}O_2$ gave the highest hydrogen yield of 6.67 mL/g [69,70]. Their follow-up studies suggested that decoration of CeO_2-ZrO_2 with 1% Gd results in further promotion of hydrogen production to 7.58 mL/g [71]. Instead of exploring the nonstoichiometric chemistry that results in partial reduction of Ce^{4+} to Ce^{3+}, Wang et al. [72] discovered that Sn-modified oxide of CeO_2-xSnO_2 (x = 0.05–0.20) was converted to thermodynamically stable $Ce_2Sn_2O_7$ during the thermal reduction step (1400 °C), which enabled deep reduction of Ce^{4+} to Ce^{3+} rather than $CeO_{2-\delta}$. When the Sn content was optimized to x = 0.15, a hydrogen production as high as 7.18 mL/g (Figure 5), 3.8 times greater than that of unmodified CeO_2, was obtained.

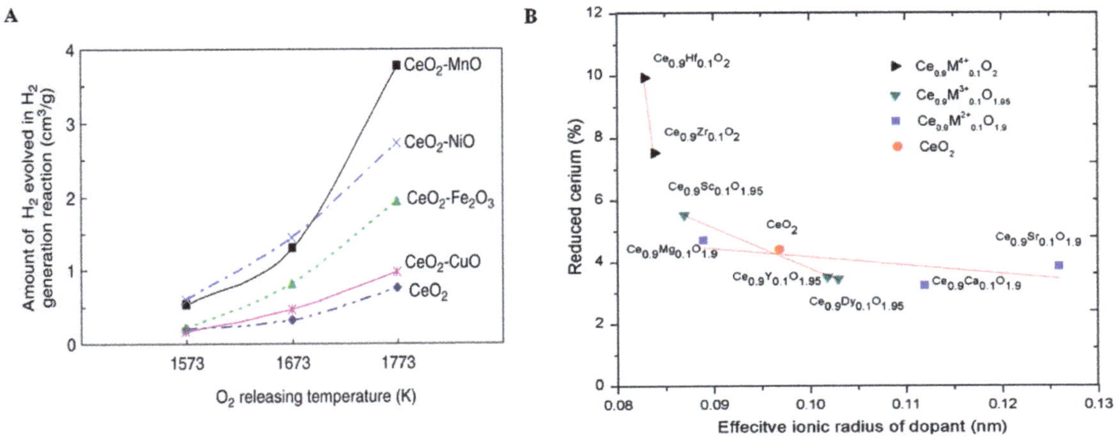

Figure 4. (**A**) Correlation between O_2-releasing temperature and the amount of hydrogen produced in the reaction with CeO_2–MO_x. Reprinted with permission from Ref. [65], copyright 2007, Elsevier. (**B**) Effect of effective ionic radius and valence state of dopants on the reduction degree of cerium. Reprinted with permission from Ref. [66], copyright 2011, Elsevier.

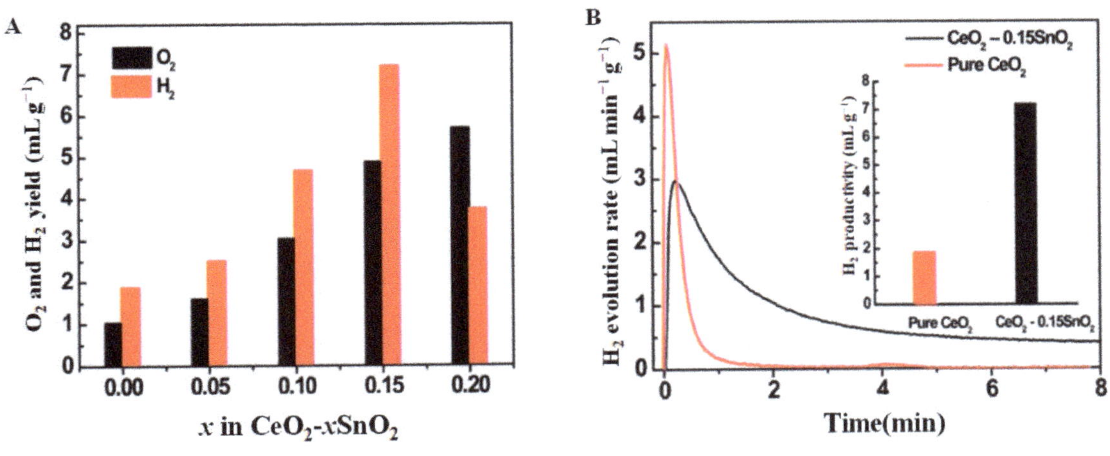

Figure 5. (**A**) Evolution of O_2 and H_2 for $CeO_{2-x}SnO_2$ (x = 0.05–0.20) during redox reactions. (**B**) H_2 evolution rate during water-splitting reaction at 800 °C for CeO_2–$0.15SnO_2$ and pure CeO_2. Adapted with permission from Ref. [72], Copyright 2017, Wiley.

Perovskite oxides (ABO_3) represent an important class of composite materials, wherein the A-site is generally occupied by rare earth, alkaline earth, or alkali metals that coordinate with 12 oxygen anions while the B-site is typically occupied by transition metals located in the octahedral interstices of the oxygen framework. These oxides hold virtue of outstanding redox properties, structure reversibility, and high-temperature stability, which render these oxides promising OCs for two-step TCWS reactions [73,74]. McDaniel et al. [75] discovered that Sr/Al doped $LaMnO_3$ exhibited exceptional oxygen mobility, which released 8 times more O_2 than CeO_2 at 1350 °C, while the temperature for lattice oxygen desorption was nearly 300 °C lower than that of CeO_2. Consequently, a hydrogen yield 9 times greater than that of CeO_2 is obtained at 1000 °C. R. Barcellos and colleagues [76] demonstrated that doping of Ce at B-site of $BaMnO_{3-\delta}$ allowed optimization of the oxygen vacancy formation energy. Among the investigated $BaCe_xMn_{1-x}O_{3-\delta}$ samples ($0 \leq x \leq 1$), $BaCe_{0.25}Mn_{0.75}O_{3-\delta}$ displayed the best water-splitting performance and yielded nearly 3 times more hydrogen than ceria when reduced at 1350 °C. Similarly, Li et al. [31] also found that introduction of Co into $LaGaO_3$ notably reduced the oxygen vacancy formation energy to ca. 3.0 eV, which was favorable for both lattice oxygen desorption and subsequent water-splitting. The best $LaGa_{0.4}Co_{0.6}O_{3-\delta}$ sample exhibited a H_2 yield of 0.478 mmol/g that was 15 times greater than CeO_2 (0.032 mmol/g) when reduced at 1350 °C and re-oxidized at 800 °C. Carter et al. [77] screened a series of $Ca_{0.5}Ce_{0.5}MO_3$ oxides (M = Sc, Ti, V, Cr, Mn, Fe, Co, and Ni) for a two-step solar thermochemical process using density functional theory (DFT) calculations. The calculation results suggested that the oxygen vacancy formation energy of $Ca_{0.5}Ce_{0.5}MnO_3$ and $Ca_{0.5}Ce_{0.5}FeO_3$ was slightly lower than that of CeO_2 (4.0~4.3 eV) and located in the range of 3.65~3.96 and 3.77~4.06 eV, respectively, which fitted well to the target window (3.2~4.1 eV) for thermochemical cycles (Figure 6). Qian et al. [78] found that the $CaTi_{0.5}Mn_{0.5}O_{3-\delta}$ was favorable for oxygen releasing and water-splitting due to its intermediate reduction enthalpy value of 200–280 KJ/mol-O (the standard enthalpy to remove one mole lattice oxygen of OC) and large entropy value of 120–180 J/mol-O·K (the standard entropy to remove one mole lattice oxygen of OC), which enables a hydrogen production of 10.0 mL/g in redox cycle between 1350 °C (thermal reduction) and 1150 °C (water-splitting) with a short cycle time of 1.5 h. All these results indicate that perovskite oxides are among the best candidates for two-step TCWS processes, and that A/B site engineering is efficient in modulating their redox properties. However, the step forward for practical application of this technology is still greatly hindered by high operating temperature and relatively low hydrogen yield due to the low oxygen redox capacity and long temperature swing time, which should be resolved ultimately by developing more effective OCs.

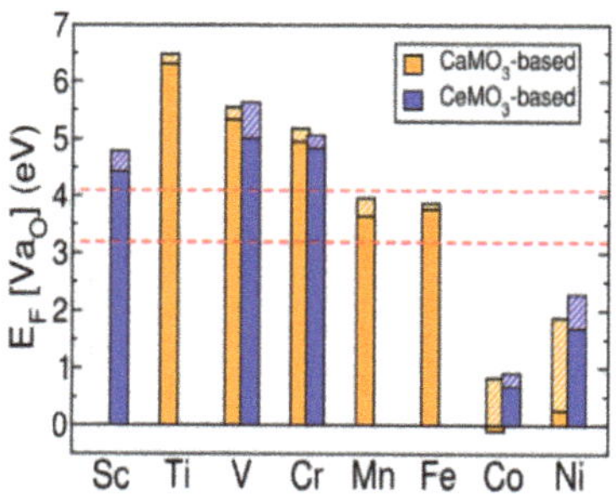

Figure 6. Oxygen vacancy formation energy of $Ca_{0.5}Ce_{0.5}MO_3$ perovskites derived from $CaMO_3$ (orange bars) and $CeMO_3$ (blue bars) oxides. Red dashed lines show the target scope for solar thermochemical hydrogen production (3.2~4.1 eV). Reprinted with permission from Ref. [77], copyright 2020, American Chemical Society.

2.2. Methane Chemical Looping Process

The two-step TCWS cycle approach has substantially decreased the reaction temperature (≤1500 °C) compared to direct thermal splitting of water, but the high reaction temperature still puts forward tough requirements for the reaction equipment. Furthermore, the continuous temperature swing and thermal shock during the cyclic reaction not only reduces the reaction efficiency but also leads to rapid decrease of redox performance, which seriously limits the commercialization of this technology. To this end, recent reports verified that introduction of reducing gas could significantly accelerate the reduction kinetics of OCs, lower the reaction temperature to below 1000 °C, and increase the amount of lattice oxygen desorption, substantially enhancing the hydrogen production efficiency [79–83].

Methane, the main component of natural gas, has been widely studied due to its abundant reserves, high hydrogen-to-carbon ratio, and strong reducibility [84,85]. The two-step TCWS cycle with methane as reducing gas is commonly referred to as methane chemical looping process (Figure 2B). The facile reduction of OCs in methane atmosphere enables a closed reaction loop at much milder conditions, which potentially reduces the energy penalty and improves the economics of this process. Compared with conventional standalone systems that generate hydrogen and electricity separately based on methane, it is assessed that chemical looping hydrogen generation with methane as reduction gas was able to save more than 16% of energy input while reducing beyond 98% of CO_2 emissions, rendering a low cost of \$32.87/MWh for H_2 production [86].

The methane conversion over OCs undergoes different processes, including partial oxidation or total combustion, when modulating the redox properties of these oxides. As for OCs with relatively low reducibility, syngas production is favored by selective methane oxidation ($CH_4 + O_L \rightarrow CO + 2\,H_2$) (Figure 2B1). The reduced OCs can be subsequently recovered by water oxidation with generation of hydrogen ($H_2O \rightarrow O_L + H_2$). The overall reaction is generally referred as chemical looping steam methane reforming (CL-SMR) [87]. Compared with traditional steam methane reforming reaction ($CH_4 + H_2O \rightarrow CO + 3\,H_2$), the CL-SMR process can obtain syngas with H_2/CO ratio of 2, which is suitable for Fischer–Tropsch synthesis and methanol. Furthermore, pure hydrogen is produced separately from syngas, bypassing the gas separation process; more importantly, through the rapid cycling of these OCs in different reactors, coking deposition on the catalyst is greatly inhibited [88–91]. When OCs with high reducibility are applied, methane is fully oxidized to CO_2 and H_2O ($CH_4 + 4\,O_L \rightarrow CO_2 + 2\,H_2O$), which is a strong exothermic reaction, and the released heat can be used for the next step water-splitting reaction, achieving heat self-sufficiency. Another dramatic difference between such a process with CL-SMR is that a third step by air oxidation is normally required to fully regenerate the OCs due to their high oxidation potential (Figure 2B2). This process (three reactors chemical looping M-TRCL) enables hydrogen production with low CO_2 emissions, since the high concentration of CO_2 can be captured by simple condensation [92].

2.2.1. Supported or Doped Iron Oxides

As for methane-driven three step CLWS, the studied OCs are mainly focused on iron oxides, which holds the virtue of low cost and environmentally friendly features [93,94]. However, developing an efficient method to avoid coke formation, which can degrade the hydrogen purity in the water-splitting step, over these OCs during methane atmosphere remains a great challenge [95]. Ku et al. [96] reported that methane dissociation and coke formation occurred over Fe_2O_3/Al_2O_3 OCs due to the generation of the $FeAl_2O_4$ phase, since the newly formed phase displayed poor oxygen mobility and rendered notably decreased oxygen capacity of OC for chemical looping reactions. To address this issue, Xiang et al. [97] proposed that loading Fe_2O_3 to $MgAl_2O_4$ support greatly suppressed the solid-phase reaction between iron oxides and the support, thereby restraining the carbon formation from CH_4 dissociation. Furthermore, they found that K-promoted Fe_2O_3/Al_2O_3 further improved the resistance towards coke deposition by decreasing the reduction activity of oxygen carrier. Li et al. [98] found that a combination of mixed ionic-electronic conductive (MIEC) support of $La_{0.8}Sr_{0.2}FeO_3$ with Fe_2O_3 improved lattice oxygen diffusion from the bulk to surface, which promoted the elimination of carbon during the methane reduction process.

Instead, when syngas is set as aimed products (methane-driven two-step CLWS), the proper OC needs to bear sufficient reactivity for selective methane conversion. There is usually a seesaw effect in reactivity and selectivity due to the proportional relationship between C–H bond activation ability of CH_4 and oxidation ability [99]. For instance, common iron oxides (Fe_2O_3 and Fe_3O_4) are reactive towards methane conversion, but CO_2 production is preferred over these OCs due to the high oxygen capacity [100,101]. To improve the selectivity of syngas at the CL-SMR process, extensive efforts have been made to break the seesaw effect. Fan et al. [102] proposed that decreasing the particle size of Fe_2O_3 to 3–5 nm by confining these oxides in mesoporous SBA-15 (Fe_2O_3@SBA-15) enabled notable enhanced syngas selectivity to above 99%. Corresponding DFT calculations demonstrated that the low coordinated Fe and O atoms of Fe_2O_3 nanoparticles promoted CH_4 activation, cleavage of Fe–O bond, and formation of CO while suppressed the further oxidation to CO_2 (Figure 7). Based on DFT + U calculations, they further analyzed the mechanism for CH_4 partial oxidation over α-Fe_2O_3 and claimed that increasing the oxygen vacancy concentration decreased the activation barriers for breaking the C–H bonds of CH_4 and Fe–C bonds, which contributed to improved reactivity and syngas selectivity (Figure 8) [103]. Recently, Wang's team [104] discovered that Y-modified Fe_2O_3/Al_2O_3 (Y: Fe = 1.5) could produce a new phase of $Y_3Fe_2Al_3O_{12}$ garnet after high temperature treatment (1200 °C), which increased the dispersion of active metal Fe, and changed the coordination environment of oxygen species. Lattice oxygen could coordinate with Fe, and also coordinate with Al and Y (Fe-O-Al, Fe-O-Y), which renders increased oxygen vacancy formation energy from 1.4 eV for bear Fe_2O_3 to 1.997 eV of $Y_3Fe_2Al_3O_{12}$, which substantially avoids excessive oxidation of CH_4 to CO_2 while maintaining CO selectivity as high as 98%.

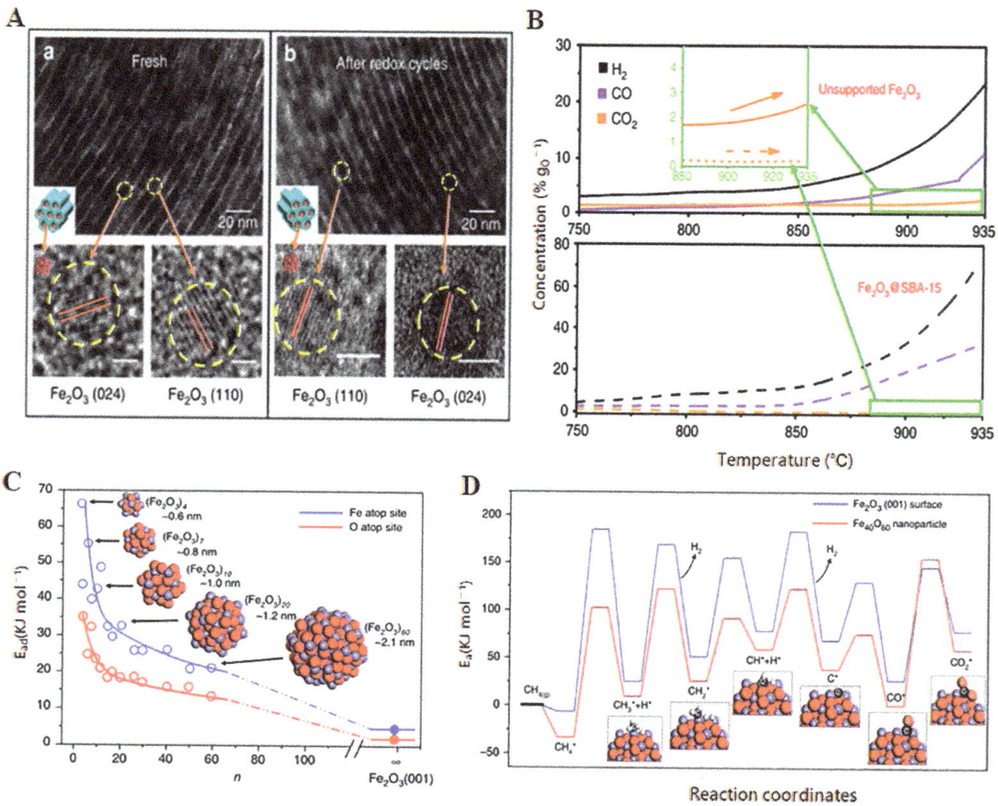

Figure 7. (**A**) Morphological characteristics of Fe_2O_3@SBA-15 OC. (**Aa**) HR–TEM images of fresh Fe_2O_3@SBA-15 and typical Fe_2O_3 nanoparticles (scale bar of 1 nm); (**Ab**) HR–TEM images of Fe_2O_3@SBA-15 after 75 cycles and typical Fe_2O_3 nanoparticles (scale bar of 5 nm); (**B**) Temperature programmed reduction profiles of Fe_2O_3@SBA-15 and unsupported Fe_2O_3; (**C**) Calculated results of CH_4 adsorption on Fe atop site and O atop site of $(Fe_2O_3)_n$ nanoparticles; (**D**) Energy evolution of CH_4 conversion on $Fe_{40}O_{60}$ nanoparticle and Fe_2O_3 (001) surface. Adapted with permission from Ref. [102], copyright 2019, Nature.

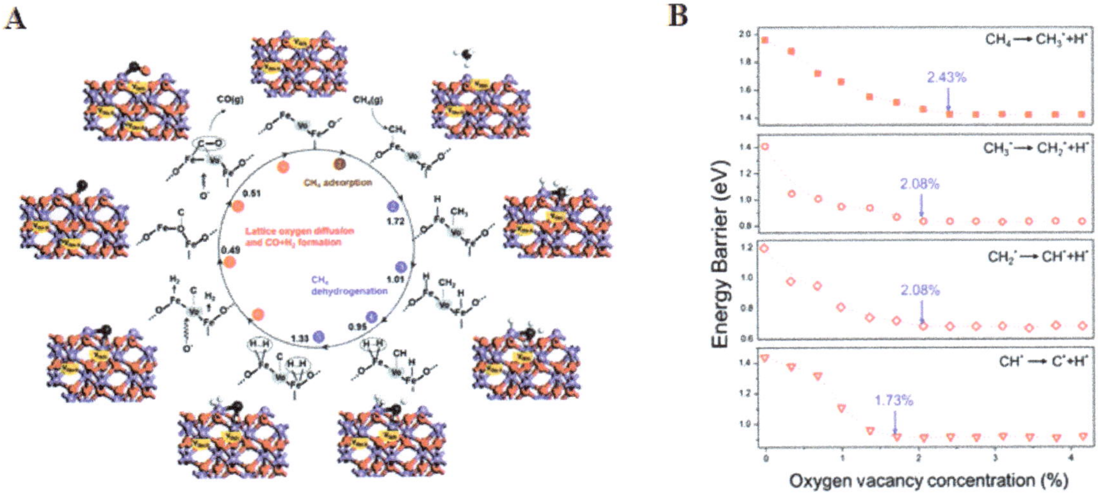

Figure 8. (**A**) Mechanism for CH_4 partial oxidation over α-Fe_2O_3 (001) surface with oxygen vacancies. Inset of the reaction loop is the activation barrier (eV) for each elementary step. V_O represents the surface oxygen vacancy. (**B**) The correlation between activation energy of C–H bond and the oxygen vacancy concentration of α-Fe_2O_3 (001) surface. Reprinted with permission from Ref. [103], copyright 2016, Royal Society of Chemistry.

2.2.2. Supported or Doped Cerium Oxides

Cerium-based OCs have been widely studied for CL-SMR in terms of their high oxygen storage capacity and reliable resistance to carbon deposition [105–107]. However, pure CeO_2 exhibits low reactivity towards methane conversion and is prone to sintering in several cyclic reactions. Therefore, great effort has been applied to improve the reactivity and stability of these OCs by doping foreign cations and constructing composite oxides. Wang et al. [108] discovered that Ni-modified (5 wt%) CeO_2-TiO_2 enabled 100% conversion of CH_4, ca. 16 times greater than that of pure CeO_2-TiO_2 (5.9%), with 85% syngas selectivity at 900 °C. A mechanism study showed that Ni species accounted for the main active centers for CH_4 activation and promoted the reduction of Ce^{4+} (CeO_2-TiO_2) to Ce^{3+} ($Ce_2Ti_2O_7$) (Figure 9B). The synergy between active Ni species and CeO_2-TiO_2/$Ce_2Ti_2O_7$ redox oxides significantly increased the water-splitting performance with hydrogen yield up to 47.0 mL/g (Figure 9). Zhu et al. [87,109,110] prepared a series of Fe-doped cerium oxides ($Ce_{1-x}Fe_xO_{2-\delta}$) for a CL-SMR process and discovered that introduction of iron notably improved the reducibility of OCs, and formation of Ce-Fe-O solid solution ($CeFeO_3$) could enhance the lattice oxygen migration, leading to increased methane conversion. As for the optimized sample (x = 0.5), conversion of methane approached 60% with syngas selectivity exceeding 80% and hydrogen productivity of 83.3 mL/g. Li et al. [111] loaded Ce-Fe-Zr-O oxides (40 wt%) onto inert MgAl-LDO (magnesium–aluminum layered double oxides) and demonstrated that the redox reactivity and stability were greatly improved due to the highly dispersed active components, accelerated reaction between Ce-Fe oxides, and restrained sintering of OCs. Experimental results showed that methane conversion decreased from 58% to 46% in the first 6 cycles and stayed steady in the following 14 cycles with CO selectivity maintained over 80% and hydrogen yield of 1.0 mmol/g. This group further studied the synergy between CeO_2 and 3DOM-$LaFeO_3$ during CL-SMR reactions and found that 10 wt% CeO_2/$LaFeO_3$ displayed the highest yields of syngas (9.94 mmol/g) and hydrogen (3.38 mmol/g), since the interaction of CeO_2-$LaFeO_3$ generated a great number of oxygen vacancies that promoted lattice oxygen diffusion and restricted the carbon deposition [112]. Zhu et al. [113] also found an obvious synergetic effect between CeO_2 and $BaFe_3Al_9O_{19}$ (BF3). The difference was that a new phase of $CeFe_xAl_{1-x}O_3$ was generated during chemical looping reactions, which resulted in promoted diffusion of lattice oxygen and increased methane conversion to ca. 90%. Wei et al. [114] investigated the influence of OCs morphology on the redox reactivity and found that CeO_2/ZrO_2 oxides calcined at low temperatures (450 °C) held stable 3DOM structure with uniform particle size, which effectively enhanced the migration of oxygen species and improved the reactivity for CH_4 partial oxidation and water-splitting. After 10 cycles at 800 °C, a hydrogen yield exceeding 38.89 mL/g was maintained.

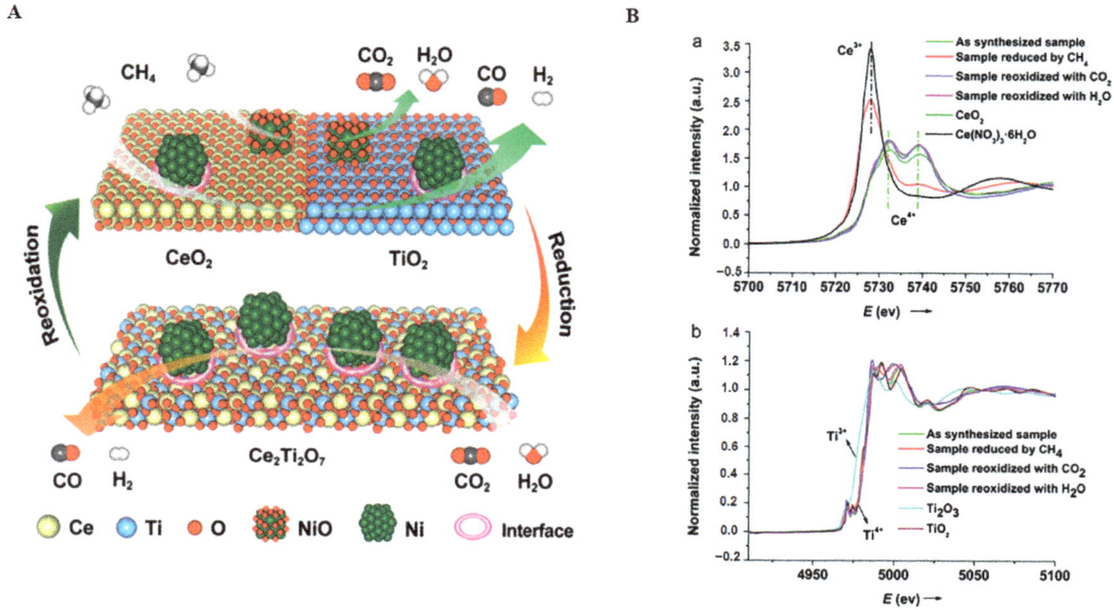

Figure 9. (**A**) Proposed mechanism for MDR–STCDS and MDR–STWS processes over $5Ni/CeO_2–TiO_2$. (**B**) Normalized XANES profiles for $5Ni/CeO_2–TiO_2$ after various steps during isothermal MDR–STCDS and MDR–STWS reactions at (**Ba**) the Ce L_{III}–edge and (**Bb**) the Ti K-edge, together with the spectra of reference compounds. Adapted with permission from Ref. [108], copyright 2019, Royal Society of Chemistry.

2.2.3. Perovskites

Perovskite oxides are promising OCs for CL-SMR due to their excellent redox activity [115,116]. Lee et al. [117] discovered that doping Fe in B-site of $LaCoO_3$ improved lattice oxygen mobility, reactivity for partial oxidation of CH_4, and adsorption and dissociation of H_2O, which enabled high CO selectivity of 92% and H_2 purity of 99.3% during the methane reduction and water-splitting steps, respectively. Gong and colleagues [118,119] discovered that substitution of Ce^{3+} for La^{3+} of $LaFeO_3$ was capable of tuning lattice oxygen activity via modulating the distortion degree of FeO_6 octahedra. When Ce/La ratio reached 1 ($La_{0.5}Ce_{0.5}FeO_3$), the oxygen carrier had the lowest oxygen vacancy formation energy and the highest oxygen mobility, which rendered the best methane-to-syngas and water-splitting performance. Furthermore, the yield of hydrogen (0.6 mmol/g) remained stable during 100 redox cycles (Figure 10), highlighting the outstanding stability of this OC. Recently, Wang et al. [120] proposed that the reactivity of lattice oxygen in $LaFe_{0.8}M_{0.2}O_3$ (M = Al, Fe, Ga, Sc) perovskite was not only influenced by B-site cations, but also greatly affected by the La cations at A-site. It was found that a decline of FeO_6 tilting degree by doping smaller cations at B-site of $LaFe_{0.8}M_{0.2}O_3$ was beneficial to weaken the La–O interaction, thereby improving the mobility of lattice oxygen and methane conversion performance. The researchers [121] further showed that doping Sr and Al at A and B sites of $LaFeO_3$, respectively, significantly promoted the deep reduction of $Fe^{4+/3+}$ cations to Fe^0 by forming Fe^0@oxides intermediate with core-shell structure. Moreover, the core-shell structure formed in methane atmosphere was able to prevent direct contact between CH_4 and Fe^0, which promoted syngas production and suppressed coke deposition over Fe^0 (Figure 11). Li's group [122] discovered that doping of Fe in $LaMnO_3$ greatly enhanced the redox performance due to lattice distortion and increased concentration of oxygen vacancy. Among the screened oxides, $La_{0.85}MnFe_{0.15}O_3$ displayed the highest CO selectivity of ~99% and hydrogen yield of 1.76 mmol/g. Similar results could be observed for Ni or Co modified $LaMnO_3$ [123]. Li et al. [124] designed an OCs of Fe_2O_3@$La_xSr_{1-x}FeO_3$ (LSF) with core-shell structure and found that LSF not only improved the diffusion of lattice oxygen in Fe_2O_3 but also reduced the direct contact between CH_4 and Fe_2O_3, which resulted in highly selective CH_4-to-syngas conversion with selectivity as high as 89%. Zhao et al. [125] found that $LaFe_{1-x}Co_xO_3$ (x = 0, 0.2, 0.4, 0.6, 0.8) with a 3DOM structure promoted the mass transfer and provided more active centers for CH_4 conversion. Moreover, Co doping increased oxygen mobility, which reduced carbon deposition and improved the purity of hydrogen. This group [126] further investigated the synergistic effect of different metals in double perovskite $La_{1.6}Sr_{0.4}FeCoO_6$, and found that CH_4 was prone to be activated on cations with high valence state (Fe^{5+}, Fe^{4+}, and Co^{3+}), while the correspondingly formed Fe^{2+}, Co^0 and the oxygen vacancies provided rich active centers for splitting of H_2O.

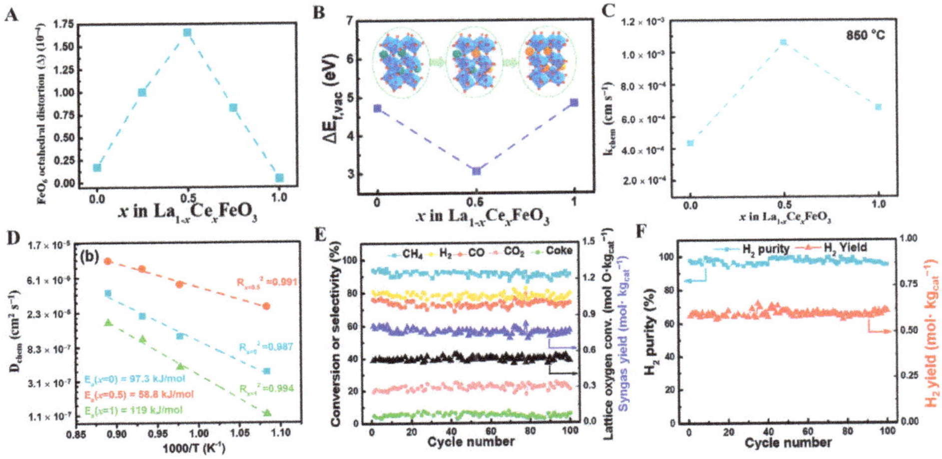

Figure 10. (**A**) Distortion degree (Δ) of FeO_6 octahedra for $La_{1-x}Ce_xFeO_3$ (x = 0, 0.25 0.5, 0.75, 1) and (**B**) calculated oxygen vacancy formation energy ($\Delta E_{f,vac}$) of $La_{1-x}Ce_xFeO_3$ (x = 0, 0.5, 1). Adapted with permission from Ref. [119], copyright 2020, American Chemical Society. (**C**) The oxygen surface exchange coefficient (k_{chem}) and (**D**) bulk diffusion coefficient (D_{chem}) of $La_{1-x}Ce_xFeO_3$ (x = 0, 0.5, 1). Redox performance of (**E**) CH_4 partial oxidation step and (**F**) the H_2O-splitting step over $La_5Ce_5FeO_3$ in CL-SRM at 925 °C. Adapted with permission from Ref. [118], copyright 2020, Elsevier.

2.3. Chemical Looping Water Gas Shift Process

As a low-cost but important platform chemical, CO can be used as a reductant-like methane to drive the CLWS process and produce high-value hydrogen (Figure 2C) [127,128]. Herein, OCs were reduced by CO ($CO + O_L \rightarrow CO_2$) and subsequently regenerated by water ($H_2O \rightarrow O_L + H_2$). This process is generally known as a chemical looping water gas shift (CL-WGS) reaction due to the same overall reaction as that of water–gas shift reaction ($CO + H_2O \rightarrow CO_2 + H_2$). Compared with CL-SMR, the application of CO as a reducing atmosphere can avoid the carbon deposition caused by CH_4 decomposition while the coke formation from the Boudouard reaction ($2CO \rightarrow C + CO_2$) is thermodynamically constrained at a temperature above 800 °C, which inhibited the contamination of hydrogen by carbon monoxide ($C + H_2O \rightarrow CO + H_2$). Furthermore, the separation of CO/CO_2

from H_2O/H_2 in chemical looping reactions also prevents the reverse water–gas shift reaction ($CO_2 + H_2 \rightarrow CO + H_2O$), leading to increased efficiency for water-splitting. Thermodynamic analysis demonstrated that the steam conversion in CL-WGS reaction reached 95% at 800 °C, which is significantly higher than the traditional WGS process [32]. Based on the ASPEN Plus simulation calculation, Zeng et al. [129] found that the thermal efficiency of H_2 production by a chemical looping process, including reduction of OCs by coal and CO (reducer reactor), partial oxidation of OC by steam oxidation to produce hydrogen (oxidizer reactor), and completely recovery of OC by air (combustor), could reach 78% for HHV (higher heating value) efficiency, which was much higher than that of hydrogen production from the traditional coal gasification method.

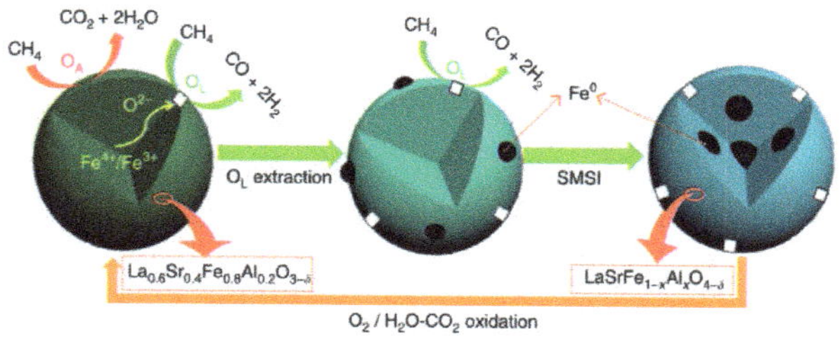

Figure 11. Proposed structural evolution of $La_{0.6}Sr_{0.4}Fe_{0.8}Al_{0.2}O_{3-\delta}$ OC during chemical looping partial oxidation of methane. White boxes represent the coordination unsaturated Fe cations. Reprinted with permission from Ref. [121], copyright 2018, Nature.

As for CL-WGS, the suitable OCs should be reactive for CO oxidation and subsequent water-splitting while possessing high redox stability. Fe_2O_3 has been widely studied due to the features of abundant reserves, favorable thermodynamic properties, and high oxygen-carrying capacity [130]. However, Fe_2O_3 is prone to agglomerate during several redox cycles, rendering rapid decline of hydrogen productivity. Recent works showed that loading of Fe_2O_3 on some inert supports, including ZrO_2 [131], Al_2O_3 [132,133], and $MgAl_2O_4$ [134], were effective in improving the specific surface area of OCs and slowing down the sintering process. Ma et al. [135] investigated in detail the effects of redox inert supports on the reactivity, redox stability, and hydrogen yield of Fe_2O_3/support OCs (support = Al_2O_3, SiO_2, $MgAl_2O_4$, and ZrO_2). It was found that the hydrogen yield of Fe_2O_3/Al_2O_3 and Fe_2O_3/SiO_2 gradually decreased with cyclic numbers, since solid phase reaction between Al_2O_3 or SiO_2 and Fe_2O_3 induced formation of less reactive $FeAl_2O_4$ and $FeSiO_3$, respectively. In contrast, when $MgAl_2O_4$ was utilized, the solid phase interaction between Fe_2O_3 and support was greatly inhibited, leading to improved redox stability and hydrogen yield of 54~93 mL/g. Yuzbasi et al. [136] reported that Na-modified Fe_2O_3/Al_2O_3 could generate a layered Na-β-Al_2O_3 phase to restrict the formation of $FeAl_2O_4$, which increased the redox stability and rendered a stable hydrogen production of 13.3 mmol/g during 15 cycles. Compared with these inert supports, introduction of MIEC support showed a much-pronounced effect in improving the efficiency of hydrogen production by enhancing migration of lattice oxygen and electron conductivity [137]. Ma et al. [138] studied the performance of $A_xCe_{1-x}O_{2-\delta}$ (A = Y, Sm and La) supported Fe_2O_3 for CL-WGS and found that Sm doping ($Fe_2O_3/Ce_{0.7}Sm_{0.2}O_{1.9}$) exhibited the best oxygen anion conductivity and oxygen mobility, which facilitated the conversion of CO and subsequent hydrogen production. In contrast, the reactivity of OCs was negatively correlated with La content due to the decreased oxygen capacity arising from formation of $LaFeO_3$ during the reaction. They further discovered that Zr and Sm co-doped $Fe_2O_3/Ce_{0.8}A_{0.2}O_{2-\delta}$ (A = Zr and Sm) displayed better CO conversion and sintering resistance, rending a hydrogen yield of 54.25 mL/g for water-splitting [139].

Iron spinels that bear features of low cost, high thermal stability, and good oxygen mobility are promising OCs for chemical looping reactions [140,141]. Huang et al. [142] found that $NiFe_2O_4/Al_2O_3$ displayed stable H_2 yield during 20 cycles of CL-WGS, since

Al_2O_3 support significantly inhibited the sintering of $NiFe_2O_4$. Kim et al. [143] also found that Al-modified $NiFe_2O_4$ exhibited good performance for water-splitting due to the formation of a spinel solid solution with Al, which could prevent densification of $NiFe_2O_4$. For the optimized sample with Al content of 3.3 wt%, maximum hydrogen productivity of 8.2 mmol/g was reached. Cui et al. [144] reported that the active Fe could completely exsolve from mixed Zn-Fe-Al spinel in CO atmosphere and resolved into spinel structure by water oxidation, enabling a hydrogen yield of 2.23 mmol/g with nearly 100% conversion of CO. More importantly, an embedded interface structure was formed for the reduced OC, which effectively limited Fe agglomeration and improved the reactivity and stability of the OCs. Based on DFT+U calculations, Xiao et al. [145] demonstrated that the reactivity of binary spinel AFe_2O_4 (A = Co, Cu, Ni, Mn) was mainly related to the concentration of oxygen vacancies (reduction level of OCs). Among these OCs, Co-modified OC was the best candidate for CL-WGS due to its suitable oxygen vacancy formation energy (2.70 eV) and lattice oxygen diffusion energy (1.03 eV), which gave an impressive yield of 11.9 mmol/g with average hydrogen production rate of 0.051 mmol/g·s at 650 °C (Figure 12). They further discovered that the oxygen vacancy formation energy and oxygen diffusion energy further decreased to 1.69 eV and 0.83 eV, respectively, after further doping of Cu in $CoFe_2O_4$ ($Cu_{0.25}Co_{0.25}Fe_{2.5}O_4$), which promoted the reduction kinetics in CO environment and increased the redox efficiency for hydrogen production. As a result, a hydrogen production as high as ~11.9 mmol/g was achieved at a temperature of 550 °C [146].

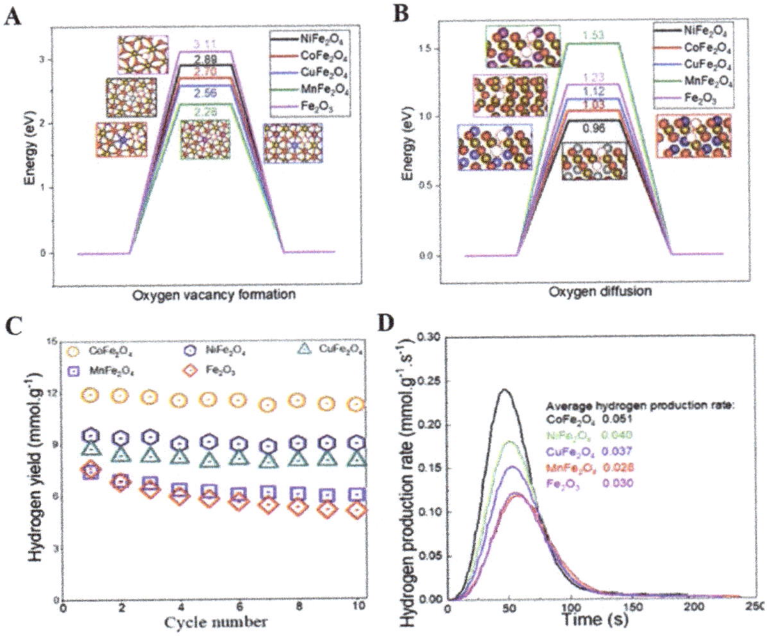

Figure 12. (**A**) The oxygen vacancy formation energy and (**B**) diffusion energy of lattice oxygen in $CoFe_2O_4$ (100), $CuFe_2O_4$ (100), $NiFe_2O_4$ (100), $MnFe_2O_4$ (100), and Fe_2O_3 (001) surfaces. O, Cu, Co, Ni, Mn, and Fe atoms are shown with red, light blue, dark blue, light grey, purple, golden spheres, respectively. (**C**) Profiles of the activation energy and (**D**) the hydrogen yield for different OCs. Reprinted with permission from Ref. [145], copyright 2019, Elsevier.

Brownmillerite ($Ca_2Fe_2O_5$) is well known for high oxygen capacity and excellent thermal stability due to the anion-deficient structure with alternating FeO_6 octahedral and FeO_4 tetrahedral layers [147,148]. Based on the thermodynamic study, Ismail et al. [149]

found that the valence state of Fe in $Ca_2Fe_2O_5$ only displayed Fe^0 and Fe^{3+} without intermediate Fe^{2+} during the redox reactions, and Fe^0 could be directly oxidized to Fe^{3+} by steam, which was suitable for a water-splitting reaction. Chan et al. [150] discovered that the steam conversion over $Ca_2Fe_2O_5$ could reach 75%, which was much higher than that of Fe_2O_3/ZrO_2 (62%). Through thermogravimetric experiments, Sun et al. [151] found that $Ca_2Fe_2O_5$ displayed much faster reaction kinetics than $CaFe_2O_4$ during the CO reduction step, rendering a higher reduction degree of 94.0% (85.5% for $CaFe_2O_4$). Guo et al. [152] employed DFT calculations to explore the mechanisms of CO adsorption, oxidation, and oxygen anion diffusion on the surface of $Ca_2Fe_2O_5$ (010). It was discovered that CO preferred to adsorb on FeO_2- and FeO-terminated surfaces rather than CaO- and OCa-terminated surfaces, while the chemisorption energy of FeO_2-terminated surfaces was considerably greater than that of FeO-terminated surfaces. In addition, it is also proposed that the oxygen anion diffusion at high temperature was the rate-limiting step because the energy barrier of oxygen diffusion was significantly greater than that of the CO oxidation process (Figure 13).

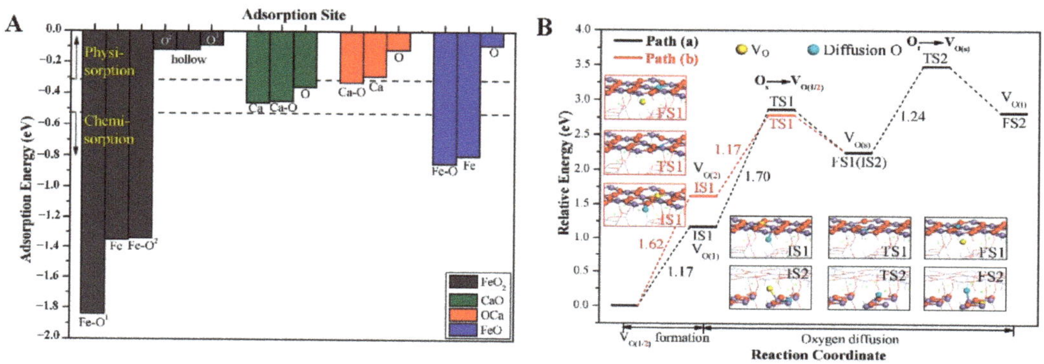

Figure 13. (**A**) Comparison of CO adsorption energies on FeO_2-, CaO-, FeO-, and OCa-terminated $Ca_2Fe_2O_5$ (0 1 0) surfaces. (**B**) The pathways and energy profiles for oxygen diffusion in $Ca_2Fe_2O_5$. Adapted with permission from Ref. [152], copyright 2020, Elsevier.

2.4. Syngas Chemical Looping Process

Syngas that can be produced via various routes, e.g., gasification of biomass, coal, and methane reforming, and represents another promising reducing agent for CLWS [153–155]. In the syngas-promoted CLWS process (Figure 2D), referred to as syngas chemical looping (SCL), syngas was utilized as the reducing gas instead of CO, while subsequent water-splitting and air oxidation (in some cases) were needed to recover the OCs. The focused OCs explored for this process are mainly iron-based oxides.

Fan et al. [33] investigated the feasibility of Ni, Cu, Cd, Co, Mn, Sn, and Fe oxides for SCL through thermodynamic analysis, and found that Fe_2O_3 exhibited the best syngas and steam conversion. The Fe_2O_3 OC displayed stable redox performance during 10 cycles at 600 °C and possessed a low attrition rate of 0.57% during the redox process in entrained flow reactor [156]. When a moving bed reactor was utilized, they found that a syngas conversion of 99.95% with Fe_2O_3 conversion close to 50% was achieved at 900 °C [157]. Dou et al. [158] showed that the syngas generated by in situ decomposition of glycerol was effective for reduction of OCs. Among the investigated OCs with different Ce/Ni/Fe molar ratios (0/0/1; 10/1/100, 10/3/100, 5/1/100, 5/2/100), OC with Ce/Ni/Fe ratio of 10/3/100 displayed the highest H_2 yield of 11.79 mmol/g at 750 °C. Mechanism studies suggested that Ni dopants promoted glycerol conversion while Ce could effectively improve the dispersion of Ni, the sintering resistance, and lattice oxygen mobility, thus substantially boosting the hydrogen yield. Sun et al. [159] investigated the performance of CeO_2-modified Fe_2O_3/Al_2O_3 in SCL process and found that CeO_2 could coordinate with Fe to generate more oxygen vacancies, which promoted the participation of lattice oxygen in the reaction and inhibited the formation of carbon or Fe_3C. The best OC with $Fe_2O_3/CeO_2/Al_2O_3$ ratio of 65/5/30 OCs achieved the best hydrogen productivity, while the overall performance only decreased by 21% during eighth cycles.

Based on thermodynamic analysis and packed bed experiments, Aston et al. [160] found that the conversion of syngas reached more than 99% during the reduction process of $NiFe_2O_4$ and $CoFe_2O_4$. Furthermore, the reduced $NiFe_2O_4$ (NiFe alloy) and $CoFe_2O_4$ (CoFe alloy) was highly reactive towards water-splitting with regeneration of spinel structure, which rendered the $NiFe_2O_4$ and $CoFe_2O_4$, bearing better redox performance than Fe_2O_3 (not recovered by water) for SCL process. He et al. [161] investigated the effect of preparation methods, including solid-state method, coprecipitation method, hydrothermal method, and sol-gel method, on the SCL performance of $NiFe_2O_4$ nanoparticles. It was found that the OC prepared by sol-gel method displayed the best hydrogen yield and highest lattice oxygen recovery degree due to its smaller particle size and porous structure. Furthermore, further mechanical mixing of $NiFe_2O_4$ and inert SiO_2 could effectively restrain the sintering of $NiFe_2O_4$, rendering outstanding redox stability at 850 °C for 20 cycles. Scheffe et al. [162] prepared m-ZrO_2 (surface area of 50 m^2/g) supported $Co_{0.85}Fe_{2.15}O_4$ via atomic layer deposition (ALD) method and found that such OC produced much more hydrogen than that over γ-Fe_2O_3/m-ZrO_2 in 7 cycles. Mechanism studies suggested that the high yield of hydrogen was ascribed to the deep reduction of Fe/Co cations into CoFe alloy, exceeding the reduction degree of γ-Fe_2O_3 (γ-Fe_2O_3→FeO), which improved the oxygen capacity and water-splitting reactivity, and thus, the hydrogen productivity (Figure 14).

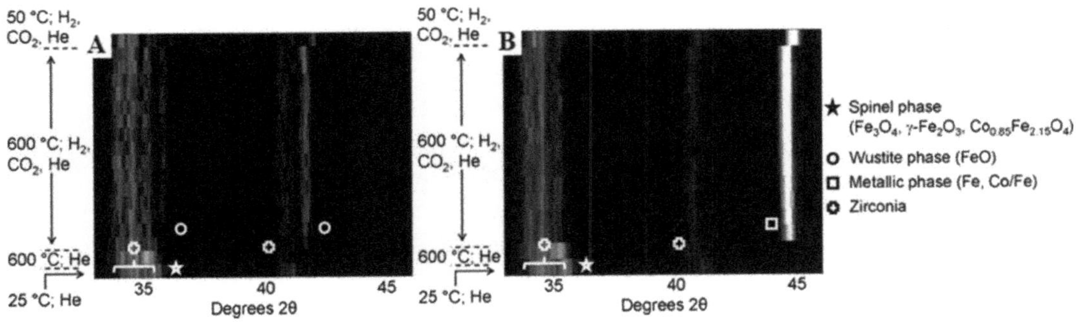

Figure 14. In situ XRD profiles of OCs reduced by 1 vol % H_2/1 vol % CO_2 in He at 600 °C. (**A**) Fe_2O_3 supported on m-ZrO_2 (ALD/20.2% mass loading) and (**B**) $Co_{0.85}Fe_{2.15}O_4$ supported on m-ZrO_2 (ALD/19% mass loading). Chemical reduction for 148 min. Adapted with permission from Ref. [162], copyright 2011, American Chemical Society.

2.5. Photo-Thermochemical Cycle

In 2016, Zhang and coworkers [34] first proposed a strategy of photo-thermochemical cycle (PTC) for CLWS, which explored the photochemical reduction method instead of thermal reduction. This method not only considerably decreased the threshold for abstracting lattice oxygen from OC, slowed down the sintering process, and improved the cyclic stability, but also enabled transformation of solar energy into chemical energy. In a typical photo-thermochemical cycle for water-splitting, the OCs were firstly reduced using ultraviolet and visible light at room temperature with releasing oxygen to generate photo-induced oxygen vacancies, and subsequently recovered by water-splitting via infrared heating to temperature of 500~600 °C with production of hydrogen.

Up to now, the most studied OCs for PTC have been restricted to pure or modified TiO_2 photocatalysts. In 2016, TiO_2 was used as OC for PTC water-splitting to produce hydrogen, and the average hydrogen production of each cycle reached 0.421 mL/g in five consecutive cyclic reactions [34]. However, TiO_2 has a narrow photoresponse range, high recombination rate of photo-induced electron-hole pairs (EHPs), and low reactivity with water. To address these problems, Xu et al. [163] prepared 0.5 wt% Fe/TiO_2 as an OC, which displayed much higher hydrogen yield of 33.36×10^{-3} mmol/g than TiO_2 (18.79×10^{-3} mmol/g). This is ascribed to the lower recombination rate of EHPs (Figure 15A) and wider photoresponse range (Figure 15B) after doping Fe (III) cations. Further DFT calculations showed that introduction of Fe cations could reduce the oxygen vacancy formation energy, improve H_2O adsorption, and enhance the desorption of generated H_2, which substantially boosted the efficiency of H_2 production (Figure 15C). Docao et al. [164] reported that doping Cu in TiO_2 could raise the distortion degree of TiO_6 octahedron, which improved the oxygen mobility and promoted water-splitting at 140 °C, with solar energy conversion efficiency reaching 0.63%.

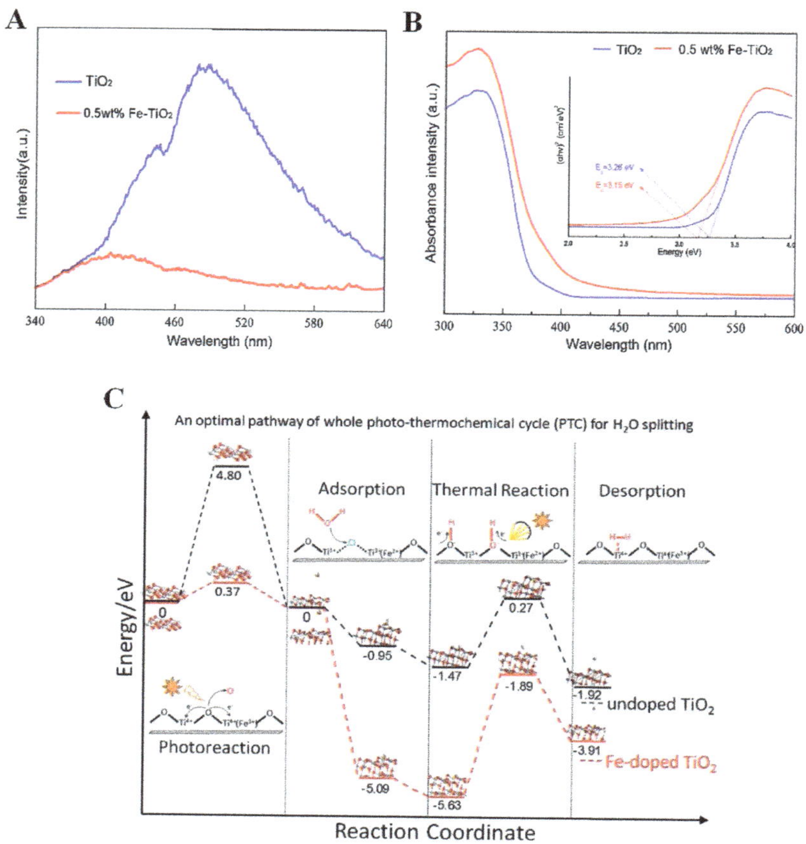

Figure 15. (**A**) PL profiles for TiO_2 and 0.5 wt% Fe-doped TiO_2. (**B**) UV–vis DRS profiles for TiO_2 and 0.5 wt% Fe-doped TiO_2. Inset shows the optical energy band gap. (**C**) The reaction pathway of H_2O splitting during PTC process. Reprinted with permission from Ref. [163], copyright 2017, Elsevier.

Wu et al. [165] modified TiO_2 OCs by doping various transition metals, including Fe, Cu, Co, Ni, and Zn, and discovered that these modified OCs showed higher oxygen vacancy concentration and wider photoresponse range than bare TiO_2. Among these OCs, Cu-doped TiO_2 displayed the highest light absorption rage and the lowest recombination rate of EHPs, which notably improved the driving force for the release of lattice oxygen and hydrogen production. Further decoration of Cu/TiO_2 by Ni could generate more oxygen vacancies and enhance the activity for water-splitting, which exhibited the maximum hydrogen output of 27.01×10^{-3} mmol/g, nearly 36.0 times higher than that of undoped

TiO_2 (0.75×10^{-3} mmol/g) (Figure 16). The follow-up research showed that 0.1~2.0 wt% of Ni-promoted Cu/TiO_2 could anchor nickel oxides (Ni_2O_3 and NiO) around the surface oxygen vacancy while the low valent state NiO showed more pronounced promotion effect towards water-splitting, which rendered the 0.5 wt% Ni-promoted Cu/TiO_2 with highest NiO content, displaying a hydrogen yield (30.6×10^{-3} mmol/g) 40 times higher than that of undoped TiO_2 [166].

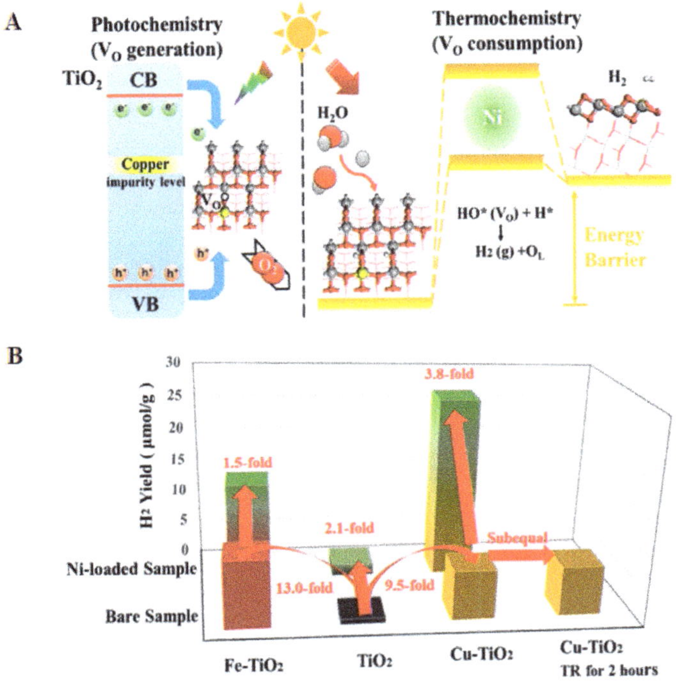

Figure 16. (**A**) Schematic of the reaction mechanism for photochemical and thermochemical processes. (**B**) The average H_2 yield of different catalysts. TR represented "thermochemical reaction". Reprinted with permission from Ref. [165], copyright 2021, Elsevier.

3. Summary and Outlook

Chemical looping water-splitting is promising for sustainable hydrogen production due to the virtue of decoupling a one-step reaction into two or three spatially separated reactions, which greatly simplifies the gas separation process and avoids the harsh conditions for direct water decomposition. To date, different processes have been developed, including a two-step thermochemical water-splitting (TCWS) cycle, methane chemical looping process, chemical looping water gas shift (CL-WGS) cycle, syngas chemical looping (SCL) process, and photo-thermochemical cycle (PTC), with attempts to reduce the energy penalty and CO_2 emissions by altering the method to abstract the lattice oxygen from the OCs, wherein the key lies in the manufacture of suitable OCs. Over the past decades, thousands of metal oxides, such as iron-based metal oxides, zinc-based metal oxides, cerium-based metal oxides, titanium-based metal oxides, and composite metal oxides (perovskite, spinel and garnets, etc.), have been prepared for chemical looping reactions. The performance summary of OCs in different processes of CLWS is summarized in Table 1.

Table 1. Performance summary of OCs in different processes in CLWS.

Hydrogen Production Processes	OCs Type	Cyclic Reaction	OCs	Reaction Temperature (°C; Red/Ox)	H_2 Yield (mmol/g)	Ref.
Two-step TCWS cycle	Ferrites ($M_xFe_{3-x}O_4$)	$M_xFe_{3-x}O_4 \rightarrow xMO + (3-x)FeO + 0.5O_2$; $xMO + (3-x)\ FeO + H_2O \rightarrow M_xFe_{3-x}O_4 + H_2$	$NiFe_2O_4$	1200/800	0.0197 *	[48]
			$Co_{0.67}Fe_{2.33}O_4/ZrO_2$	1300~1400/1000	0.357~0.535	[49]
			$NiFe_2O_4$	1400/1000	~0.446	[50]
			$NiFe_2O_4$/m-ZrO_2	1400/1000	0.535~0.714	[50]
	ZnO_2	$ZnO \rightarrow Zn + O_2$; $Zn + H_2O \rightarrow ZnO + H_2$	ZnO	-	-	
	Cerium oxide ($Ce_xM_{1-x}O_2$)	$Ce_xM_{1-x}O_2 \rightarrow Ce_xM_{1-x}O_{2-\delta} + 0.5\delta\ O_2$; $Ce_xM_{1-x}O_{2-\delta} + H_2O \rightarrow \delta H_2O \rightarrow Ce_xM_{1-x}O_2 + \delta H_2$	CeO_2	1350/800	0.032	[31]
			CeO_2	1500/500	0.142 *	[66]
			$Ce_{0.9}Hf_{0.1}O_2$	1500/500	0.2 *	[66]
			$Ce_{0.9}Sc_{0.1}O_{1.95}$	1500/500	0.181 *	[66]
			$Ce_{0.75}Zr_{0.25}O_2$	1400/1050	0.297	[69]
			1% Gd/CeO_2-ZrO_2	1400/1050	0.338 *	[71]
			CeO_2-0.15SnO_2	1400/800	0.321 *	[72]
			CeO_2	1350/850	0.05	[76]
			CeO_2	1500/1000	0.22	[76]
	ABO_3	$ABO_3 \leftrightarrow ABO_{3-\delta} + 0.5\delta O_2$; $ABO_{3-\delta} + \delta H_2O \rightarrow ABO_3 + \delta H_2$	$LaGa_{0.4}Co_{0.6}O_{3-\delta}$	1350/800	0.478	[31]
			$Sr_{0.4}La_{0.6}Mn_{0.6}Al_{0.4}O_3$	1350/1000	0.307	[75]
			$Sr_{0.4}La_{0.6}Mn_{0.6}Al_{0.4}O_{3-\delta}$	1350/850	0.194	[76]
			$BaCe_{0.25}Mn_{0.75}O_{3-\delta}$	1350/850	0.140	[76]
			$CaTi_{0.5}Mn_{0.5}O_{3-\delta}$	1350/1150	0.446	[78]
Methane chemical looping process: Three reactors chemical looping (M-TRCL)	Iron oxides (Fe_2O_3)	$Fe_2O_3 + CH_4 \rightarrow FeO + 3CO_2 + 2H_2O$; $3FeO + H_2O \rightarrow Fe_3O_4 + H_2$; $2Fe_3O_4 + 0.5O_2 \rightarrow 3Fe_2O_3$	K promoted-Fe_2O_3/Al_2O_3	850/850	-	[97]
			$Fe_2O_3/La_{0.8}Sr_{0.2}FeO_3$	900/900	-	[98]
Methane chemical looping process: CL-SMR	Cerium oxide ($Ce_xM_{1-x}O_2$)	$Ce_xM_{1-x}O_2 + \delta CH_4 \rightarrow Ce_xM_{1-x}O_{2-\delta} + \delta CO + 2\delta H_2$; $Ce_xM_{1-x}O_{2-\delta} + \delta H_2O \rightarrow Ce_xM_{1-x}O_2 + \delta H_2$	$Ce_{0.5}Fe_{0.5}O_{2-\delta}$	850/850	3.71	[87]
			5Ni/CeO_2-TiO_2	900/900	2.098 *	[108]
			CeO_2-Fe_2O_3 (Ce/Fe = 7/3)	850/700	1.116	[109]
			40 wt% Ce-Fe-Zr-O/MgAl-LDO	850/850	1.0	[111]
	ABO_3	$ABO_3 + \delta CH_4 \rightarrow ABO_{3-\delta} + \delta CO + 2\delta H_2$; $ABO_{3-\delta} + \delta H_2O \rightarrow ABO_3 + \delta H_2$	CeO_2-$LaFeO_3$	800/800	3.38	[112]
			CeO_2/ZrO_2	800/800	1.736	[114]
			$LaCo_{0.6}Fe_{0.4}O_3$	700/700	2.22	[117]
			$La_{0.5}Ce_{0.5}FeO_3$	925/925	0.6	[118]
			$La_{0.85}MnFe_{0.15}O_3$	850/850	1.76	[122]
CL-WGS	Iron oxides (Fe_2O_3)	$Fe_2O_3 + 3CO \rightarrow 2Fe + 3CO_2$; $3Fe + 4H_2O \rightarrow Fe_3O_4 + 4H_2$; $2Fe_3O_4 + 0.5O_2 \rightarrow 3Fe_2O_3$	Fe_2O_3/SiO_2	900/900	~0.745 *	[135]
			Fe_2O_3/Al_2O_3	900/900	~0.982 *	[135]
			Fe_2O_3/YSZ	900/900	~2.232 *	[135]
			Fe_2O_3/ZrO_2	900/900	~2.977 *	[135]
			$Fe_2O_3/MgAl_2O_4$	900/900	2.41~4.152 *	[135]
			Na-modified Fe_2O_3/Al_2O_3	800/800	~13.3	[136]

Table 1. *Cont.*

Hydrogen Production Processes	OCs Type	Cyclic Reaction	OCs	Reaction Temperature (°C; Red/Ox)	H_2 Yield (mmol/g)	Ref.
SCL process	Iron spinels ($M_xFe_{3-x}O_4$)	$M_xFe_{3-x}O_4 + (2-x)H_2 + CO \rightarrow xMO + (3-x)Fe + CO_2 + (3-x)H_2O$; $Fe + H_2O \rightarrow Fe_3O_4 + H_2$; $xMO + (3-x)/3Fe_3O_4 + 0.5xO_2 \rightarrow M_xFe_{3-x}O_4$	Fe_2O_3/ZrO_2	850/850	~7.143 *	[137]
			Fe_2O_3/CeO_2	850/850	~5.80 *	[137]
			$Fe_2O_3/Ce_{0.6}Sm_{0.15}Zr_{0.25}O_{1.925}$	850/850	2.422 *	[139]
			3.3 wt% Al-modified $NiFe_2O_4$	800/800	8.2	[143]
			$ZnFeAlO_x$	850/850	2.23	[144]
			$CoFe_2O_4$	650/650	11.9	[145]
			$Cu_{0.25}Co_{0.25}Fe_{2.5}O_4$	550/550	~11.9	[146]
	Brownmillerite ($Ca_2Fe_2O_5$)	$Ca_2Fe_2O_5 + 3CO \rightarrow 2CaO + 2Fe + 3CO_2$; $2CaO + 2Fe + 3H_2O \rightarrow Ca_2Fe_2O_5 + 3H_2$	$Ca_2Fe_2O_5$	900/900	-	[149]
	Iron oxides (Fe_2O_3)	$Fe_2O_3 + 2H_2 + CO \rightarrow 2Fe + CO_2 + 2H_2O$ $3Fe + 4H_2O \rightarrow Fe_3O_4 + 4H_2$; $2Fe_3O_4 + 0.5O_2 \rightarrow 3Fe_2O_3$	Fe-Ce-Ni based OCs (Ce/Ni/Fe ratio of 10/3/100)	750/750	11.79	[158]
			Fe_2O_3	600/600	1.14	[162]
	Iron spinels ($M_xFe_{3-x}O_4$)	$M_xFe_{3-x}O_4 + CO \rightarrow xMO + (3-x)Fe + CO_2$; $xMO + (3-x)FeO + H_2O \rightarrow M_xFe_{3-x}O_4 + H_2$;	$Co_{0.85}Fe_{2.15}O_4/m\text{-}ZrO_2$	600/600	3.944	[162]
PTC	TiO_2	$2TiO_2 \rightarrow Ti_2O_3 + 0.5O_2$; $Ti_2O_3 + H_2O \rightarrow 2TiO_2 + H_2$ $(hv + TiO_2 \rightarrow e^- + h^+$; $e^- + Ti^{4+} \rightarrow Ti^{3+}$; $h^+ + O_2 \rightarrow O_2 + Vo$; $Vo + H_2O \rightarrow Vo(H_2O)$ $Ti^{3+}\text{-}Vo\ (H_2O)\text{-}Ti^{3+} \rightarrow Ti^{4+}\text{-}O\text{-}Ti^{4+} + H_2)$	TiO_2	room temperature/500-600	0.0188 *	[34]
			0.5 wt% Fe/TiO_2	room temperature/600	0.0334	[163]
			TiO_2	room temperature/350	0.75×10^{-3}	[165]
			Fe/TiO_2	room temperature/350	9.73×10^{-3}	[165]
			Cu/TiO_2	room temperature/350	7.13×10^{-3}	[165]
			Ni-promoted Cu/TiO_2	room temperature/350	0.027	[165]
			0.5% Ni-promoted Cu/TiO_2	room temperature/350	0.0306	[166]

* The unit of raw hydrogen production data is "mL/g", which can be converted into "mmol/g" by the formula PV = nRT with hypothesis of STP condition.

For the two-step TCWS cycle, various OCs, such as iron oxides, zinc oxides, cerium oxides, and perovskite have been widely studied. Among them, cerium oxides have attracted particular attention due to their high structural stability and water-splitting conversion. However, the relatively low reduction degree during the redox cycle, rendering a low hydrogen yield (0.72~7.58 mL/g), greatly hampered its practical applications. Recent work showed that perovskite oxides are promising candidates for two-step TCWS with hydrogen yields of up to 3.13 to 10.71 mL/g, since their redox properties can be facilely modulated by tuning the A/B sites. This gives a clue that constructing composite materials to adjust the redox potential suitable for oxygen desorption and water-splitting should be the key for improving the hydrogen productivity. Furthermore, the sintering problems, thermal shocks, and long single redox cycle time, induced by application of high thermal reduction temperature and low water-splitting temperature, represents another huge obstacle for commercialization of this process. Therefore, introducing support to stabilize the OCs and exploring isothermal redox reactions would be effective in improving the process efficiency for hydrogen production.

Compared to the two-step TCWS cycle, introducing reducing gas, such as methane, carbon monoxide, and syngas, to reduce the oxygen carrier is capable of notably decreasing the reaction temperature to below 1000 °C while enhancing the available oxygen capacity, which significantly decreases the energy consumption, slows down the sintering of OCs, and improve the yield of hydrogen to 13.44~267.63 mL/g. As for methane-driven reduction, valuable syngas with H_2/CO ratio of two for Fischer–Tropsch synthesis and methanol production is produced when a suitable OC is selected. Upon OCs with high reducibility applied, the reducing gaseous can be totally combusted with generation of high concentration CO_2 (and H_2O). All these processes greatly inhibit the side reactions and reduce the burden for gas separation and CO_2 caption, rendering improved efficiency and lowered cost for hydrogen production. Among the investigated OCs, iron-based oxides are among the most studied materials due to the virtues of low-cost, environmentally benign features with the high capacity to donate lattice oxygen by varying the valence state of Fe cations. Herein, the relatively low reactivity of iron cations towards conversion of these reducing agents and general trade-off between selectivity and hydrogen yield represent the main challenge for widespread application of these techniques. For CL-SMR with the aim of methane partial oxidation, the reactivity for C–H bond cleavage can be improved by introducing reactive promoters (e.g., Ni and Cu), creating oxygen vacancies, and engineering the morphology of OCs to provide more active centers. To break the trade-off between selectivity and hydrogen productivity, the key lies in a balance between oxygen supply and methane decomposition, which sustainably restrains the coke formation and increases the oxygen capacity for water-splitting. Based on available results, constructing OCs with core-shell structure, composite oxides, and poisoning active Fe^0 by forming an alloy or covering with oxides are effective methods to inhibit the carbon deposition. When total combustion of reducing agents (processes of methane-driven three step CLWS, CL-WGS, and SCL) is desired, the reduction degree of OC and hydrogen productivity can be improved by introducing MIEC support, alkali promoters, and some active metals (e.g., Co and Cu), which contribute to promote oxygen mobility and oxygen capacity.

In the case of PTC cycle, the reduction of OCs was realized at ambient temperature with hydrogen production at low temperature (140~600 °C), which greatly suppressed the sintering of OCs. More importantly, the solar energy can be well converted to chemical energy, implementing production of green hydrogen. Up to now, the studied OCs are still limited to bare or modified titanium oxide. Although much progress has been achieved in restraining the recombination rate of EHPs and enhancing the reactivity towards water-splitting, which resulted in improved redox performance, the hydrogen yield (0.0168~0.685 mL/g) is still lower than the above CLWS process, because the photo-driven reduction mainly drives the surface lattice oxygen desorption of OCs, and the amount of desorbed oxygen desorption is limited. In addition to exploring advanced OCs that release more oxygen in the photothermal reduction process, it is anticipated that a combination of suitable reducing gas and photothermal reduction would be promising to significantly enhance the efficiency of hydrogen production.

Over the past 30 years, great effort has been made to develop various OCs by engineering the composition of oxides, including simple metal oxides (cerium oxides, iron oxides, titanium oxides, etc.), composite metal oxides (e.g., perovskite, spinel and garnets), and mixed oxides (e.g., supported OCs, Fe_2O_3-CeO_2, and CeO_2-TiO_2), which greatly improved the redox performance in CLWS processes [31,48,58,97,104,108,110,125,145,167]. B In addition, modulating the microstructure during the redox reactions also exerts a big influence on the performance due to the modulating of surface active centers, concentration of oxygen vacancies, or metal–oxygen bond strength. For example, Li et al. [124] prepared a kind of Fe_2O_3@$La_xSr_{1-x}FeO_{3-\delta}$ (LSF) OC with core-shell structure and showed that the LSF shell with high methane-to-syngas selectivity could reduce the contact between methane and Fe_2O_3, which enables a high CO selectivity above 98%, while the Fe_2O_3 core can donate oxygen to the LSF shell that greatly enhances the oxygen capacity of this OC. Qian et al. [78] found that the crystal phase of $CaTi_{0.5}Mn_{0.5}O_{3-\delta}$ OCs could change from an orthogonal phase to a cubic phase during the thermal reduction process, which greatly promoted the desorption of lattice oxygen. Furthermore, the OCs with cubic phase integrate the virtues of high entropy and medium enthalpy, which improves the reactivity for hydrogen production via water-splitting. Wang et al. [121] demonstrated that a specific Fe^0@ oxide mediate with core-shell structure was formed when $La_{0.6}Sr_{0.4}Fe_{0.8}Al_{0.2}O_{3-\delta}$ was reduced by methane, which notably reduced the direct contact between methane and generated Fe^0 species, restrained the coke deposition, and improved the syngas productivity. Cui et al. [144] indicated that the active component (Fe) of $ZnFeAlO_x$ OCs reversibly exsolved and dissolved from the interface structure of Zn-Fe-Al-O mixed spinel during a CL-WGS process, which can effectively hinder the migration and agglomeration of iron ions and avoid sintering of materials. These results clearly indicate that adjusting the structure changes during redox reactions represents an effective method to modulate the redox performance for CLWS processes.

At present, selection of OCs for CLWS reactions mainly relies on screening method. This is mainly ascribed to the harsh reaction conditions and dynamic structural evolution during redox reactions, which poses a huge challenge for comprehensively understanding the reaction mechanism and designing advanced OCs for CLWS reactions. Future studies should pay more attention to establish a more precise structure–function relationship with the help of in situ characterization, theoretical calculations, and thermodynamic analysis to provide a theoretical basis and development direction for the design of new efficient long-life OCs. Furthermore, according to the pioneering studies, the research focus of OCs is gradually transferred from simple metal oxides to composite oxides (e.g., perovskite) and mixed oxides due to the feasibility of modulating the redox properties by altering the composition of OCs or synergy between different oxides, which bypasses the shortcomings of single metal oxides, and improves the performance of hydrogen production. Therefore, exploring composite oxides to precisely control the metal–oxygen bond strength and mixed oxides to integrate the advantages of different oxides would be an effective strategy for further improving the redox performance of OCs.

Author Contributions: Conceptualization, writing—original draft preparation, W.C.; writing—review and editing, Y.H. (Yue Hu) and W.X.; supervision, writing—review and editing, C.H.; investigation, H.C.; investigation, J.H.; resources, Y.H. (Yujia Han); supervision, writing—review and editing, Y.Z.; resources, X.M.; supervision, X.W. All authors have read and agreed to the published version of the manuscript.

Funding: National Natural Science Foundation of China (No. 21706254, 21978239, and 22178337), Joint Fund of the Yulin University and the Dalian National Laboratory for Clean Energy (No. 2021012), and the Cyrus Tang Foundation are gratefully acknowledged.

Conflicts of Interest: The authors declare no conflict of interest.

References

1. Bockris, J. The origin of ideas on a Hydrogen Economy and its solution to the decay of the environment. *Int. J. Hydrogen Energy* **2002**, *27*, 731–740. [CrossRef]
2. Safari, F.; Dincer, I. A review and comparative evaluation of thermochemical water splitting cycles for hydrogen production. *Energy Convers. Manag.* **2020**, *205*, 112182. [CrossRef]
3. Alves, H.J.; Bley Junior, C.; Niklevicz, R.R.; Frigo, E.P.; Frigo, M.S.; Coimbra-Araújo, C.H. Overview of hydrogen production technologies from biogas and the applications in fuel cells. *Int. J. Hydrogen Energy* **2013**, *38*, 5215–5225. [CrossRef]
4. Dincer, I.; Acar, C. Review and evaluation of hydrogen production methods for better sustainability. *Int. J. Hydrogen Energy* **2015**, *40*, 11094–11111. [CrossRef]
5. Mahant, B.; Linga, P.; Kumar, R. Hydrogen Economy and Role of Hythane as a Bridging Solution: A Perspective Review. *Energy Fuels* **2021**, *35*, 15424–15454. [CrossRef]
6. Navarro, R.M.; Pena, M.A.; Fierro, J.L. Hydrogen production reactions from carbon feedstocks: Fossil fuels and biomass. *Chem. Rev.* **2007**, *107*, 3952–3991. [CrossRef]
7. Go, K.S.; Son, S.R.; Kim, S.D.; Kang, K.S.; Park, C.S. Hydrogen production from two-step steam methane reforming in a fluidized bed reactor. *Int. J. Hydrogen Energy* **2009**, *34*, 1301–1309. [CrossRef]
8. Lee, K.B.; Beaver, M.G.; Caram, H.S.; Sircar, S. Novel Thermal-Swing Sorption-Enhanced Reaction Process Concept for Hydrogen Production by Low-Temperature Steam–Methane Reforming. *Ind. Eng. Chem. Res.* **2007**, *46*, 5003–5014. [CrossRef]
9. Wang, J.; Sakanishi, K.; Saito, I.; Takarada, T.; Morishita, K. High-Yield Hydrogen Production by Steam Gasification of HyperCoal (Ash-Free Coal Extract) with Potassium Carbonate: Comparison with Raw Coal. *Energy Fuels* **2005**, *19*, 2114–2120. [CrossRef]
10. Wang, Z.; Li, L.; Zhang, G. Life cycle greenhouse gas assessment of hydrogen production via chemical looping combustion thermally coupled steam reforming. *J. Clean. Prod.* **2018**, *179*, 335–346. [CrossRef]
11. Li, J.; Cheng, W. Comparative life cycle energy consumption, carbon emissions and economic costs of hydrogen production from coke oven gas and coal gasification. *Int. J. Hydrogen Energy* **2020**, *45*, 27979–27993. [CrossRef]
12. Zhang, H.; Sun, Z.; Hu, Y.H. Steam reforming of methane: Current states of catalyst design and process upgrading. *Renew. Sust. Energy Rev.* **2021**, *149*, 111330. [CrossRef]
13. Siefert, N.S.; Shekhawat, D.; Litster, S.; Berry, D.A. Steam–Coal Gasification Using CaO and KOH for in Situ Carbon and Sulfur Capture. *Energy Fuels* **2013**, *27*, 4278–4289. [CrossRef]
14. Kodama, T.; Gokon, N. Thermochemical cycles for high-temperature solar hydrogen production. *Chem. Rev.* **2007**, *107*, 4048–4077. [CrossRef]
15. Kogan, A. Direct solar thermal splitting of water and on-site separation of the products-II. Experimental feasibility study. *Int. J. Hydrogen Energy* **1998**, *23*, 89–98. [CrossRef]
16. Bilgen, E.; Ducarroir, M.; Foex, M.; Sibieude, F.; Trombe, F. Use of solar energy for direct and two-step water decomposition cycles. *Int. J. Hydrogen Energy* **1977**, *2*, 251–257. [CrossRef]
17. Marwat, M.A.; Humayun, M.; Afridi, M.W.; Zhang, H.; Abdul Karim, M.R.; Ashtar, M.; Usman, M.; Waqar, S.; Ullah, H.; Wang, C.; et al. Advanced Catalysts for Photoelectrochemical Water Splitting. *ACS Appl. Energy Mater.* **2021**, *4*, 12007–12031. [CrossRef]
18. Zhou, R.; Li, D.; Qu, B.; Sun, X.; Zhang, B.; Zeng, X.C. Rutile $TiO_2(011)$-2 $\times$ 1 Reconstructed Surfaces with Optical Absorption over the Visible Light Spectrum. *ACS Appl. Mater. Interfaces* **2016**, *8*, 27403–27410. [CrossRef] [PubMed]
19. Anpo, M.; Li, J.; Liu, W.; Qi, W.; Song, F. Hydrogen production technology by electrolysis of water and its application in renewable energy consumption. In Proceedings of the 3rd International Conference on Energy Resources and Sustainable Development (ICERSD 2020), Incheon, Republic of Korea, 13–14 January 2020.
20. Laguna-Bercero, M.A. Recent advances in high temperature electrolysis using solid oxide fuel cells: A review. *J. Power Sources* **2012**, *203*, 4–16. [CrossRef]
21. Xue, F.; Su, J.; Li, P.; Zhang, Y. Application of Proton Exchange Membrane Electrolysis of Water Hydrogen Production Technology in Power Plant. In Proceedings of the 3rd International Conference on Air Pollution and Environmental Engineering, Xi'an, China, 28–29 September 2020.
22. Megía, P.J.; Vizcaíno, A.J.; Calles, J.A.; Carrero, A. Hydrogen Production Technologies: From Fossil Fuels toward Renewable Sources. A Mini Review. *Energy Fuels* **2021**, *35*, 16403–16415. [CrossRef]
23. Li, D.; Xu, R.; Li, X.; Li, Z.; Zhu, X.; Li, K. Chemical Looping Conversion of Gaseous and Liquid Fuels for Chemical Production: A Review. *Energy Fuels* **2020**, *34*, 5381–5413. [CrossRef]
24. Moghtaderi, B. Review of the Recent Chemical Looping Process Developments for Novel Energy and Fuel Applications. *Energy Fuels* **2011**, *26*, 15–40. [CrossRef]
25. De Vos, Y.; Jacobs, M.; Van Der Voort, P.; Van Driessche, I.; Snijkers, F.; Verberckmoes, A. Development of Stable Oxygen Carrier Materials for Chemical Looping Processes—A Review. *Catalysts* **2020**, *10*, 926. [CrossRef]
26. Funk, J.E.; Reinstrom, R.M. Energy Requirements in Production of Hydrogen from Water. *Ind. Eng. Chem. Process* **1966**, *5*, 336–342. [CrossRef]
27. Krenzke, P.T.; Fosheim, J.R.; Davidson, J.H. Solar fuels via chemical-looping reforming. *Sol. Energy* **2017**, *156*, 48–72. [CrossRef]
28. Kang, K.-S.; Kim, C.-H.; Bae, K.-K.; Cho, W.-C.; Kim, S.-H.; Park, C.-S. Oxygen-carrier selection and thermal analysis of the chemical-looping process for hydrogen production. *Int. J. Hydrogen Energy* **2010**, *35*, 12246–12254. [CrossRef]

29. Kim, Y.; Lim, H.S.; Kim, H.S.; Lee, M.; Lee, J.W.; Kang, D. Carbon dioxide splitting and hydrogen production using a chemical looping concept: A review. *J. CO_2 Util.* **2022**, *63*, 102139. [CrossRef]
30. Zhao, L.; Dou, B.; Zhang, H.; Wang, Z. Oxygen carriers for chemical-looping water splitting to hydrogen production: A critical review. *Carbon Capture Sci. Technol.* **2021**, *1*, 100006. [CrossRef]
31. Chen, Z.; Jiang, Q.; Cheng, F.; Tong, J.; Yang, M.; Jiang, Z.; Li, C. Sr- and Co-doped $LaGaO_{3-\delta}$ with high O_2 and H_2 yields in solar thermochemical water splitting. *J. Mater. Chem. A* **2019**, *7*, 6099–6112. [CrossRef]
32. de Leeuwe, C.; Hu, W.; Evans, J.; von Stosch, M.; Metcalfe, I.S. Production of high purity H_2 through chemical-looping water–gas shift at reforming temperatures—The importance of non-stoichiometric oxygen carriers. *Chem. Eng. J.* **2021**, *423*, 130174. [CrossRef]
33. Gupta, P.; Velazquez-Vargas, L.G.; Fan, L.-S. Syngas Redox (SGR) Process to Produce Hydrogen from Coal Derived Syngas. *Energy Fuels* **2007**, *21*, 2900–2908. [CrossRef]
34. Zhang, Y.; Chen, J.; Xu, C.; Zhou, K.; Wang, Z.; Zhou, J.; Cen, K. A novel photo-thermochemical cycle of water-splitting for hydrogen production based on TiO_{2-x}/TiO_2. *Int. J. Hydrogen Energy* **2016**, *41*, 2215–2221. [CrossRef]
35. Li, F.; Fan, L.-S. Clean coal conversion processes—Progress and challenges. *Energy Environ. Sci.* **2008**, *1*, 248–267. [CrossRef]
36. Adanez, J.; Abad, A.; Garcia-Labiano, F.; Gayan, P.; de Diego, L.F. Progress in Chemical-Looping Combustion and Reforming technologies. *Prog. Energy Combust. Sci.* **2012**, *38*, 215–282. [CrossRef]
37. Zhu, X.; Imtiaz, Q.; Donat, F.; Müller, C.R.; Li, F. Chemical looping beyond combustion—A perspective. *Energy Environ. Sci.* **2020**, *13*, 772–804. [CrossRef]
38. Zeng, L.; Cheng, Z.; Fan, J.A.; Fan, L.-S.; Gong, J. Metal oxide redox chemistry for chemical looping processes. *Nat. Rev. Chem.* **2018**, *2*, 349–364. [CrossRef]
39. Qasim, M.; Ayoub, M.; Ghazali, N.A.; Aqsha, A.; Ameen, M. Recent Advances and Development of Various Oxygen Carriers for the Chemical Looping Combustion Process: A Review. *Ind. Eng. Chem. Res.* **2021**, *60*, 8621–8641. [CrossRef]
40. Tian, M.; Wang, C.; Han, Y.; Wang, X. Recent Advances of Oxygen Carriers for Chemical Looping Reforming of Methane. *ChemCatChem* **2021**, *13*, 1615–1637. [CrossRef]
41. Li, D.; Xu, R.; Gu, Z.; Zhu, X.; Qing, S.; Li, K. Chemical-Looping Conversion of Methane: A Review. *Energy Technol.* **2019**, *8*, 1900925. [CrossRef]
42. Voitic, G.; Hacker, V. Recent advancements in chemical looping water splitting for the production of hydrogen. *RSC Adv.* **2016**, *6*, 98267–98296. [CrossRef]
43. Luo, M.; Yi, Y.; Wang, S.; Wang, Z.; Du, M.; Pan, J.; Wang, Q. Review of hydrogen production using chemical-looping technology. *Renew. Sustain. Energy Rev.* **2018**, *81*, 3186–3214. [CrossRef]
44. Abanades, S. Metal Oxides Applied to Thermochemical Water-Splitting for Hydrogen Production Using Concentrated Solar Energy. *ChemEngineering* **2019**, *3*, 63. [CrossRef]
45. Muhich, C.L.; Ehrhart, B.D.; Al-Shankiti, I.; Ward, B.J.; Musgrave, C.B.; Weimer, A.W. A review and perspective of efficient hydrogen generation via solar thermal water splitting. *Wiley Interdiscip. Rev. Energy Environ.* **2016**, *5*, 261–287. [CrossRef]
46. Nakamura, T. Hydrogen production from water utilizing solar heat at high temperatures. *Sol. Energy* **1977**, *19*, 467–475. [CrossRef]
47. Steinfeld, A.; Sanders, S.; Palumbo, R. Design aspects of solar thermochemical engineering—A case study: Two-step water-splitting cycle using the Fe3O4/FeO redox system. *Sol. Energy* **1999**, *65*, 43–53. [CrossRef]
48. Han, S.B.; Kang, T.B.; Joo, O.S.; Jung, K.D. Water splitting for hydrogen production with ferrites. *Sol. Energy* **2007**, *81*, 623–628. [CrossRef]
49. Kodama, T.; Kondoh, Y.; Yamamoto, R.; Andou, H.; Satou, N. Thermochemical hydrogen production by a redox system of ZrO_2-supported Co(II)-ferrite. *Sol. Energy* **2005**, *78*, 623–631. [CrossRef]
50. Kodama, T.; Gokon, N.; Yamamoto, R. Thermochemical two-step water splitting by ZrO_2-supported $Ni_xFe_{3-x}O_4$ for solar hydrogen production. *Sol. Energy* **2008**, *82*, 73–79. [CrossRef]
51. Ishihara, H.; Kaneko, H.; Yokoyama, T.; Fuse, A.; Hasegawa, N.; Tamaura, Y. Hydrogen Production Through Two-Step Water Splitting Using YSZ (Ni,Fe) System for Solar Hydrogen Production. In Proceedings of the ASME 2005 International Solar Energy Conference, Orlando, FL, USA, 6–12 August 2005.
52. Miller, J.E.; Allendorf, M.D.; Diver, R.B.; Evans, L.R.; Siegel, N.P.; Stuecker, J.N. Metal oxide composites and structures for ultra-high temperature solar thermochemical cycles. *J. Mater. Sci.* **2008**, *43*, 4714–4728. [CrossRef]
53. Palumbo, R.; Léde, J.; Boutin, O.; Elorza Ricart, E.; Steinfeld, A.; Möller, S.; Weidenkaff, A.; Fletcher, E.A.; Bielicki, J. The production of Zn from ZnO in a high-temperature solar decomposition quench process—I. The scientific framework for the process. *Chem. Eng. Sci.* **1998**, *53*, 2503–2517. [CrossRef]
54. Weidenkaff, A.; Reller, A.W.; Wokaun, A.; Steinfeld, A. Thermogravimetric analysis of the ZnO/Zn water splitting cycle. *Thermochim. Acta* **2000**, *359*, 69–75. [CrossRef]
55. Ernst, F.; Steinfeld, A.; Pratsinis, S. Hydrolysis rate of submicron Zn particles for solar H_2 synthesis. *Int. J. Hydrogen Energy* **2009**, *34*, 1166–1175. [CrossRef]
56. Lv, M.; Zhou, J.; Yang, W.; Cen, K. Thermogravimetric analysis of the hydrolysis of zinc particles. *Int. J. Hydrogen Energy* **2010**, *35*, 2617–2621. [CrossRef]
57. Abanades, S.; Flamant, G. Thermochemical hydrogen production from a two-step solar-driven water-splitting cycle based on cerium oxides. *Sol. Energy* **2006**, *80*, 1611–1623. [CrossRef]

58. Chueh, W.C.; Falter, C.; Abbott, M.; Scipio, D.; Furler, P.; Haile, S.M.; Steinfeld, A. High-flux solar-driven thermochemical dissociation of CO_2 and H_2O using nonstoichiometric ceria. *Science* **2010**, *330*, 1797–1801. [CrossRef] [PubMed]
59. Costa Oliveira, F.A.; Barreiros, M.A.; Haeussler, A.; Caetano, A.P.F.; Mouquinho, A.I.; Oliveira e Silva, P.M.; Novais, R.M.; Pullar, R.C.; Abanades, S. High performance cork-templated ceria for solar thermochemical hydrogen production via two-step water-splitting cycles. *Sustain. Energy Fuels* **2020**, *4*, 3077–3089. [CrossRef]
60. Singh, P.; Hegde, M.S. $Ce_{0.67}Cr_{0.33}O_{2.11}$: A New Low-Temperature O_2 Evolution Material and H_2 Generation Catalyst by Thermochemical Splitting of Water. *Chem. Mater.* **2009**, *22*, 762–768. [CrossRef]
61. Le Gal, A.; Abanades, S.; Flamant, G. CO_2 and H_2O Splitting for Thermochemical Production of Solar Fuels Using Nonstoichiometric Ceria and Ceria/Zirconia Solid Solutions. *Energy Fuels* **2011**, *25*, 4836–4845. [CrossRef]
62. Chen, Z.; Jiang, Q.; An, H.; Zhang, J.; Hao, S.; Li, X.; Cai, L.; Yu, W.; You, K.; Zhu, X.; et al. Platinum Group Metal Catalyst (RuO_x, PtO_x, and IrO_x)-Decorated Ceria-Zirconia Solid Solution as High Active Oxygen Carriers for Solar Thermochemical CO_2 Splitting. *ACS Catal.* **2022**, *12*, 7719–7736. [CrossRef]
63. Petkovich, N.D.; Rudisill, S.G.; Venstrom, L.J.; Boman, D.B.; Davidson, J.H.; Stein, A. Control of Heterogeneity in Nanostructured $Ce_{1-x}Zr_xO_2$ Binary Oxides for Enhanced Thermal Stability and Water Splitting Activity. *J. Phys. Chem. C* **2011**, *115*, 21022–21033. [CrossRef]
64. Gokon, N.; Suda, T.; Kodama, T. Oxygen and hydrogen productivities and repeatable reactivity of 30-mol%-Fe-, Co-, Ni-, Mn-doped $CeO_{2-\delta}$ for thermochemical two-step water-splitting cycle. *Energy* **2015**, *90*, 1280–1289. [CrossRef]
65. Kaneko, H.; Miura, T.; Ishihara, H.; Taku, S.; Yokoyama, T.; Nakajima, H.; Tamaura, Y. Reactive ceramics of CeO_2–MO_x (M = Mn, Fe, Ni, Cu) for H_2 generation by two-step water splitting using concentrated solar thermal energy. *Energy* **2007**, *32*, 656–663. [CrossRef]
66. Meng, Q.-L.; Lee, C.-i.; Ishihara, T.; Kaneko, H.; Tamaura, Y. Reactivity of CeO_2-based ceramics for solar hydrogen production via a two-step water-splitting cycle with concentrated solar energy. *Int. J. Hydrogen Energy* **2011**, *36*, 13435–13441. [CrossRef]
67. Lee, C.-i.; Meng, Q.-L.; Kaneko, H.; Tamaura, Y. Solar Hydrogen Productivity of Ceria–Scandia Solid Solution Using Two-Step Water-Splitting Cycle. *J. Sol. Energy Eng.* **2013**, *135*, 011002. [CrossRef]
68. Le Gal, A.; Abanades, S.; Bion, N.; Le Mercier, T.; Harlé, V. Reactivity of Doped Ceria-Based Mixed Oxides for Solar Thermochemical Hydrogen Generation via Two-Step Water-Splitting Cycles. *Energy Fuels* **2013**, *27*, 6068–6078. [CrossRef]
69. Le Gal, A.; Abanades, S. Catalytic investigation of ceria-zirconia solid solutions for solar hydrogen production. *Int. J. Hydrogen Energy* **2011**, *36*, 4739–4748. [CrossRef]
70. Abanades, S.; Legal, A.; Cordier, A.; Peraudeau, G.; Flamant, G.; Julbe, A. Investigation of reactive cerium-based oxides for H_2 production by thermochemical two-step water-splitting. *J. Mater. Sci.* **2010**, *45*, 4163–4173. [CrossRef]
71. Le Gal, A.; Abanades, S. Dopant Incorporation in Ceria for Enhanced Water-Splitting Activity during Solar Thermochemical Hydrogen Generation. *J. Phys. Chem. C* **2012**, *116*, 13516–13523. [CrossRef]
72. Ruan, C.; Tan, Y.; Li, L.; Wang, J.; Liu, X.; Wang, X. A novel $CeO_{2-x}SnO_2/Ce_2Sn_2O_7$ pyrochlore cycle for enhanced solar thermochemical water splitting. *AIChE J.* **2017**, *63*, 3450–3462. [CrossRef]
73. Zhu, X.; Li, K.; Neal, L.; Li, F. Perovskites as Geo-inspired Oxygen Storage Materials for Chemical Looping and Three-Way Catalysis: A Perspective. *ACS Catal.* **2018**, *8*, 8213–8236. [CrossRef]
74. Takacs, M.; Hoes, M.; Caduff, M.; Cooper, T.; Scheffe, J.R.; Steinfeld, A. Oxygen nonstoichiometry, defect equilibria, and thermodynamic characterization of $LaMnO_3$ perovskites with Ca/Sr A-site and Al B-site doping. *Acta Mater.* **2016**, *103*, 700–710. [CrossRef]
75. McDaniel, A.H.; Miller, E.C.; Arifin, D.; Ambrosini, A.; Coker, E.N.; O'Hayre, R.; Chueh, W.C.; Tong, J. Sr- and Mn-doped $LaAlO_{3-\delta}$ for solar thermochemical H_2 and CO production. *Energy Environ. Sci.* **2013**, *6*, 2424–2428. [CrossRef]
76. Barcellos, D.R.; Sanders, M.D.; Tong, J.; McDaniel, A.H.; O'Hayre, R.P. $BaCe_{0.25}Mn_{0.75}O_{3-\delta}$—A promising perovskite-type oxide for solar thermochemical hydrogen production. *Energy Environ. Sci.* **2018**, *11*, 3256–3265. [CrossRef]
77. Sai Gautam, G.; Stechel, E.B.; Carter, E.A. Exploring Ca–Ce–M–O (M = 3d Transition Metal) Oxide Perovskites for Solar Thermochemical Applications. *Chem. Mater.* **2020**, *32*, 9964–9982. [CrossRef]
78. Qian, X.; He, J.; Mastronardo, E.; Baldassarri, B.; Yuan, W.; Wolverton, C.; Haile, S.M. Outstanding Properties and Performance of $CaTi_{0.5}Mn_{0.5}O_{3-\delta}$ for Solar-Driven Thermochemical Hydrogen Production. *Matter* **2021**, *4*, 688–708. [CrossRef]
79. Chuayboon, S.; Abanades, S.; Rodat, S. High-Purity and Clean Syngas and Hydrogen Production From Two-Step CH_4 Reforming and H_2O Splitting Through Isothermal Ceria Redox Cycle Using Concentrated Sunlight. *Front. Energy Res.* **2020**, *8*, 00128. [CrossRef]
80. Wang, L.; Ma, T.; Chang, Z.; Li, H.; Fu, M.; Li, X. Solar fuels production via two-step thermochemical cycle based on Fe_3O_4/Fe with methane reduction. *Sol. Energy* **2019**, *177*, 772–781. [CrossRef]
81. He, F.; Li, F. Perovskite promoted iron oxide for hybrid water-splitting and syngas generation with exceptional conversion. *Energy Environ. Sci.* **2015**, *8*, 535–539. [CrossRef]
82. Chuayboon, S.; Abanades, S.; Rodat, S. Syngas production via solar-driven chemical looping methane reforming from redox cycling of ceria porous foam in a volumetric solar reactor. *Chem. Eng. J.* **2019**, *356*, 756–770. [CrossRef]
83. Welte, M.; Warren, K.; Scheffe, J.R.; Steinfeld, A. Combined Ceria Reduction and Methane Reforming in a Solar-Driven Particle-Transport Reactor. *Ind. Eng. Chem. Res.* **2017**, *56*, 10300–10308. [CrossRef] [PubMed]

84. Saha, D.; Grappe, H.A.; Chakraborty, A.; Orkoulas, G. Postextraction Separation, On-Board Storage, and Catalytic Conversion of Methane in Natural Gas: A Review. *Chem. Rev.* **2016**, *116*, 11436–11499. [CrossRef] [PubMed]
85. Olivos-Suarez, A.I.; Szécsényi, À.; Hensen, E.J.M.; Ruiz-Martinez, J.; Pidko, E.A.; Gascon, J. Strategies for the Direct Catalytic Valorization of Methane Using Heterogeneous Catalysis: Challenges and Opportunities. *ACS Catal.* **2016**, *6*, 2965–2981. [CrossRef]
86. He, Y.; Zhu, L.; Li, L.; Liu, G. Hydrogen and Power Cogeneration Based on Chemical Looping Combustion: Is It Capable of Reducing Carbon Emissions and the Cost of Production? *Energy Fuels* **2020**, *34*, 3501–3512. [CrossRef]
87. Zhu, X.; Wei, Y.; Wang, H.; Li, K. Ce–Fe oxygen carriers for chemical-looping steam methane reforming. *Int. J. Hydrogen Energy* **2013**, *38*, 4492–4501. [CrossRef]
88. Zhu, Y.; Liu, W.; Sun, X.; Ma, X.; Kang, Y.; Wang, X.; Wang, J. La-hexaaluminate for synthesis gas generation by Chemical Looping Partial Oxidation of Methane Using CO_2 as Sole Oxidant. *AIChE J.* **2018**, *64*, 550–563. [CrossRef]
89. Shen, Q.; Huang, F.; Tian, M.; Zhu, Y.; Li, L.; Wang, J.; Wang, X. Effect of Regeneration Period on the Selectivity of Synthesis Gas of Ba-Hexaaluminates in Chemical Looping Partial Oxidation of Methane. *ACS Catal.* **2018**, *9*, 722–731. [CrossRef]
90. Krenzke, P.T.; Davidson, J.H. Thermodynamic Analysis of Syngas Production via the Solar Thermochemical Cerium Oxide Redox Cycle with Methane-Driven Reduction. *Energy Fuels* **2014**, *28*, 4088–4095. [CrossRef]
91. Wang, X.; Du, X.; Yu, W.; Zhang, J.; Wei, J. Coproduction of Hydrogen and Methanol from Methane by Chemical Looping Reforming. *Ind. Eng. Chem. Res.* **2019**, *58*, 10296–10306. [CrossRef]
92. Zhang, F.; Zhu, L.; Wang, Y.; Sun, L. Exergy analysis on the process for three reactors chemical looping hydrogen generation. *Int. J. Hydrogen Energy* **2020**, *45*, 24322–24332. [CrossRef]
93. De Vos, Y.; Jacobs, M.; Van Der Voort, P.; Van Driessche, I.; Snijkers, F.; Verberckmoes, A. Sustainable iron-based oxygen carriers for Chemical Looping for Hydrogen Generation. *Int. J. Hydrogen Energy* **2019**, *44*, 1374–1391. [CrossRef]
94. Go, K.; Son, S.; Kim, S. Reaction kinetics of reduction and oxidation of metal oxides for hydrogen production. *Int. J. Hydrogen Energy* **2008**, *33*, 5986–5995. [CrossRef]
95. Kang, K.-S.; Kim, C.-H.; Cho, W.-C.; Bae, K.-K.; Woo, S.-W.; Park, C.-S. Reduction characteristics of $CuFe_2O_4$ and Fe_3O_4 by methane; $CuFe_2O_4$ as an oxidant for two-step thermochemical methane reforming. *Int. J. Hydrogen Energy* **2008**, *33*, 4560–4568. [CrossRef]
96. Ku, Y.; Wu, H.-C.; Chiu, P.-C.; Tseng, Y.-H.; Kuo, Y.-L. Methane combustion by moving bed fuel reactor with Fe_2O_3/Al_2O_3 oxygen carriers. *Appl. Energy* **2014**, *113*, 1909–1915. [CrossRef]
97. Zhu, M.; Chen, S.; Ma, S.; Xiang, W. Carbon formation on iron-based oxygen carriers during CH_4 reduction period in Chemical Looping Hydrogen Generation process. *Chem. Eng. J.* **2017**, *325*, 322–331. [CrossRef]
98. Galinsky, N.L.; Huang, Y.; Shafiefarhood, A.; Li, F. Iron Oxide with Facilitated O^{2-} Transport for Facile Fuel Oxidation and CO_2 Capture in a Chemical Looping Scheme. *ACS Sustain. Chem. Eng.* **2013**, *1*, 364–373. [CrossRef]
99. Fung, V.; Polo-Garzon, F.; Wu, Z.; Jiang, D.-e. Exploring perovskites for methane activation from first principles. *Catal. Sci. Technol.* **2018**, *8*, 702–709. [CrossRef]
100. Mattisson, T.; Lyngfelt, A. Capture of CO_2 using chemical-looping combustion. In Proceedings of the First Biennial Meeting of the Scandinavian-928 Nordic Section of the Combustion Institute, Göteborg, Sweden, 18–20 April 2001.
101. Moghtaderi, B. Hydrogen enrichment of fuels using a novel miniaturised chemical looping steam reformer. *Chem. Eng. Res. Des.* **2012**, *90*, 19–25. [CrossRef]
102. Liu, Y.; Qin, L.; Cheng, Z.; Goetze, J.W.; Kong, F.; Fan, J.A.; Fan, L.S. Near 100% CO selectivity in nanoscaled iron-based oxygen carriers for chemical looping methane partial oxidation. *Nat. Commun.* **2019**, *10*, 5503. [CrossRef] [PubMed]
103. Cheng, Z.; Qin, L.; Guo, M.; Xu, M.; Fan, J.A.; Fan, L.S. Oxygen vacancy promoted methane partial oxidation over iron oxide oxygen carriers in the chemical looping process. *Phys. Chem. Chem. Phys.* **2016**, *18*, 32418–32428. [CrossRef] [PubMed]
104. Kang, Y.; Tian, M.; Huang, C.; Lin, J.; Hou, B.; Pan, X.; Li, L.; Rykov, A.I.; Wang, J.; Wang, X. Improving Syngas Selectivity of Fe_2O_3/Al_2O_3 with Yttrium Modification in Chemical Looping Methane Conversion. *ACS Catal.* **2019**, *9*, 8373–8382. [CrossRef]
105. Li, D.; Li, K.; Xu, R.; Wang, H.; Tian, D.; Wei, Y.; Zhu, X.; Zeng, C.; Zeng, L. $Ce_{1-x}Fe_xO_{2-\delta}$ catalysts for catalytic methane combustion: Role of oxygen vacancy and structural dependence. *Catal. Today* **2018**, *318*, 73–85. [CrossRef]
106. Han, Y.; Tian, M.; Wang, C.; Kang, Y.; Kang, L.; Su, Y.; Huang, C.; Zong, T.; Lin, J.; Hou, B.; et al. Highly Active and Anticoke Ni/CeO_2 with Ultralow Ni Loading in Chemical Looping Dry Reforming via the Strong Metal–Support Interaction. *ACS Sustain. Chem. Eng.* **2021**, *9*, 17276–17288. [CrossRef]
107. Murray, E.P.; Tsai, T.; Barnett, S.A. A direct-methane fuel cell with a ceria-based anode. *Nature* **1999**, *400*, 649–651. [CrossRef]
108. Ruan, C.; Huang, Z.-Q.; Lin, J.; Li, L.; Liu, X.; Tian, M.; Huang, C.; Chang, C.-R.; Li, J.; Wang, X. Synergy of the catalytic activation on Ni and the CeO_2–$TiO_2/Ce_2Ti_2O_7$ stoichiometric redox cycle for dramatically enhanced solar fuel production. *Energy Environ. Sci.* **2019**, *12*, 767–779. [CrossRef]
109. Zhu, X.; Wang, H.; Wei, Y.; Li, K.; Cheng, X. Hydrogen and syngas production from two-step steam reforming of methane over CeO_2-Fe_2O_3 oxygen carrier. *J. Rare Earths* **2010**, *28*, 907–913. [CrossRef]
110. Zhu, X.; Li, K.; Wei, Y.; Wang, H.; Sun, L. Chemical-Looping Steam Methane Reforming over a CeO_2–Fe_2O_3 Oxygen Carrier: Evolution of Its Structure and Reducibility. *Energy Fuels* **2014**, *28*, 754–760. [CrossRef]
111. Yuan, J.; Zhao, Y.; Xu, H.; Lu, C.; Yang, K.; Zhu, X.; Li, K. Layered Mg-Al spinel supported Ce-Fe-Zr-O oxygen carriers for chemical looping reforming. *Chin. J. Chem. Eng.* **2020**, *28*, 2668–2676. [CrossRef]

112. Zheng, Y.; Li, K.; Wang, H.; Tian, D.; Wang, Y.; Zhu, X.; Wei, Y.; Zheng, M.; Luo, Y. Designed oxygen carriers from macroporous $LaFeO_3$ supported CeO_2 for chemical-looping reforming of methane. *Appl. Catal. B* **2017**, *202*, 51–63. [CrossRef]
113. Cheng, Z.; Zhang, L.; Jin, N.; Zhu, Y.; Chen, L.; Yang, Q.; Yan, M.; Ma, X.; Wang, X. Effect of calcination temperature on the performance of hexaaluminate supported CeO_2 for chemical looping dry reforming. *Fuel Process. Technol.* **2021**, *218*, 106873. [CrossRef]
114. Zheng, Y.; Wei, Y.; Li, K.; Zhu, X.; Wang, H.; Wang, Y. Chemical-looping steam methane reforming over macroporous CeO_2–ZrO_2 solid solution: Effect of calcination temperature. *Int. J. Hydrogen Energy* **2014**, *39*, 13361–13368. [CrossRef]
115. Rydén, M.; Leion, H.; Mattisson, T.; Lyngfelt, A. Combined oxides as oxygen-carrier material for chemical-looping with oxygen uncoupling. *Appl. Energy* **2014**, *113*, 1924–1932. [CrossRef]
116. Long, Y.; Yang, K.; Gu, Z.; Lin, S.; Li, D.; Zhu, X.; Wang, H.; Li, K. Hydrogen generation from water splitting over polyfunctional perovskite oxygen carriers by using coke oven gas as reducing agent. *Appl. Catal. B* **2022**, *301*, 120778. [CrossRef]
117. Lee, M.; Lim, H.S.; Kim, Y.; Lee, J.W. Enhancement of highly-concentrated hydrogen productivity in chemical looping steam methane reforming using Fe-substituted $LaCoO_3$. *Energy Convers. Manag.* **2020**, *207*, 112507. [CrossRef]
118. Zhang, X.; Su, Y.; Pei, C.; Zhao, Z.-J.; Liu, R.; Gong, J. Chemical looping steam reforming of methane over Ce-doped perovskites. *Chem. Eng. Sci.* **2020**, *223*, 115707. [CrossRef]
119. Zhang, X.; Pei, C.; Chang, X.; Chen, S.; Liu, R.; Zhao, Z.J.; Mu, R.; Gong, J. FeO_6 Octahedral Distortion Activates Lattice Oxygen in Perovskite Ferrite for Methane Partial Oxidation Coupled with CO_2 Splitting. *J. Am. Chem. Soc.* **2020**, *142*, 11540–11549. [CrossRef] [PubMed]
120. Xia, X.; Chang, W.; Cheng, S.; Huang, C.; Hu, Y.; Xu, W.; Zhang, L.; Jiang, B.; Sun, Z.; Zhu, Y.; et al. Oxygen Activity Tuning via FeO_6 Octahedral Tilting in Perovskite Ferrites for Chemical Looping Dry Reforming of Methane. *ACS Catal.* **2022**, *12*, 7326–7335. [CrossRef]
121. Huang, C.; Wu, J.; Chen, Y.-T.; Tian, M.; Rykov, A.I.; Hou, B.; Lin, J.; Chang, C.-R.; Pan, X.; Wang, J.; et al. In situ encapsulation of iron(0) for solar thermochemical syngas production over iron-based perovskite material. *Commun. Chem.* **2018**, *1*, 55. [CrossRef]
122. Wang, Y.; Zheng, Y.; Wang, Y.; Wang, H.; Zhu, X.; Wei, Y.; Wang, Y.; Jiang, L.; Yang, Z.; Li, K. Evaluation of Fe substitution in perovskite $LaMnO_3$ for the production of high purity syngas and hydrogen. *J. Power Sources* **2020**, *449*, 227505. [CrossRef]
123. Yin, X.; Wang, S.; Wang, B.; Shen, L. Perovskite-type $LaMn_{1-x}B_xO_{3+\delta}$ (B = Fe, CO and Ni) as oxygen carriers for chemical looping steam methane reforming. *Chem. Eng. J.* **2021**, *422*, 128751. [CrossRef]
124. Shafiefarhood, A.; Galinsky, N.; Huang, Y.; Chen, Y.; Li, F. Fe_2O_3@$La_xSr_{1-x}FeO_3$ Core-Shell Redox Catalyst for Methane Partial Oxidation. *ChemCatChem* **2014**, *6*, 790–799. [CrossRef]
125. Zhao, K.; Chen, J.; Li, H.; Zheng, A.; Huang, Z.; Wei, G.; Zhao, Z.; Wang, X. Investigation of the relationship between electronic properties and reactivity of 3DOM $LaFe_{1-x}Co_xO_3$ for methane reforming to produce syngas. *Int. J. Energy Res.* **2019**, *43*, 7120–7134. [CrossRef]
126. Zhao, K.; Zheng, A.; Li, H.; He, F.; Huang, Z.; Wei, G.; Shen, Y.; Zhao, Z. Exploration of the mechanism of chemical looping steam methane reforming using double perovskite-type oxides $La_{1.6}Sr_{0.4}FeCoO_6$. *Appl. Catal. B* **2017**, *219*, 672–682. [CrossRef]
127. Wang, S.; Guan, B.Y.; Lou, X.W.D. Construction of $ZnIn_2S_4$-In_2O_3 Hierarchical Tubular Heterostructures for Efficient CO_2 Photoreduction. *J. Am. Chem. Soc.* **2018**, *140*, 5037–5040. [CrossRef] [PubMed]
128. Bahmanpour, A.M.; Héroguel, F.; Kılıç, M.; Baranowski, C.J.; Artiglia, L.; Röthlisberger, U.; Luterbacher, J.S.; Kröcher, O. Cu–Al Spinel as a Highly Active and Stable Catalyst for the Reverse Water Gas Shift Reaction. *ACS Catal.* **2019**, *9*, 6243–6251. [CrossRef]
129. Zeng, L.; He, F.; Li, F.; Fan, L.-S. Coal-Direct Chemical Looping Gasification for Hydrogen Production: Reactor Modeling and Process Simulation. *Energy Fuels* **2012**, *26*, 3680–3690. [CrossRef]
130. Bohn, C.D.; Cleeton, J.P.; Müller, C.R.; Chuang, S.Y.; Scott, S.A.; Dennis, J.S. Stabilizing Iron Oxide Used in Cycles of Reduction and Oxidation for Hydrogen Production. *Energy Fuels* **2010**, *24*, 4025–4033. [CrossRef]
131. Liu, W.; Dennis, J.S.; Scott, S.A. The Effect of Addition of ZrO_2 to Fe_2O_3 for Hydrogen Production by Chemical Looping. *Ind. Eng. Chem. Res.* **2012**, *51*, 16597–16609. [CrossRef]
132. Kierzkowska, A.M.; Bohn, C.D.; Scott, S.A.; Cleeton, J.P.; Dennis, J.S.; Müller, C.R. Development of Iron Oxide Carriers for Chemical Looping Combustion Using Sol–Gel. *Ind. Eng. Chem. Res.* **2010**, *49*, 5383–5391. [CrossRef]
133. Chen, S.; Shi, Q.; Xue, Z.; Sun, X.; Xiang, W. Experimental investigation of chemical-looping hydrogen generation using Al_2O_3 or TiO_2-supported iron oxides in a batch fluidized bed. *Int. J. Hydrogen Energy* **2011**, *36*, 8915–8926. [CrossRef]
134. Hafizi, A.; Rahimpour, M. Inhibiting Fe–Al Spinel Formation on a Narrowed Mesopore-Sized $MgAl_2O_4$ Support as a Novel Catalyst for H_2 Production in Chemical Looping Technology. *Catalysts* **2018**, *8*, 27. [CrossRef]
135. Ma, S.; Chen, S.; Soomro, A.; Xiang, W. Effects of supports on hydrogen production and carbon deposition of Fe-based oxygen carriers in chemical looping hydrogen generation. *Int. J. Hydrogen Energy* **2017**, *42*, 11006–11016. [CrossRef]
136. Yüzbasi, N.S.; Armutlulu, A.; Huthwelker, T.; Abdala, P.M.; Müller, C.R. Na-β-Al_2O_3 stabilized Fe_2O_3 oxygen carriers for chemical looping water splitting: Correlating structure with redox stability. *J. Mater. Chem. A* **2022**, *10*, 10692–10700. [CrossRef] [PubMed]
137. Ma, S.; Chen, S.; Soomro, A.; Xiang, W. Effects of CeO_2, ZrO_2, and Al_2O_3 Supports on Iron Oxygen Carrier for Chemical Looping Hydrogen Generation. *Energy Fuels* **2017**, *31*, 8001–8013. [CrossRef]

138. Ma, S.; Chen, S.; Zhao, Z.; Soomro, A.; Zhu, M.; Hu, J.; Wu, M.; Xiang, W. Enhanced Hydrogen Generation for Fe_2O_3/CeO_2 Oxygen Carrier via Rare-Earth (Y, Sm, and La) Doping in Chemical Looping Process. *Energy Fuels* **2018**, *32*, 11362–11374. [CrossRef]
139. Ma, S.; Chen, S.; Ge, H.; Song, T.; Lu, P.; Xiang, W. Synergistic Effects of the Zr and Sm Co-doped Fe_2O_3/CeO_2 Oxygen Carrier for Chemical Looping Hydrogen Generation. *Energy Fuels* **2020**, *34*, 10256–10267. [CrossRef]
140. Do, J.Y.; Son, N.; Park, N.-K.; Kwak, B.S.; Baek, J.-I.; Ryu, H.-J.; Kang, M. Reliable oxygen transfer in $MgAl_2O_4$ spinel through the reversible formation of oxygen vacancies by Cu^{2+}/Fe^{3+} anchoring. *Appl. Energy* **2018**, *219*, 138–150. [CrossRef]
141. Liu, F.; Wu, F.; Liu, J.; Li, Y.; Yang, Y. Experimental and theoretical insights into the mechanism of spinel $CoFe_2O_4$ reduction in CO chemical looping combustion. *Fuel* **2021**, *293*, 120473. [CrossRef]
142. Huang, Z.; Deng, Z.; Chen, D.; Wei, G.; He, F.; Zhao, K.; Zheng, A.; Zhao, Z.; Li, H. Exploration of Reaction Mechanisms on Hydrogen Production through Chemical Looping Steam Reforming Using $NiFe_2O_4$ Oxygen Carrier. *ACS Sustain. Chem. Eng.* **2019**, *7*, 11621–11632. [CrossRef]
143. Kim, Y.; Lim, H.S.; Lee, M.; Kim, M.; Kang, D.; Lee, J.W. Enhanced Morphological Preservation and Redox Activity in Al-Incorporated $NiFe_2O_4$ for Chemical Looping Hydrogen Production. *ACS Sustain. Chem. Eng.* **2021**, *9*, 14800–14810. [CrossRef]
144. Cui, D.; Li, M.; Qiu, Y.; Ma, L.; Zeng, D.; Xiao, R. Improved hydrogen production with 100% fuel conversion through the redox cycle of $ZnFeAlO_x$ oxygen carrier in chemical looping scheme. *Chem. Eng. J.* **2020**, *400*, 125769. [CrossRef]
145. Li, M.; Du, Y.; Qiu, Y.; Ma, L.; Cui, D.; Zhang, S.; Tippayawong, N.; Zeng, D.; Xiao, R. The use of ferrites as highly active oxygen storage materials for chemical looping hydrogen production under intermediate temperature. *Int. J. Hydrogen Energy* **2019**, *44*, 28638–28648. [CrossRef]
146. Zeng, D.; Qiu, Y.; Li, M.; Cui, D.; Ma, L.; Lv, Y.; Zhang, S.; Xiao, R. Ternary Mixed Spinel Oxides as Oxygen Carriers for Chemical Looping Hydrogen Production Operating at 550 °C. *ACS Appl. Mater. Interfaces* **2019**, *11*, 44223–44232. [CrossRef]
147. Hirabayashi, D.; Yoshikawa, T.; Mochizuki, K.; Suzuki, K.; Sakai, Y. Formation of brownmillerite type calcium ferrite ($Ca_2Fe_2O_5$) and catalytic properties in propylene combustion. *Catal. Lett.* **2006**, *110*, 155–160. [CrossRef]
148. Shah, V.; Mohapatra, P.; Fan, L.-S. Thermodynamic and Process Analyses of Syngas Production Using Chemical Looping Reforming Assisted by Flexible Dicalcium Ferrite-Based Oxygen Carrier Regeneration. *Energy Fuels* **2020**, *34*, 6490–6500. [CrossRef]
149. Ismail, M.; Liu, W.; Chan, M.S.C.; Dunstan, M.T.; Scott, S.A. Synthesis, Application, and Carbonation Behavior of $Ca_2Fe_2O_5$ for Chemical Looping H_2 Production. *Energy Fuels* **2016**, *30*, 6220–6232. [CrossRef]
150. Chan, M.S.C.; Liu, W.; Ismail, M.; Yang, Y.; Scott, S.A.; Dennis, J.S. Improving hydrogen yields, and hydrogen:steam ratio in the chemical looping production of hydrogen using $Ca_2Fe_2O_5$. *Chem. Eng. J.* **2016**, *296*, 406–411. [CrossRef]
151. Sun, Z.; Chen, S.; Hu, J.; Chen, A.; Rony, A.H.; Russell, C.K.; Xiang, W.; Fan, M.; Darby Dyar, M.; Dklute, E.C. $Ca_2Fe_2O_5$: A promising oxygen carrier for CO/CH_4 conversion and almost-pure H_2 production with inherent CO_2 capture over a two-step chemical looping hydrogen generation process. *Appl. Energy* **2018**, *211*, 431–442. [CrossRef]
152. Feng, Y.; Wang, N.; Guo, X.; Zhang, S. Reaction mechanism of $Ca_2Fe_2O_5$ oxygen carrier with CO in chemical looping hydrogen production. *Appl. Surf. Sci.* **2020**, *534*, 147583. [CrossRef]
153. You, S.; Ok, Y.S.; Chen, S.S.; Tsang, D.C.W.; Kwon, E.E.; Lee, J.; Wang, C.H. A critical review on sustainable biochar system through gasification: Energy and environmental applications. *Bioresour. Technol.* **2017**, *246*, 242–253. [CrossRef] [PubMed]
154. Hu, Z.; Miao, Z.; Chen, H.; Wu, J.; Wu, W.; Ren, Y.; Jiang, E. Chemical looping gasification of biochar to produce hydrogen-rich syngas using Fe/Ca-based oxygen carrier prepared by coprecipitation. *J. Energy Inst.* **2021**, *94*, 157–166. [CrossRef]
155. Bracciale, M.P.; Damizia, M.; De Filippis, P.; de Caprariis, B. Clean Syngas and Hydrogen Co-Production by Gasification and Chemical Looping Hydrogen Process Using MgO-Doped Fe_2O_3 as Redox Material. *Catalysts* **2022**, *12*, 1273. [CrossRef]
156. Li, F.; Kim, H.R.; Sridhar, D.; Wang, F.; Zeng, L.; Chen, J.; Fan, L.S. Syngas Chemical Looping Gasification Process: Oxygen Carrier Particle Selection and Performance. *Energy Fuels* **2009**, *23*, 4182–4189. [CrossRef]
157. Li, F.; Zeng, L.; Velazquez-Vargas, L.G.; Yoscovits, Z.; Fan, L.-S. Syngas chemical looping gasification process: Bench-scale studies and reactor simulations. *AlChE J.* **2010**, *56*, 2186–2199. [CrossRef]
158. Luo, C.; Dou, B.; Zhang, H.; Liu, D.; Zhao, L.; Chen, H.; Xu, Y. Co-production of hydrogen and syngas from chemical looping water splitting coupled with decomposition of glycerol using Fe-Ce-Ni based oxygen carriers. *Energy Convers. Manag.* **2021**, *238*, 114166. [CrossRef]
159. Sun, S.; Zhao, M.; Cai, L.; Zhang, S.; Zeng, D.; Xiao, R. Performance of CeO_2-Modified Iron-Based Oxygen Carrier in the Chemical Looping Hydrogen Generation Process. *Energy Fuels* **2015**, *29*, 7612–7621. [CrossRef]
160. Aston, V.J.; Evanko, B.W.; Weimer, A.W. Investigation of novel mixed metal ferrites for pure H_2 and CO_2 production using chemical looping. *Int. J. Hydrogen Energy* **2013**, *38*, 9085–9096. [CrossRef]
161. Liu, S.; He, F.; Huang, Z.; Zheng, A.; Feng, Y.; Shen, Y.; Li, H.; Wu, H.; Glarborg, P. Screening of $NiFe_2O_4$ Nanoparticles as Oxygen Carrier in Chemical Looping Hydrogen Production. *Energy Fuels* **2016**, *30*, 4251–4262. [CrossRef]
162. Scheffe, J.R.; Allendorf, M.D.; Coker, E.N.; Jacobs, B.W.; McDaniel, A.H.; Weimer, A.W. Hydrogen Production via Chemical Looping Redox Cycles Using Atomic Layer Deposition-Synthesized Iron Oxide and Cobalt Ferrites. *Chem. Mater.* **2011**, *23*, 2030–2038. [CrossRef]
163. Xu, C.; Zhang, Y.; Chen, J.; Lin, J.; Zhang, X.; Wang, Z.; Zhou, J. Enhanced mechanism of the photo-thermochemical cycle based on effective Fe-doping TiO_2 films and DFT calculations. *Appl. Catal. B* **2017**, *204*, 324–334. [CrossRef]

164. Docao, S.; Koirala, A.R.; Kim, M.G.; Hwang, I.C.; Song, M.K.; Yoon, K.B. Solar photochemical–thermal water splitting at 140 °C with Cu-loaded TiO_2. *Energy Environ. Sci.* **2017**, *10*, 628–640. [CrossRef]
165. Wu, Q.; Li, Z.; Zhang, X.; Huang, W.; Ni, M.; Cen, K.; Zhang, Y. Enhanced defect-water hydrogen evolution method for efficient solar utilization: Photo-thermal chemical coupling on oxygen vacancy. *Chem. Eng. J.* **2021**, *408*, 127248. [CrossRef]
166. Wu, Q.; Li, Z.; Zhang, X.; Xu, C.; Ni, M.; Cen, K.; Zhang, Y. Oxygen-vacancy-anchoring NiO loading towards efficient hydrogen evolution via photo-thermal coupling reaction. *J. Energy Chem.* **2021**, *61*, 77–87. [CrossRef]
167. Zeng, D.; Qiu, Y.; Zhang, S.; Ma, L.; Li, M.; Cui, D.; Zeng, J.; Xiao, R. Synergistic effects of binary oxygen carriers during chemical looping hydrogen production. *Int. J. Hydrogen Energy* **2019**, *44*, 21290–21302. [CrossRef]

catalysts MDPI

Article

Ce-Doped $LaMnO_3$ Redox Catalysts for Chemical Looping Oxidative Dehydrogenation of Ethane

Jingwei Wang [1,2], Xiaocen Liang [1,2], Zifan Xing [1,2], Haitao Chen [1,2], Yang Li [3], Da Song [1,2] and Fang He [1,2,*]

[1] College of Chemistry and Bioengineering, Guilin University of Technology, Guilin 541006, China
[2] Guangxi Key Laboratory of Electrochemical and Magneto-Chemical Functional Materials, College of Chemistry and Bioengineering, Guilin University of Technology, Guilin 541004, China
[3] College of Chemistry and Engineering, Northeast Petroleum University, Daqing 163318, China
* Correspondence: hefang@glut.edu.cn

Abstract: As a novel reaction mode of oxidative dehydrogenation of ethane to ethylene, the chemical looping oxidative dehydrogenation (CL-ODH) of ethane to ethylene has attracted much attention. Instead of using gaseous oxygen, CL-ODH uses lattice oxygen in an oxygen carrier or redox catalyst to facilitate the ODH reaction. In this paper, a perovskite type redox catalyst $LaMnO_{3+\delta}$ was used as a substrate, Ce^{3+} with different proportions was introduced into its A site, and its CL-ODH reaction performance for ethane was studied. The results showed that the ratio of Mn^{4+}/Mn^{3+} on the surface of Ce-modified samples decreased significantly, and the lattice oxygen species in the bulk phase increased; these were the main reasons for improving ethylene selectivity. $La_{0.7}Ce_{0.3}MnO_3$ showed the best performance during the ODH reaction and showed good stability in twenty redox cycle tests.

Keywords: chemical looping oxidative dehydrogenation; $LaMnO_3$; Ethane; perovskite type redox catalyst

1. Introduction

Ethylene is one of the most commonly used feedstocks in the production of numerous chemical intermediates and polymers, and is one of the most produced chemicals in the world [1–3]. The global ethylene market was estimated to be 158 million tons in 2020, and is expected to reach 207.4 million tons in 2027 [4]. At present, steam cracking is mainly carried out by petroleum raw materials, such as petroleum brain and condensate, to achieve ethylene production in industry [5]. Recently, with the increase in shale gas production, there has been extensive attention paid to the use of the underutilized ethane, extracted from shale gas, to produce ethylene [6,7].

In industry, the steam cracking process for ethylene production from ethane is shown in Equation (1) [8]. This reaction is a strongly endothermic reaction with a high reaction temperature (higher than 1000 °C), which inevitably leads to high energy consumption (approximately 16 GJ/[t C_2H_4]), serious carbon deposition, and produces a large number of environmentally unfriendly by-products such as CO_2 (1−1.2 t CO_2/[t C_2H_4]) and NO_x [9,10].

$$C_2H_6\,(g) \Leftrightarrow C_2H_4\,(g) + H_2(g) \quad \Delta H_{298K} = +136\ \text{kJ/mol} \tag{1}$$

In order to avoid the disadvantages of the traditional steam cracking process, ethylene production via the oxidative dehydrogenation (ODH) of ethane (Equation (2) is proposed as an alternative method. Compared to steam cracking technology, ODH is an exothermic reaction; therefore, the reaction temperature is lower (300–700 °C), and the energy consumption of tons of ethylene can be reduced by 35%. In addition, due to the participation of gaseous oxygen in the reaction, the hydrogen generated by the cleavage of ethane in the reaction Equation (2) will undergo a combustion reaction (Equation (3); this can eliminate the thermodynamic barrier to achieve a high ethane conversion rate and make the reaction kinetics possess lower Gibbs free energy [8,11,12].

$$C_2H_6\ (g) + \frac{1}{2}O_2\ (g) \rightarrow C_2H_4\ (g) + H_2O\ (g) \quad \Delta H_{298K} = -150\ kJ/mol \tag{2}$$

$$2H_2\ (g) + O_2\ (g) \rightarrow 2H_2O\ (g) \quad \Delta H_{298K} = -286\ kJ/mol \tag{3}$$

In the past 20 years, a number of ODH catalyst systems have been reported, such as NiO [13], MO-V-O [14], Sn-doped Pt [15], La/Sr/Nd [16], and Mo/V/Te/Nb/O [17]. It has been reported that the ethylene yield of NiO and NiO catalysts, modified with Zr, Ti, Mo, W, and V in the ODH reaction of ethane, ranges from 35% to 38%, among which Ni-W-O and Ni-Ti-O catalysts show the highest ethylene yield [13]. The Mo-V-containing catalysts achieved a high ethylene selectivity of 76% under the condition of 30% ethane conversion [14]. The Mo/V/Te/Nb/O catalyst can achieve 78% ethylene yield [17]. Despite the ODH reaction possessing many potential advantages, it can not replace the steam cracking process in the industry, because there are still some problems that need to be solved; these include the following: (1) In order for pure oxygen or oxygen-enriched air to participate in the ethane ODH reaction, it needs to be in an air separation plant; this no doubt increases the cost of investment and the extent of air separation oxygen in the energy intensive process. (2) The co-feeding of ethane and oxygen has safety hazards. If inert gas is added to dilute the feedstock gas, it will increase the additional costs and bring unnecessary trouble to the separation of subsequent products [18]. (3) With the participation of gas-phase oxygen, the product is more likely to be deeply oxidized to CO_x, which reduces ethylene selectivity [11].

To solve these problems, the chemical looping was combined with the ODH reaction of ethane to form the chemical looping oxidative dehydrogenation of ethane, as shown in Scheme 1 [19–21]. This scheme uses lattice oxygen in an oxygen carrier (also known as a redox catalyst) instead of gas oxygen, to participate in the oxidative dehydrogenation reaction, without needing air separation; thus, the feed mixture of ethane and oxygen is avoided, which inhibits the deep oxidation of the product and minimizes explosion hazards from co-feeding oxygen with ethane. In the CL-ODH scheme, the reaction process between fuel and the oxygen carrier is divided into two steps: In the fuel reactor, ethane is partially oxidized into ethylene by an oxygen carrier (MeO_x); meanwhile, the oxygen carrier is reduced to a reduced state. In the air reactor, the reduced oxygen carrier (MeO_{x-1}) is reoxidized by air to restore its lattice oxygen, completing the redox cycle [22]. Previous studies show that in the chemical looping oxidative dehydrogenation reaction, ethane is firstly cleaved to produce ethylene and hydrogen, and the produced hydrogen is selectively oxidized on the surface of oxygen carrier, breaking the thermodynamic equilibrium and making the reaction go in the direction of ethylene formation [23]. In the ODH reaction, the adsorbed oxygen on the surface of the oxygen carrier will be rapidly consumed to form oxygen vacancies, and the internal lattice oxygen needs to migrate to the surface to fill the oxygen vacancies.. Therefore, the lattice oxygen mobility of the oxygen carrier plays a crucial role in the success of the CL-ODH reaction [24]. In the CL-ODH of ethane, the key to this process is to design oxygen carriers with high lattice oxygen mobility, which can selectively burn hydrogen and inhibit the deep oxidation of products.

Recently, perovskite-type oxygen carriers have attracted much attention as redox catalysts due to their unique structure and high stability. The general formula of perovskite-type oxide is ABO_3, where A is an alkaline earth or rare earth metal cation and B is a transition metal cation. To date, a number of studies have been conducted on the chemical looping oxidative dehydrogenation of ethane using perovskite-type oxygen carriers. In these previous studies, perovskite-type oxides are widely used in chemical looping reactions due to their excellent redox properties and high oxygen mobility. Tian et al. [10] studied that Ce-modified $SrFeO_{3-\delta}$ achieved up to 29% ethane conversion and 82% ethylene selectivity at 725 °C. In this study, the modification of Ce promoted the formation of surface oxygen vacancies and increased the resistance of lattice oxygen diffusion to the surface, which was the reason for the increased reactivity and ethylene selectivity. Gao et al. [25] found that the $La_xSr_{2-x}FeO_{4-\delta}$ catalyst, modified with Li and K, had an ethane conversion of 60% and ethylene selectivity of 86%o at 700 °C. It is very important to design and develop efficient redox catalysts to replace gaseous oxygen, inhibit the deep oxidation of ethane, and improve ethylene selectivity for the chemical looping oxidative dehydrogenation of ethane. It was also found that Mn-based oxygen carriers, such as $CaMnO_3$, Mg_6MnO_8, $SrMnO_2$, and Mn_7SiO_{12}, had good performances in the CL-ODH of ethane, and the yield of Mn-based oxygen carriers, modified by Na_2WO_4, could reach 68.2% [23,24,26–28]. These results indicated that the Mn-based oxygen carrier is a promising oxygen carrier for the oxidative dehydrogenation of ethane. Among the various perovskite-type oxides, $LaMnO_3$ exhibited a higher thermal stability and better activity, so it is considered to be a more meaningful carrier for the CL-ODH of ethane to produce ethylene [11]. It is generally believed that when different cations are introduced into the A or B sites of perovskite oxides, the lattice structure of perovskite will not change fundamentally, but its reactivity can be improved by regulating the number of oxygen vacancies, regulating the oxygen vacancy symmetry, and oxide lattice distortion [29]. Recently, T. Rajkumar et al. [30] found that in the CeO_2/MnO_x system, through the mixing of CeO_2 and MnO_x, the oxidation states of Ce^{3+} and Ce^{4+} in the CeO_2 can be shifted to improve the oxygen vacancy, oxygen mobility and oxygen storage capacity in the lattice; this can change the catalytic activity of the CeO_2. Therefore, in this paper, by introducing Ce^{3+} cations as heterogeneous ions into the A site of $LaMnO_3$ to adjust the local chemical environment of the oxygen vacancy, the influence of heterogeneous ions on ethane cleavage and the selective hydrogen burning mechanism in the CL-ODH of ethane were studied. Different proportions of Ce^{3+} were introduced into the A site of the perovskite-type oxide $LaMnO_3$ to carry out the CL-ODH reaction. The doping of Ce^{3+} can form asymmetric oxygen vacancies (some studies have shown that asymmetric oxygen vacancies easily adsorb oxygen in an oxidizing atmosphere, and desorb oxygen in a reducing atmosphere), actively regulate the local chemical environment of lattice oxygen and oxygen vacancies, and achieve the effect of increasing ethylene selectivity [29].

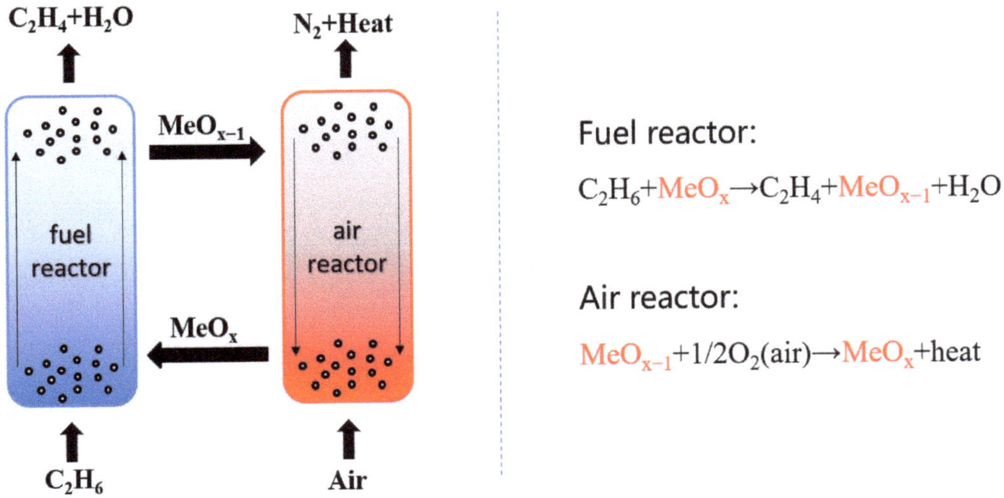

Scheme 1. A schematic view for the concept of ethane CL-ODH.

2. Results and Discussions

2.1. Structural Characterization

The lattice structure, crystal size and lattice parameters of the prepared LMO, 0.1LCMO, 0.2LCMO, 0.3LCMO and 0.4LCMO samples were characterized by XRD; the results are shown in Figure 1 and Table 1. The appearance of a peak indicates a good crystallinity. For the samples without Ce doping, there was no secondary phase. According to the JCPDS card (PDF # 82-1152), the sample shows the characteristic peaks of a hexagonal $LaMnO_3$ phase. The main peaks are at $2\theta = 21.69°$, 31.25°, 31.57°, 38.87°, 45.54° and 56.8°, corresponding to (012), (110), (104), (202), (024), and (214) planes, respectively. When Ce is doped into $LaMnO_3$, there will be a second phase, and a orthorhombic perovskite is formed. The secondary phase has a main peak at 2θ of 27.34°, 31.88°, 46.22° and 55.12°, which is assigned to the CeO_2 (PDF # 89-8436) phase. Moreover, it can be clearly observed that with the gradual increase in the Ce doping amount, the characteristic peak intensity of the parent $LaMnO_3$ gradually weakens, while the peak intensity of the CeO_2 gradually increases [31]. The presence of the CeO_2 phase was attributed to the surface enrichment of atomic cerium and it not being incorporated into the $LaMnO_3$ perovskite framework during the calcination process. Therefore, the $La_{1-x}Ce_xMnO_{3+\delta}$ samples were considered to be multiple oxides, containing Ce-doped perovskite and cerium oxide [32]. The lattice structure, crystal size and lattice parameters of each sample are listed in Table 1. It can be seen that the total crystal size of $La_{1-x}Ce_xMnO_{3+\delta}$ samples was smaller than that of $LaMnO_{3+\delta}$, indicating that the doping of Ce degraded the crystallinity of perovskite crystal. The lattice parameters of $La_{1-x}Ce_xMnO_{3+\delta}$ decreased due to Ce doping. The lattice shrinks of $La_{1-x}Ce_xMnO_{3+\delta}$ crystal were attributed to the substitution of La^{3+} (ionic radii is 1.032 Å) by Ce^{3+} (ionic radii is 1.02 Å), though there is a certain amount of Ce^{4+} (ionic radii is 0.87 Å) in the $La_{1-x}Ce_xMnO_{3+\delta}$ perovskite [33].

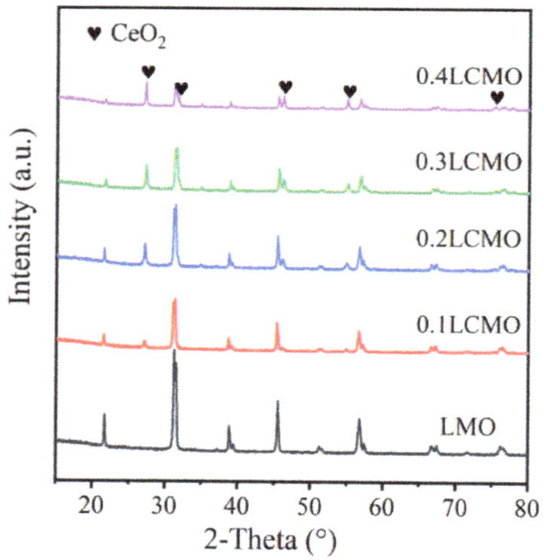

Figure 1. XRD patterns of as-prepared redox catalysts.

Table 1. Lattice structure, crystal size and lattice parameters of each fresh sample.

Samples	Lattice Structure	Crystal Size (nm)	a (Å)	b (Å)	c (Å)	Vol (Å^3)
LMO	hexagonal	21.93	5.525	5.525	13.361	407.853
0.1LCMO	orthorhombic	21.02	5.517	5.517	13.356	406.530
0.2LCMO	orthorhombic	19.56	5.516	5.516	13.348	406.130
0.3LCMO	orthorhombic	19.15	5.517	5.517	13.351	406.368
0.4LCMO	orthorhombic	18.23	5.521	5.521	13.358	407.171

In order to observe the micromorphology and element distribution of the samples more intuitively, the fresh samples were characterized by SEM-EDS, as shown in Figures 2 and 3. It can be clearly seen that all samples have a similar porous structure, and all elements are uniformly dispersed in the prepared samples. From the results of the BET experiment, as shown in Table 2, the specific surface area of the samples is between 3.1–5.1 m^2/g, and the specific surface area of $LaMnO_3$ without Ce doping is the smallest.

Figure 2. SEM images of fresh LMO (**a**), 0.1LCMO (**b**), 0.2LCMO (**c**), 0.3LCMO (**d**), 0.4LCMO (**e**).

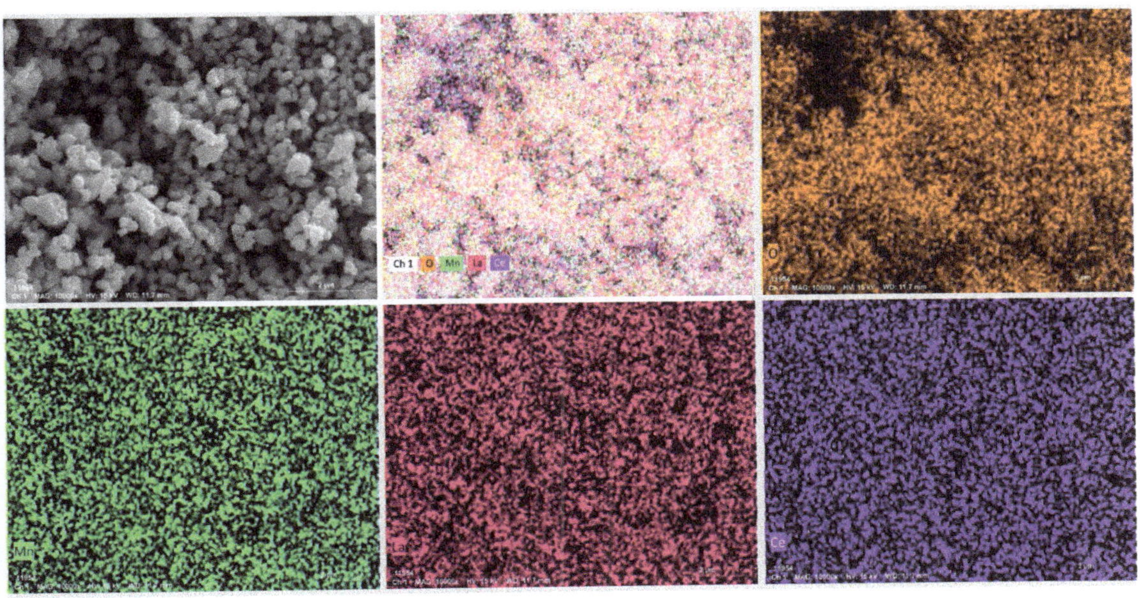

Figure 3. EDS spectra of fresh 0.3LCMO.

2.2. XPS

In order to further explore the influence of Ce-doping on the performance of $LaMnO_3$ in ODH reactions of ethane, XPS technology was used to test the types of manganese ions and oxygen species near the surface of the oxygen carrier. It is known that the manganese ions, located in the B sites of perovskite, directly affect the reactivity of perovskite;

meanwhile, the properties of the near-surface oxygen species greatly affect the ethylene selectivity of oxygen carriers for ethane conversion. Therefore, the O 1s and Mn 2p spectra of each sample were analyzed in detail, and the results are shown in Figure 4.

Table 2. Structural information of the fresh catalysts.

	S_{BET} (m^2/g)	V_{pore} (cm^3/g)	Average Pore Diameter (nm)
LCMO	3.1	0.01	9.2
0.1LCMO	5.1	0.02	14.7
0.2LCMO	3.4	0.02	17.1
0.3LCMO	4.7	0.04	26.4
0.4LCMO	5.0	0.03	15.8

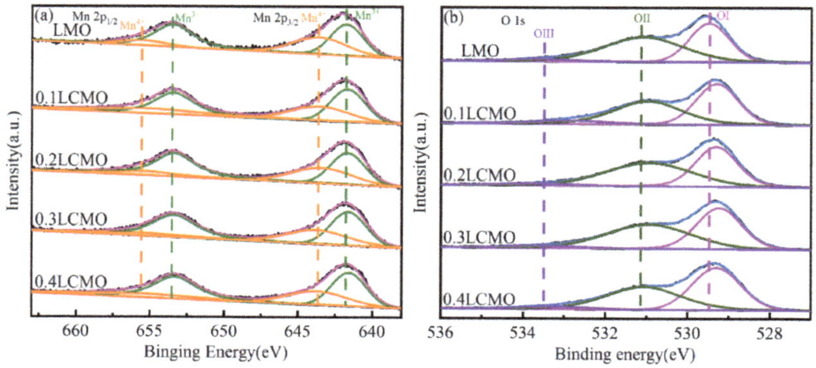

Figure 4. XPS spectra for (**a**) and (**b**) of the fresh samples.

As shown in Figure 4a, the Mn 2p is featured with a double-peaked spectrum, assigned to Mn $2p_{1/2}$ and Mn $2p_{3/2}$, respectively, due to energy level splitting. For each sample, two peaks with binding energies of around 641.5~641.8 eV and 643.4~643.8 eV were obtained via the deconvolution of Mn $2p_{3/2}$ spectra, corresponding to Mn^{3+} and Mn^{4+} species, respectively [34–36]. This is consistent with the analysis of the binding energy location of Mn^{3+} and Mn^{4+} species in Yin et al.'s study [34]. According to the detailed data of Mn $2p_{3/2}$ shown in Table 3, as Ce is introduced into the oxygen carrier, the characteristic peak of manganese shows a trend reflecting atransfer from a high binding energy to a low binding energy, and the value of Mn^{4+}/Mn^{3+} is obviously reduced. Considering the potential incorporation of Ce^{4+} with a higher valence than La^{3+}, which causes the distortion of the perovskite structure and the imbalance of internal charges, it is necessary to compensate by converting Mn^{4+} into Mn^{3+}, so more active sites are generated and more oxygen vacancies are formed [10]. Combined with the CL-ODH results, the changing trend in ethylene selectivity is negatively correlated with the changing trend in Mn^{4+}/Mn^{3+} value, indicating that the formation of Mn^{3+} is conducive to the selective production of ethylene from ethane.

Table 3. Molar proportion of surface element for each fresh sample was measured by XPS.

	Mn $2p_{3/2}$			O 1s			
	Mn^{3+}	Mn^{4+}	Mn^{4+}/Mn^{3+}	OI	OII	OIII	OI/(OII+OIII)
LCMO	0.53	0.47	0.90	0.40	0.55	0.05	0.67
0.1LCMO	0.58	0.42	0.74	0.47	0.44	0.09	0.88
0.2LCMO	0.54	0.46	0.85	0.45	0.52	0.03	0.81
0.3LCMO	0.55	0.45	0.81	0.46	0.50	0.04	0.85
0.4LCMO	0.58	0.42	0.71	0.49	0.46	0.05	0.96

For the deconvolution of O 1s spectrum, two main peaks and one secondary peak can be observed in all samples; these are the lattice oxygen species (O^{2-}, O I) near 529.2~529.5 eV, surface electrophilic oxygen species (O II) near 530.9~531.1 eV and the hydroxide/carbonate species ($OH^-/CO_3{}^{2-}$, O III) near 533.1~533.9 eV [37,38]. It is generally believed that both lattice oxygen and surface electrophilic oxygen species can improve the conversion rate of ethane in CL-ODH reactions. The difference is that lattice oxygen can promote the selective conversion of ethane to ethylene, whereas the electrophilic oxygen species is usually non-selective and prone to causing the deep oxidation of ethane [24,28]. It can be seen from the O 1s in Figure 4b, and the data in Table 3, that the relative content of the lattice oxygen of $LaMnO_3$ is obviously less than that of the doped sample, consistent with its worst selectivity in the CL-ODH process. The higher proportion of the lattice oxygen species in 0.1LCMO and 0.4LCMO is consistent with their relatively high ethane selectivity. As far as 0.2LCMO and 0.3LCMO are concerned, due to the similar relative amount of active oxygen substances (i.e., lattice oxygen and surface electrophilic oxygen), their reaction activities are almost the same (97% vs. 96%). As for the 0.3LCMO sample, it has a higher proportion of lattice oxygen, thereby giving a better reaction performance in CL-ODH.

2.3. H_2-TPR and O_2−TPD

In order to understand the reducibility of oxygen carriers in the CL-ODH reaction, a H_2-TPR test was conducted, and the H_2-TPR curve of the sample is shown in Figure 5a. It can be observed that all samples have two obvious reduction peaks, which are located at 300–620 °C and 650–850 °C, respectively. The reduction peak at low temperatures may originate from the consumption of adsorbed oxygen on the surface and the reduction of Mn^{4+} to Mn^{3+}, while the reduction peak at high temperatures is corresponding to the removal of lattice oxygen from the reduction of Mn^{3+} to Mn^{2+} [34,39]. It can be observed that the initial reduction peak position of the modified oxygen carrier obviously moves to the lower temperature, indicating that the addition of Ce improves the oxygen mobility of $LaMnO_3$ and the reducibility of the sample. It is generally believed that the samples with high reducibility have higher activity, which is consistent with the CL-ODH reaction results [25].

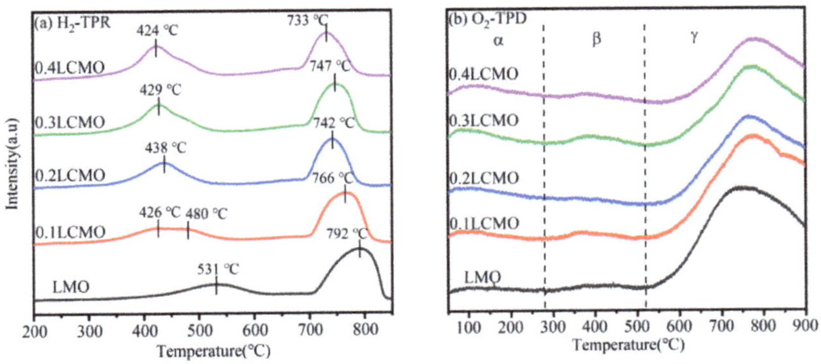

Figure 5. Results of the oxygen carriers.

In order to study the relative amounts of various oxygen species in the oxygen carrier, an O_2-TPD test was conducted on the fresh oxygen carrier. Figure 5b shows the O_2-TPD curve for each sample. According to the previous literature published to date, the curve is usually divided into three peaks, labeled as α, β and γ [40,41]. The α peak is in the temperature range below 280 °C, and the appearance of this decomposition absorption peak is attributed to the desorption of surface chemisorbed oxygen ($O_2{}^-$, $O_2{}^{2-}$, O^-). The β desorption peak, in the range of 280 °C–520 °C, is caused by the surface lattice oxygen

(O^{2-}) species. The higher temperature range of γ is related to the bulk lattice oxygen of perovskite crystals [40–43]. It can be noticed that the peaks in the α and β regions of the doped samples have significant changes, especially for the sample 0.3LCMO; this gives a significantly enhanced β peak region, which indicates that the introduction of cerium is conducive to the formation of more oxygen vacancies and improves the sample oxygen mobility. At elevated temperatures, it can be seen that the peak value of the γ region is shifted to a higher temperature due to the doping of cerium, which indicates that the oxidation activity of modified samples is worse at high temperatures. Therefore, the modification of the $LaMnO_3$ substrate is more conducive to the optimization of the reaction performance at low temperatures.

2.4. Ethane CL-ODH Reaction Resting

In the continuous cycle pulse experiment, due to the existence of adsorbed oxygen on the surface of the oxygen carrier, non-selective oxidation may occur at the beginning of the pulse, resulting in the poor performance of the sample. The continuous pulse experiment results of the 0.2LCMO sample at 700 °C are shown in Figure 6. It can be observed that in the first two pulses, the non selective oxygen species are rapidly consumed, and the sample performance tends to be stable without too much negative impact [44]. The ethane conversion, ethylene selectivity and ethylene yield data of different samples obtained in the CL-ODH step, at various temperatures (650 °C, 675 °C and 700 °C), are shown in Figure 7. The carbon balance calculations for all tests were in the range of 93102%. Perovskite-type oxygen carriers exhibit a satisfactory redox performance in the CL-ODH reaction at 700 °C. It can be seen from Figure 7c that, compared with the oxygen carrier without Ce, the reactivity of the modified oxygen carrier was almost improved, and it obtained a higher ethylene selectivity. These results show that the introduction of Ce is effective in improving the CL-ODH reaction performance of the $LaMnO_3$ substrate. Among all oxygen carriers, 0.3LCMO exhibited the highest activity for ethane conversion and moderate ethylene selectivity. While 0.1LCMO had a higher ethylene selectivity, the activity was the worst. Once the Ce-doping content was above 0.3, the activity or ethylene selectivity of the Ce-doped samples was further improved with the Ce content. Therefore, 0.3LCMO is considered to have the most suitable Ce doping ratio.

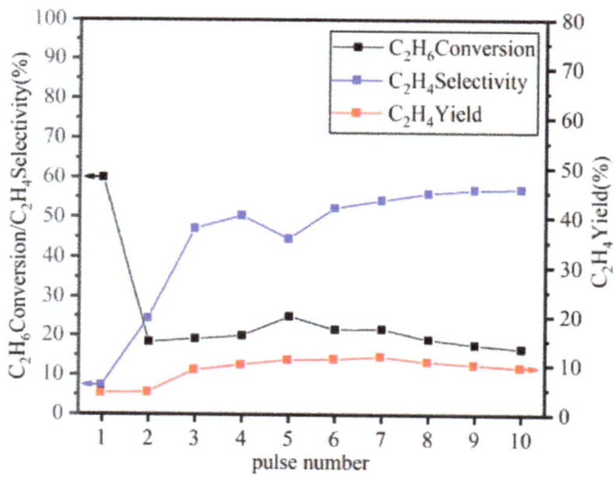

Figure 6. The 0.2LCMO continuous pulse experiment at 700 °C (GHSV = 1200 mL/h·g).

2.5. Cyclic Stability

The excellent stability and recyclability of perovskite-type catalysts at high temperatures are critical for multiple redox cycles in the utilization of chemical looping. Therefore, the 0.3LCMO sample with the best reaction performance was subjected to 20 consecutive cycles of redox tests, and the reaction results are shown in Figure 8. In the ethane ODH step, the conversion of ethane fluctuated slightly between 35–40%, while ethylene selectivity remained in the range of 25–30%, and the yield did not decline significantly after 20 cycles. The XRD results (Figure 9) can further confirm the structural stability of the 0.3LCMO sample. The sample after the re-oxidation cycle obtained the same XRD patterns as the fresh sample, and the characteristic peaks after successive redox cycles were dominated by $LaMnO_3$ characteristic peaks. In addition, it is noted that the sample showed weak peaks at $2\theta = 34.9°$ and $2\theta = 40.5°$ after cycling, which corresponded to MnO [34]. The SEM-EDS results (Figures 10 and 11) show that the structure of the circulating sample remains good after the cycle, and all elements are evenly distributed in the oxygen carrier.

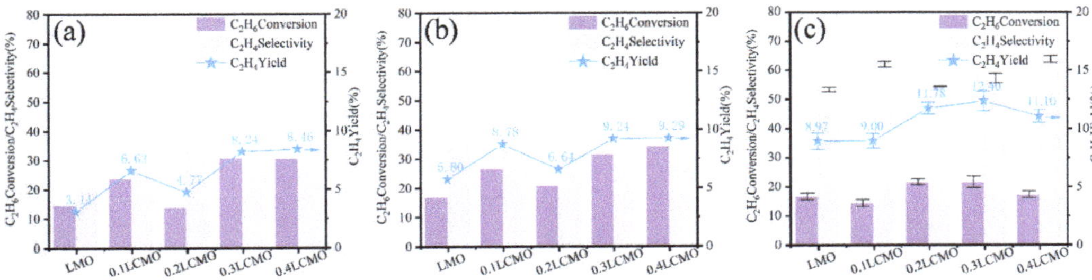

Figure 7. The ethane ODH reaction performances of different oxygen carriers at (**a**) 650 °C, (**b**) 675 °C and (**c**) 700 °C (GHSV = 1200 mL/h·g).

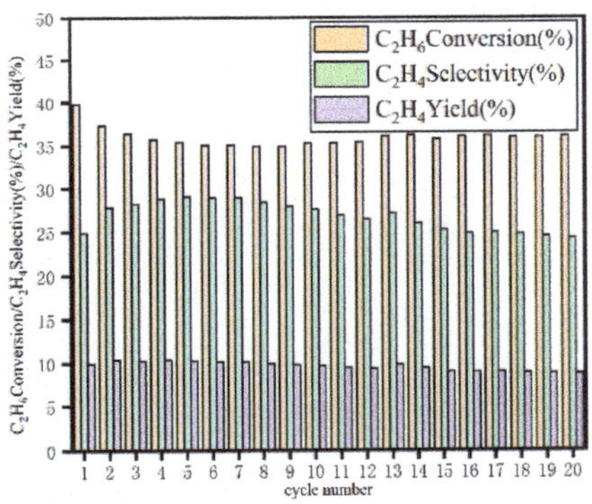

Figure 8. CL-ODH cycle for 0.3LCMO redox catalyst (GHSV = 1200 mL/h·g).

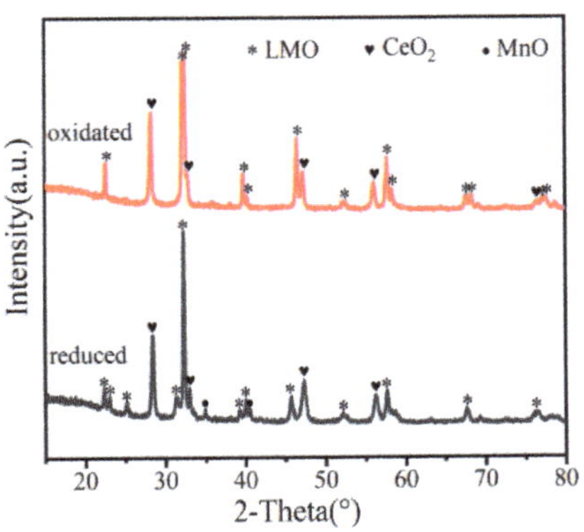

Figure 9. XRD patterns of 0.3LCMO after 20 continuous cycles test and regeneration.

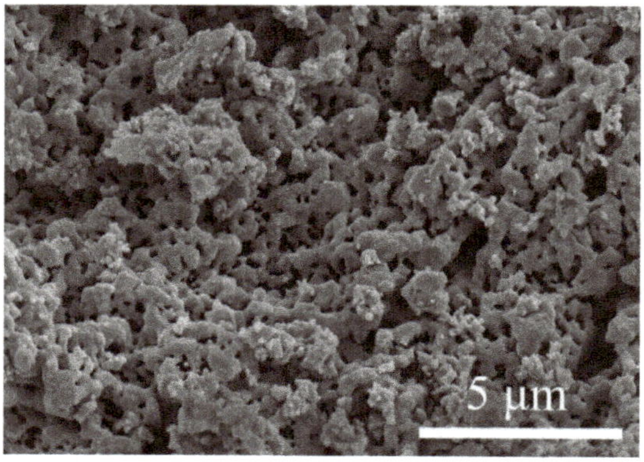

Figure 10. SEM images of 0.3LCMO after circulation.

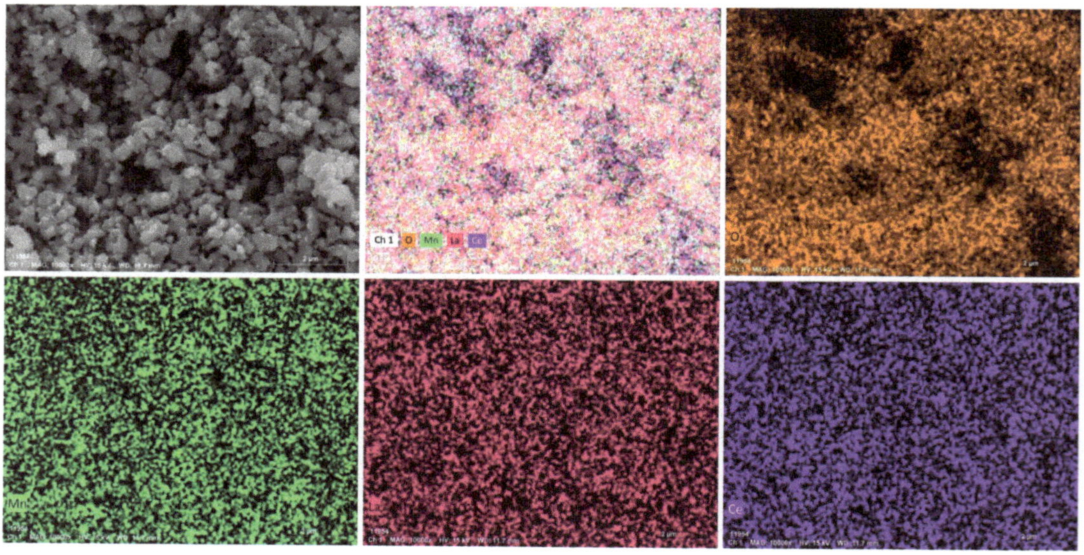

Figure 11. EDS spectra of 0.3LCMO after circulation.

3. Experimental Section

3.1. Redox catalyst Preparation

The $La_{1-x}Ce_xMnO_{3+\delta}$ (x = 0, 0.1, 0.2, 0.3, 0.4) oxygen carriers used in this work were all prepared by a simple sol–gel method described elsewhere [45–48]. Firstly, La $(NO_3)_3{\cdot}6H_2O$ (Aladdin, 99.99% purity), Ce $(NO_3)_3{\cdot}6H_2O$ (Aladdin, 99.95% purity) and Mn $(NO_3)_2$ (Rhawn,50 wt.% in solution) were dissolved in deionized water with a certain molar ratio (e.g. $La_{0.9}Ce_{0.1}MnO_{3+\delta}$, $La(NO_3)_3{\cdot}6H_2O$: $Ce(NO_3)_3{\cdot}6H_2O$: Mn $(NO_3)_2$ = 0.9:0.1:1), and the total nitrate concentration reached 1 mol/L. Then citric acid ($C_6H_8O_7{\cdot}H_2O$, Aladdin, ≥99.5% purity) was added to the solution, and the molar ratio of citric acid to total nitrate ions was 3:1. After the mixture was stirred in a water bath at 60 °C for 30 min, glycol solution ($C_2H_6O_2$, Aladdin, ≥99.5% purity) was added, and the molar ratio of glycol solution to citric acid was 2:1. The solution was heated to 80 °C under constant stirring (approximately 24 h) until a sol-like viscous substance formed. The gel was dried in a blast drying oven at 120 °C for 24 h. Finally, the sample was decomposed at 350 °C for 2 h and then calcinated at 1000 °C for 8 h in a muffle furnace; the heating rate of the calcination was 10 °C/min.

Eventually, five samples were attained, and they were labeled as LMO, 0.1LCMO, 0.2LCMO, 0.3LCMO, and 0.4LCMO, respectively. All samples were ground and sieved into the size range of 0.45–0.9 mm for reaction testing.

3.2. CL-ODH Experiments

The reaction performance of ethane CL-ODH was tested in a micro fixed-bed quartz tube reactor under atmospheric pressure; the length of the catalyst bed was 300 nm. In this experiment, the length of the quartz tube was 440 mm, the outer diameter was 8 mm, and the inner diameter was 6 mm. The amount of redox catalyst was 1.5 g. The appropriate amount of quartz wool was used to fix the catalyst on its upper and lower sides, but also to reduce the gas flow in the heated reaction area.

The whole fixed-bed reaction was divided into two stages: reduction and oxidation. In the reduction stage, the fixed-bed reactor acted as a fuel reactor and was heated to the reduction temperature in pure argon at 10 °C/min. Then, the atmosphere was switched to 40% ethane/balance nitrogen, and the air flow rate was 30 mL/min. After that, the argon gas was blown at the same speed and gas bags were used to collect all the tail gas at this stage. Next was the oxidation stage; here, the fixed-bed reactor acted as an air reactor, using 21% oxygen/balanced nitrogen to oxidize the catalyst; argon was also used for line purging.

Before the performance test, in order to explore the potential oxygen capacity of the prepared oxygen carrier, a continuous pulse experiment was carried out. After the reactor was filled with reaction gas (40% ethane/balance nitrogen) for 1 min, it was purged with argon for 2 min to complete a single pulse. Then, it was pulsed ten times continuously to observe the best performance time. All samples adopted the performance data at the same time. Each experiment was repeated three times.

In order to test the cycle stability of the materials, redox cycle experiments were carried out for the materials with the best performance. The reaction temperature was set to 700 °C, reaction gas was injected for 1 min in the reduction phase, and it was purged with argon for 2 min. Then, 21% O_2 (N_2 as balance gas) was used in the oxidation stage, and the flow rate was 20 mL /min for 4 min.The reactor was purged for 4 min by 30 mL/min argon between each redox cycle.

A Shimadzu GC-2014 gas chromatograph equipped with a flame ionization detector (FID, for hydrocarbon analysis) and two thermal conductivity detectors (TCD, for CO/CO_2 and H_2 analysis, respectively) was used to test the exhaust of the phone. According to the carbon balance, the C_2H_6 conversion, C_2H_4 selectivity, and C_2H_4 yield can be calculated using the following formulas:

$$\text{Conversion of ethane}(\%) = \frac{\text{introduced ethane} - \text{residual ethane}}{\text{introduced ethane}} \times 100\% \quad (4)$$

$$\text{Selectivity of ethylene}(\%) = \frac{\text{produced ethylene}}{\text{converted ethane}} \times 100\% \quad (5)$$

$$\text{Yield of ethylene}(\%) = \frac{\text{produced ethylene}}{\text{introduced ethane}} \times 100\% \quad (6)$$

3.3. Characterization Techniques

X-ray diffraction (XRD) was used to analyze the crystal structure of the samples. The test conditions were the following: Cu Kα radiation, a wavelength of λ = 1.5406 Å, a scanning speed of 15°/min, a target voltage of 40 kV, a target current of 40 mA, and a scanning angle of 15–80°. The specific surface area of the catalyst was measured by the JW-BK112 automatic physical adsorption analyzer using the Brunauer Emmett Teller (BET) theory. Before the test, the samples were dried in a vacuum at 300 °C for 3 h to completely remove the adsorbed water vapor and other impurities. Then, the pretreated samples were tested at low temperatures with liquid nitrogen (−196 °C) to obtain the adsorption and analytical curves. SU-5000 field emission scanning electron microscopy (SEM) was used to observe the microstructure of the catalyst under the conditions of 3 kV accelerating voltage and 5 mm working distance. At the same time, an XFlash6110 energy-dispersive spectrum analyzer (EDS) was used to analyze the element content of the catalyst. It is worth noting that in an SEM test, the sample needs to be treated with gold spraying before observation.

A hydrogen temperature programmed reduction (H_2-TPR) experiment was performed, using a PCA-1200 full-automatic temperature program of Beijing Builder Electronic Technology Company. First, the sample (0.01~0.015 g) was heated to 300 °C at 10 °C/min under 30 mL/min nitrogen atmosphere and kept constant for 1 h to remove moisture and adsorbed gas, and then cooled to room temperature naturally. In the reduction process, 5% H_2/Ar gas (the gas flow rate is 30 mL/min) was used and the temperature was increased to 900 °C (the heating rate iwas 10 °C/min) to obtain the H_2-TPR curve. Oxygen temperature programmed desorption (O_2-TPD) experiments were carried out using the same chemisorption instrument; in this experiment, the sample quality was 0.1~0.15 g, and the gas flow rate of the gas used was 30 mL/min. The pretreatment process of the samples was the same as that of the H_2-TPR experiment. After that, the pretreated samples were adsorbed in a pure oxygen environment for 1 h at 60 °C, and then switched to a helium atmosphere. After the TCD signal was stable, the temperature rate was increased to 900 °C at 10 °C/min, and signal data were recorded. The X-ray photoelectron spectroscopy (XPS) test used the ESCALAB 250 Xi multifunctional and high-performance surface analyzer of the US Thermo-Fisher Scientific Company. The experimental conditions were Al Kα rays (hν = 1486.6 eV) as the excitation source, and a power of 150 W. With channel energies of 20 eV and 100 eV, the total spectrum and the single element spectrum of C, La, Mn, Ce and O were obtained, respectively.

4. Conclusions

In this study, $LaMnO_3$ redox catalysts, modified with different proportions of cerium, were prepared by the sol–gel method for the CL-ODH reaction of ethane. Their properties were characterized by XRD, SEM, BET, XPS, H_2-TPR, O_2-TPD, etc. XRD shows that a CeO_2 phase is formed in Ce doped $LaMnO_3$, and the introduction of Ce^{4+} will cause the lattice distortion of calcium titanium. From the XPS results, it can be concluded that Ce^{4+} will promote the conversion of Mn^{4+} to Mn^{3+}, which is conducive to improving the ethylene conversion rate. The increase in active oxygen species plays a positive role in the reaction activity of the samples. Among all the catalysts, 0.3LCMO showed the best performance in the ODH reaction process, obtaining a 22% ethane conversion and 57% ethylene selectivity; this was improved compared with the substrate $LaMnO_3$, and showed a higher stability in the redox cycle experiment. This study makes a new attempt to further the design and development of perovskite-type redox catalysts; it also lays a foundation for the development of high-performance oxygen carriers for the ethane CL-ODH reaction.

Author Contributions: Conceptualization, J.W. and X.L.; methodology, J.W.; software, J.W.; validation, J.W., X.L. and F.H.; formal analysis, Z.X. and F.H.; investigation, J.W.; resources, H.C.; data curation, J.W.; writing—original draft preparation, J.W.; writing—review and editing, D.S. and F.H.; visualization, Y.L.; supervision, X.L. and D.S.; project administration, Y.L.; funding acquisition, F.H. All authors have read and agreed to the published version of the manuscript.

Funding: The authors gratefully acknowledge the financial support from the National Natural Science Foundation of China (22179027), Guangxi Natural Science Foundation (2018GXNSFDA281005, 2021GXNSFAA075036).

Data Availability Statement: Not applicable.

Conflicts of Interest: The authors declare no conflict of interest.

References

1. Dai, Y.H.; Gao, X.; Wang, Q.; Wan, X.Y.; Zhou, C.M.; Yang, Y.H. Recent Progress in Heterogeneous Metal and Metal Oxide Catalysts for Direct Dehydrogenation of Ethane and Propane. *Chem. Soc. Rev.* **2021**, *50*, 5590–5630. [CrossRef] [PubMed]
2. Mishanin, I.I.; Bogdan, T.V.; Koklin, A.E.; Bogdan, V.I. Design of Highly Selective Heterogeneous Catalyst for CO_2-mediated Ethane Oxidative Dehydrogenation Based on Nonoxidative Catalysis in Stainless-steel Reactor. *Chem. Eng. J.* **2022**, *446*, 137184. [CrossRef]
3. Najari, S.; Saeidi, S.; Concepcion, P.; Dionysiou, D.D.; Bhargava, S.K.; Lee, A.F.; Wilson, K. Oxidative Dehydrogenation of Ethane: Catalytic and Mechanistic Aspects and Future Trends. *Chem. Soc. Rev.* **2021**, *50*, 4564–4605. [CrossRef] [PubMed]
4. Tian, Y.; Westmoreland, P.R.; Li, F.X. $CaMn_{0.9}Ti_{0.1}O_3$ Based Redox Catalysts for Chemical Looping-Oxidative Dehydrogenation of Ethane: Effects of Na_2MoO_4 Promoter and Degree of Reduction on the Reaction Kinetics. *Catal. Today* **2022**. [CrossRef]
5. Amghizar, I.; Vandewalle, L.A.; Van Geem, K.M.; Marin, G.B. New Trends in Olefin Production. *Engineering* **2017**, *3*, 171–178. [CrossRef]
6. Sattler, J.J.H.B.; Ruiz-Martinez, J.; Santillan-Jimenez, E.; Weckhuysen, B.M. Catalytic Dehydrogenation of Light Alkanes on Metals and Metal Oxides. *Chem. Rev.* **2014**, *114*, 10613–10653. [CrossRef]
7. Zhang, R.H.; Wang, H.; Tang, S.Y.; Liu, C.J.; Dong, F.; Yue, H.R.; Liang, B. Photocatalytic Oxidative Dehydrogenation of Ethane Using CO_2 as a Soft Oxidant over Pd/TiO_2 Catalysts to C_2H_4 and Syngas. *ACS Catal.* **2018**, *8*, 9280–9286. [CrossRef]
8. Luongo, c.; Donat, F.; Krödel, M.; Cormos, C.; Müller, C.R. Experimental Data Supported Techno-economic Assessment of the Oxidative Dehydrogenation of Ethane through Chemical Looping with Oxygen Uncoupling. *Renew. Sust. Energ. Rev.* **2021**, *149*, 111403. [CrossRef]
9. Neal, L.M.; Yusuf, S.; Sofranko, J.A.; Li, F.X. Oxidative Dehydrogenation of Ethane: A Chemical Looping Approach. *Energy Technol.* **2016**, *4*, 1200–1208. [CrossRef]
10. Tian, X.; Zheng, C.H.; Zhao, H.B. Ce-modified $SrFeO_{3-\delta}$ for Ethane Oxidative Dehydrogenation Coupled with CO_2 Splitting via a Chemical Looping Scheme. *Appl. Catal. B-Environ.* **2022**, *303*, 120894. [CrossRef]
11. Ding, W.X.; Zhao, K.; Jiang, S.C.; Zhao, Z.L.; Cao, Y.; He, F. Alkali-metal Enhanced $LaMnO_3$ Perovskite Oxides for Chemical Looping Oxidative Dehydrogenation of Ethane. *Appl. Catal. A-Gen.* **2021**, *609*, 117910. [CrossRef]
12. Maffia, G.J.; Gaffney, A.M.; Mason, O.M. Techno-Economic Analysis of Oxidative Dehydrogenation Options. *Top. Catal.* **2016**, *59*, 1573–1579. [CrossRef]
13. Zhu, H.B.; Dong, H.L.; Laveille, P.; Saih, Y.; Caps, V.; Basset, J. Metal Oxides Modified NiO Catalysts for Oxidative Dehydrogenation of Ethane to Ethylene. *Catal. Today* **2014**, *228*, 58–64. [CrossRef]
14. Solsona, B.; Dejoz, A.; Garcia, T.; Concepción, P.; Lopez Nieto, J.M.; Vázquez, M.I.; Navarro, M.T. Molybdenum–vanadium Supported on Mesoporous Alumina Catalysts for the Oxidative Dehydrogenation of Ethane. *Catal. Today* **2006**, *117*, 228–233. [CrossRef]
15. Yokoyama, C.; Bharadwaj, S.S.; Schmidt, L.D. Platinum-tin and Platinum-copper Catalysts for Autothermal Oxidative Dehydrogenation of Ethane to Ethylene. *Catal. Lett.* **1996**, *38*, 181–188. [CrossRef]
16. Dai, H.X.; Ng, C.F.; Au, C.T. Perovskite-Type Halo-oxide $La_{1-x}Sr_xFeO_{3-\delta}X_\sigma$ (X=F, Cl) Catalysts Selective for the Oxidation of Ethane to Ethene. *J. Catal.* **2000**, *189*, 52–62. [CrossRef]
17. Védrine, J.C.; Fechete, I. Heterogeneous Partial Oxidation Catalysis on Metal Oxides. *CR Chim.* **2016**, *19*, 1203–1225. [CrossRef]
18. Baroi, C.; Gaffney, A.M.; Fushimi, R. Process Economics and Safety Considerations for the Oxidative Dehydrogenation of Ethane Using the M1 Catalyst. *Catal. Today* **2017**, *298*, 138–144. [CrossRef]
19. Mendiara, T.; García-Labiano, F.; Abad, A.; Gayán, P.; de Diego, L.F.; Izquierdo, M.T.; Adánez, J. Negative CO_2 Emissions through the Use of Biofuels in Chemical Looping Technology: A Review. *Appl. Energ.* **2018**, *232*, 657–684. [CrossRef]
20. Novotný, P.; Yusuf, S.; Li, F.X.; Lamb, H.H. Oxidative Dehydrogenation of Ethane Using MoO_3/Fe_2O_3 Catalysts in a Cyclic Redox Mode. *Catal. Today* **2018**, *317*, 50–55. [CrossRef]
21. Zhao, K.; He, F.; Huang, Z.; Wei, G.Q.; Zheng, A.Q.; Liu, H.B.; Zhao, Z.L. Perovskite-type Oxides $LaFe_{1-x}Co_xO_3$ for Chemical Looping Steam Methane Reforming to Syngas and Hydrogen Co-production. *Appl. Energ.* **2016**, *168*, 193–203. [CrossRef]
22. Zhao, H.B.; Tian, X.; Ma, J.C.; Su, M.Z.; Wang, B.W.; Mei, D.F. Development of Tailor-made Oxygen Carriers and Reactors for Chemical Looping Processes at Huazhong University of Science & Technology. *Int. J. Greenh. Gas Con.* **2020**, *93*, 102898.
23. Dudek, R.B.; Gao, Y.F.; Zhang, J.S.; Li, F.X. Manganese-containing Redox Catalysts for Selective Hydrogen Combustion under a Cyclic Redox Scheme. *AICHE J.* **2018**, *64*, 3141–3150. [CrossRef]
24. Yusuf, S.; Neal, L.; Bao, Z.H.; Wu, Z.L.; Li, F.X. Effects of Sodium and Tungsten Promoters on Mg_6MnO_8-Based Core–Shell Redox Catalysts for Chemical Looping—Oxidative Dehydrogenation of Ethane. *ACS Catal.* **2019**, *9*, 3174–3186. [CrossRef]
25. Gao, Y.F.; Neal, L.M.; Li, F.X. Li-Promoted $La_xSr_{2-x}FeO_{4-\delta}$ Core–Shell Redox Catalysts for Oxidative Dehydrogenation of Ethane under a Cyclic Redox Scheme. *ACS Catal.* **2016**, *6*, 7293–7302. [CrossRef]
26. Tian, Y.; Dudek, R.B.; Westmoreland, P.R.; Li, F.X. Effect of Sodium Tungstate Promoter on the Reduction Kinetics of $CaMn_{0.9}Fe_{0.1}O_3$ for Chemical Looping-Oxidative Dehydrogenation of Ethane. *Chem. Eng. J.* **2020**, *398*, 125583. [CrossRef]
27. Yusuf, S.; Neal, L.; Haribal, V.; Lamb, H.H.; Li, F.X. Manganese Silicate Based Redox Catalysts for Greener Ethylene Production via Chemical Looping-Oxidative Dehydrogenation of Ethane. *Appl. Catal. B-Environ.* **2018**, *232*, 77–85. [CrossRef]
28. Yusuf, S.; Neal, L.M.; Li, F.X. Effect of Promoters on Manganese-Containing Mixed Metal Oxides for Oxidative Dehydrogenation of Ethane via a Cyclic Redox Scheme. *ACS Catal.* **2017**, *7*, 5163–5173. [CrossRef]

29. Zhang, L.; Xu, W.B.; Wu, J.; Hu, Y.; Huang, C.D.; Zhu, Y.Y.; Tian, M.; Kang, Y.; Pan, X.L.; Su, Y.; et al. Identifying the Role of A-Site Cations in Modulating Oxygen Capacity of Iron-Based Perovskite for Enhanced Chemical Looping Methane-to-Syngas Conversion. *ACS Catal.* **2020**, *10*, 9420–9430. [CrossRef]
30. Rajkumar, T.; Sápi, A.; Ábel, M.; Kiss, J.; Szenti, I.; Baán, K.; Gómez-Pérez, J.F.; Kukovecz, Á.; Kónya, Z. Surface Engineering of CeO_2 Catalysts: Differences Between Solid Solution Based and Interfacially Designed $Ce_{1-x}M_xO_2$ and MO/CeO_2 (M=Zn, Mn) in CO_2 Hydrogenation Reaction. *Catal. Lett.* **2021**, *151*, 3477–3491. [CrossRef]
31. Ahmad, J.; Ahmad, U.; Bukhari, S.H. Synthesis and Optical Properties of $La_{1-X}Ce_XMnO_3$ Studied by Infrared Reflectivity Measurements. *Chin. J. Phy.* **2018**, *56*, 1439–1448. [CrossRef]
32. Zhang, C.H.; Hua, W.C.; Wang, C.; Guo, Y.L.; Guo, Y.; Lua, G.Z.; Baylet, A.; Giroir-Fendler, A. The Effect of A-site Substitution by Sr, Mg and Ce on the Catalytic Performance of LaMnO3 Catalysts for the Oxidation of Vinyl Chloride Emission. *Appl. Catal. B-Environ.* **2013**, *134–135*, 310–315. [CrossRef]
33. Suárez-Vázquez, S.I.; Moreno-Román, E.J.; Zanella, R.; Cruz-López, A.; García-Goméz, C.; Nieto-Márquez, A.; Gil, S. Insight into the Surface Reaction Mechanism of Toluene Oxidation over a Composite $CeO_x/La_{1-x}Ce_xMnO_3$ Catalyst using DRIFTS. *Chem. Eng. Sci.* **2022**, *259*, 117831. [CrossRef]
34. Yin, X.L.; Wang, S.; Wang, B.Y.; Shen, L.H. Perovskite-type $LaMn_{1-x}B_xO_{3+\delta}$ (B = Fe, CO and Ni) as Oxygen Carriers for Chemical Looping Steam Methane Reforming. *Chem. Eng. J.* **2021**, *422*, 128751. [CrossRef]
35. Wang, Y.J.; Zheng, Y.E.; Wang, Y.H.; Wang, H.; Zhu, X.; Wei, Y.G.; Wang, Y.M.; Jiang, L.H.; Yang, Z.Y.; Li, K.Z. Evaluation of Fe Substitution in Perovskite $LaMnO_3$ for the Production of High Purity Syngas and Hydrogen. *J. Power Sources* **2020**, *449*, 227505. [CrossRef]
36. Zhao, K.; He, F.; Huang, Z.; Wei, G.Q.; Zheng, A.Q.; Li, H.B.; Zhao, Z.L. Perovskite-type $LaFe_{1-x}Mn_xO_3$ (x = 0, 0.3, 0.5, 0.7, 1.0) Oxygen Carriers for Chemical-looping Steam Methane Reforming: Oxidation Activity and Resistance to Carbon Rormation. *Korean J. Chem. Eng.* **2017**, *34*, 1651–1660. [CrossRef]
37. Jiang, S.C.; Ding, W.X.; Zhao, K.; Huang, Z.; Wei, G.Q.; Feng, Y.Y.; Lv, Y.J.; He, F. Enhanced Chemical Looping Oxidative Coupling of Methane by Na-doped $LaMnO_3$ Redox Catalysts. *Fuel* **2021**, *299*, 120932. [CrossRef]
38. Ponce, S.; Peña, M.A.; Fierro, J.L.G. Surface Properties and Catalytic Performance in Methane Combustion of Sr-substituted Lanthanum Manganites. *Appl. Catal. B-Environ.* **2000**, *24*, 193–205. [CrossRef]
39. Zhang, C.H.; Zeng, K.; Wang, C.; Liu, X.H.; Wu, G.L.; Wang, Z.; Wang, D. $LaMnO_3$ Perovskites via a Facile Nickel Substitution Strategy for Boosting Propane Combustion Performance. *Ceram. Int.* **2020**, *46*, 6652–6662. [CrossRef]
40. Najjar, H.; Lamonier, J.-F.; Mentré, O.; Giraudon, J.-M.; Batis, H. Optimization of the Combustion Synthesis towards Efficient $LaMnO_{3+y}$ Catalysts in Methane Oxidation. *Appl. Catal. B-Environ.* **2011**, *106*, 149–159. [CrossRef]
41. Yang, X.Q.; Yu, X.L.; Jing, M.Z.; Song, W.Y.; Liu, J.; Ge, M.F. Defective $MnxZr_{1-x}O_2$ Solid Solution for the Catalytic Oxidation of Toluene: Insights into the Oxygen Vacancy Contribution. ACS Appl. *Mater. Inter.* **2019**, *11*, 730–739. [CrossRef] [PubMed]
42. Zhang, C.H.; Wang, C.; Hua, W.H.; Guo, Y.L.; Lu, G.Z.; Gil, S.; Giroir-Fendler, A. Relationship between Catalytic Deactivation and PhysicoChemical Properties of $LaMnO_3$ Perovskite Catalyst during Catalytic Oxidation of Vinyl Chloride. *Appl. Catal. B-Environ.* **2016**, *186*, 173–183. [CrossRef]
43. Zhu, J.J.; Zhao, Z.; Xiao, D.H.; Li, J.; Yang, X.G.; Wu, Y. Study of $La_{2-x}Sr_xCuO_4$ (x = 0.0, 0.5, 1.0) Catalysts for NO+CO Reaction from the Measurements of O_2-TPD, H_2-TPR and Cyclic Voltammetry. *J. Mol. Catal. A-Chem.* **2005**, *238*, 35–40. [CrossRef]
44. Wang, T.; Gao, Y.F.; Liu, Y.Z.; Song, M.H.; Liu, J.J.; Guo, Q.J. Core-shell $Na_2WO_4/CuMn_2O_4$ Oxygen Carrier with High Oxygen Capacity for Chemical Looping Oxidative Dehydrogenation of Ethane. *Fuel* **2021**, *303*, 121286. [CrossRef]
45. Galinsky, N.; Mishra, A.; Zhang, J.; Li, F.X. $Ca_{1-x}A_xMnO_3$ (A = Sr and Ba) Perovskite Based Oxygen Carriers for Chemical Looping with Oxygen Uncoupling (CLOU). *Appl. Energ.* **2015**, *157*, 358–367. [CrossRef]
46. He, F.; Li, X.A.; Zhao, K.; Huang, Z.; Wei, G.Q.; Li, H.B. The Use of $La_{1-x}Sr_xFeO_3$ Perovskite-type Oxides as Oxygen Carriers in Chemical-looping Reforming of Methane. *Fuel* **2013**, *108*, 465–473. [CrossRef]
47. Song, D.; Lin, Y.; Zhao, K.; Huang, Z.; He, F.; Xiong, Y. Migration Mechanism of Lattice Oxygen: Conversion of CO_2 to CO Using $NiFe_2O_4$ Spinel Oxygen Carrier in Chemical Looping Reactions. *Catalysts* **2022**, *12*, 1181. [CrossRef]
48. Zhang, C.H.; Guo, Y.L.; Guo, J.; Lu, G.Z.; Boreave, A.; Retailleau, L.; Baylet, A.; Giroir-Fendler, A. $LaMnO_3$ Perovskite Oxides Prepared by Different Methods for Catalytic Oxidation of Toluene. *Appl. Catal. B-Environ.* **2014**, *148–149*, 490–498. [CrossRef]

catalysts

MDPI

Article

Limestone Calcination Kinetics in Microfluidized Bed Thermogravimetric Analysis (MFB-TGA) for Calcium Looping

Dan Li, Yang Wang and Zhenshan Li *

Key Laboratory for Thermal Science and Power Engineering of Ministry of Education, Department of Energy and Power Engineering, Tsinghua University, Beijing 100084, China

* Correspondence: lizs@mail.tsinghua.edu.cn; Tel.: +86-10-62789955

Abstract: Limestone calcination is an important part of calcium looping (CaL) technology and is critical to the design and operation optimization of fluidized bed reactors. However, obtaining a method of measuring the fast calcination kinetics in a fluidizing environment with isothermal conditions is still a challenge in the field of calcium looping. We address this challenge in this work and develop a new method of obtaining limestone calcination kinetics by injecting limestone particles into the hot fluidizing sands in a microfluidized bed thermogravimetric analysis (MFB-TGA) with a mass measurement resolution of 1 mg. The calcination characteristics of limestone are investigated at different particle sizes (150–1250 μm), temperatures (750–920 °C), and CO_2 concentrations (0–30 vol.%). The experimental data measured from MFB-TGA were analyzed using a detailed model including surface reaction and intraparticle and external diffusion. The results show that the kinetics of limestone calcination measured by MFB-TGA are faster than those measured via regular TGA. This particle-injecting method of MFB-TGA provides a new experimental idea for measuring fast calcination kinetics occurring inside fluidized bed reactors and provides guidance on the application of CaL technology.

Keywords: limestone calcination; microfluidized bed thermogravimetric analysis (MFB-TGA); calcination kinetics; model

1. Introduction

Calcium looping technology is a promising technology based on the calcination/carbonation reaction of calcium carbonate ($CaCO_3$) [1,2]. It can be widely applied in carbon capture, adsorption-enhanced steam methane reforming (SE-SMR), integrated CO_2 capture and conversion, and thermochemical heat storage [3–7]. CaL technology typically involves the use of dual interconnected circulating fluidized bed (CFB) reactors [8,9]. In the first reactor, lime (CaO) reacts with CO_2 in the temperature range of 600–700 °C to form $CaCO_3$. The $CaCO_3$ is then subsequently transported to the second CFB reactor for calcination. The regenerated CaO is then looped back to the first reactor. CaL technology has the following advantages [10,11]: abundant and inexpensive natural limestone with high theoretical CO_2 capture capacity and relatively low energy consumption; high energy storage density and great heat recovery potential due to high-temperature carbonation reaction. The $CaCO_3$ calcination stage, as a major part of the CaL technology, has a significant influence on the design and operation of the regeneration reactor, system integration and optimization, and the sintering of the CaO absorbent [12,13].

$CaCO_3$ calcination is a complex physical and chemical process in which reaction, heat transfer, and mass transfer are mutually coupled [14]. Researchers have studied the influence mechanism of $CaCO_3$ calcination, and it is generally accepted that with a decrease in particle size, the influence of heat transfer on calcination kinetics weakens and the influence of chemical reaction enhances [15–17]. For very small particles (<90 μm), the chemical reaction is the controlling step [17]. For particles >6 mm, heat transfer becomes a key process [18]. For the particle size between these values, both chemical reaction and mass transfer control the calcination kinetics [19]. Several models have been developed in the past to describe the calcination behavior, including the shrinking core model [20], the grain model [21], the uniform transformation model [17], and the random pore model [15]. The calcination of $CaCO_3$ particles with low porosity has been typically described using the shrinking core model, which assumes that there is an obvious reaction interface between the nonporous $CaCO_3$ and the generated CaO [19,20,22]. However, the diffusion of generated CO_2 through the CaO layer and external diffusion processes in the atmosphere are generally ignored. Diffusion resistance has an important influence on the decomposition of kinetics, especially for larger particle sizes [23]. In addition, due to several factors affecting the calcination of $CaCO_3$, such as particle size, temperature, CO_2 concentration, and experimental apparatus, there is no consensus on the calcination mechanism of $CaCO_3$ particles, and the obtained kinetic parameters are markedly different. It has been reported [24–26] that the activation energy of $CaCO_3$ calcination is in the range of 100–230 kJ/mol. Furthermore, researchers [19,27–29] have found that the CO_2 concentration in the atmosphere has a significant effect on the $CaCO_3$ calcination rate. The effect of CO_2 partial pressure on the calcination is normally described with the form of $(1 - P/P_{eq})^n$ (where P is the CO_2 partial pressure in the atmosphere and P_{eq} is the CO_2 equilibrium partial pressure), and the value of the exponent n ranges from 0.5 to 2 [19,30,31]. Some researchers adopt surface elementary reactions to explain the chemical reaction kinetics of $CaCO_3$ calcination in CO_2 atmosphere [19,32,33]. The broad activation energy and CO_2 partial pressure ranges reveal that calcination experiments and modeling under fluidized bed conditions are of great significance for the future scaling up of CaL technology. Regardless of the concept used for describing the calcination kinetics, the calcination data must be measured based on an accurate experimental method.

In order to obtain the kinetics of $CaCO_3$ calcination, there are mainly the following methods: (i) Thermogravimetric analyzer (TGA) is a method commonly adopted by researchers [22,25,29,34–36]. However, it is still a challenge to obtain calcination results at constant temperatures [36], especially at high temperatures. In addition, the heating rate of particles is usually lower than 100 K/min, and the diffusion resistance is serious in TGA instruments, which is not conducive to the development of kinetics models. The solid sample in the crucible of TGA is in a packed state, which is somewhat different from the fluidization in a fluidized reactor. (ii) Fluidized bed reactor based on gas signal measurement [27]. The vigorous movement of solid particles means excellent heat and mass transfer, which is a natural advantage of fluidized bed reactors for kinetic studies. The difficulty of the method lies in the capture of minor change in the gas concentration signal of rapid reaction, especially when the CO_2 produced by the sample is very little at high CO_2 concentrations [37]. Moreover, the blend of gas species affects the original signal and causes errors when there are variations in the concentration of multiple constituent gases in the experiment. (iii) Tube furnace [38] and high-temperature gas–solid suspension furnace [39,40]; the particle heating rate is fast, but the sample size is smaller than ~70 μm, and the residence time is shorter than 5 s. At the same time, the gas and particles inside the drop tube furnace are in entrained flow, while the gas inside the fluidized bed is divided into a bubble phase and an emulsion phase. The particles inside the fluidized bed will form a dense phase. Therefore, the calcination kinetics obtained from drop tube furnace cannot be accurately applied for the development of a fluidized bed reactor. (iv) In situ XRD method [32]; the change in crystal structure of $CaCO_3$ particles during calcination is analyzed using X-ray diffraction, which can be combined with TGA to obtain in situ particle conversion.

Limestone calcination for calcium looping occurs in a fluidizing state; therefore, it is necessary to measure the calcination kinetic data from the fluidized bed reactor. The aim of the work is to obtain the limestone calcination characteristics in the fluidized bed condition and to describe the calcination kinetics to guide the design and operation scale-up of the fluidized bed reactor. First, based on the previously constructed MFB-TGA with a mass measurement resolution of 1 mg and strong heat and mass transfer properties [41], a new method of injecting the limestone particles into the MFB-TGA is proposed and developed to measure the calcination kinetics under the fluidization and isothermal conditions with various particle sizes, temperatures, and CO_2 partial pressures under constant temperature conditions. Then, a shrinking core model is established considering the effects of CO_2 concentration and mass transfer rate on the calcination process. Finally, the relative importance of surface reactions, intraparticle diffusion, and external gas diffusion is analyzed and discussed based on experimental data and established models.

2. Results

2.1. Comparison of MFB-TGA with Regular TGA

TGA is a commonly used commercial instrument for measuring the kinetics of limestone calcination [22,25,29,35]. In the TGA experiment, a 5–10 mg sample is placed in a crucible and heated at a heating rate of 50 K/min to obtain the weightless curvature of the limestone. It can be seen from Figure 1 that the limestone particles are decomposed during the heating process. The decomposition characteristics of limestone particles under constant temperature conditions are difficult to obtain, which is of great significance for the establishment of the decomposition models and the determination of kinetics. Furthermore, the conversion of limestone particles in TGA is slower compared to MFB-TGA due to the gas diffusion resistance. Thus, a smaller value of calculated chemical reaction rate constant could be obtained from TGA results. Correspondingly, the controlling step may be wrongly determined as a chemical reaction. It still remains a challenge for the TGA method to eradicate the mass transfer effects and maintain isothermal decomposition, especially for rapid reactions [41,42]. MFB-TGA couples the accurate and real-time quality monitoring characteristics of TGA and the motion state of particle fluidization with good heat and mass transfer effects [27]. The CO_2 produced by the calcination of the 60–70 mg limestone sample is limited. Coupled with the fast external diffusion rate, the effect of mass transfer is significantly reduced. Details on the reduction of mass transfer effects by MFB-TGA can be found in previous works [41,43].

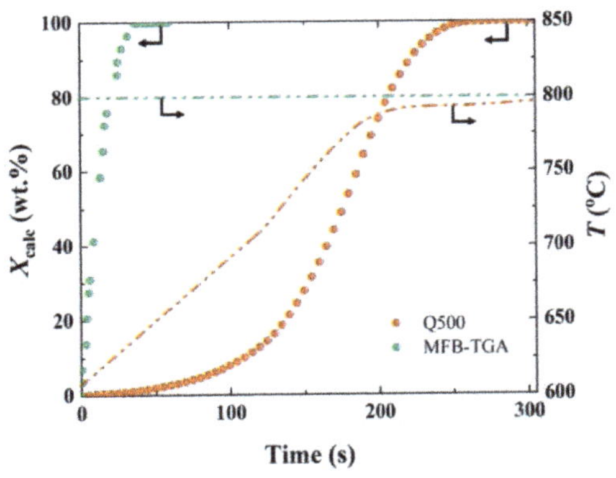

Figure 1. Comparison of MFB-TGA experimental results to TGA (800 °C, 0 vol.% CO_2).

2.2. Effect of Temperature on the Calcination Kinetics

Temperature has a significant effect on the calcination performance of limestone in a broad range. Calcination experiments of limestone with particle sizes of 150 μm are conducted at temperatures from 750 to 920 °C under N_2 atmosphere in MFB-TGA apparatus. The model results are compared with the experimental data of MFB-TGA at various temperatures shown in Figure 2a. The conversion of limestone particles increases rapidly from initial to full conversion with time, especially at high temperature. Under 900 °C, it only takes ~10 s for particle conversion to achieve ~100%. Even at 750 °C, the particle conversion reaches more than 90% in ~60 s, which is significantly faster than the calcination rate reported in other works [23,25,44]. It is mainly due to the fact that the limestone particles are injected into the quartz reactor in the MFB-TGA experiments. The particles achieve the target temperature and CO_2 concentration in an instant for calcination. In the state of vigorous fluidization, the particles are well blended with the inert bed material, and the temperature of the particles are basically kept consistent with the inert bed material.

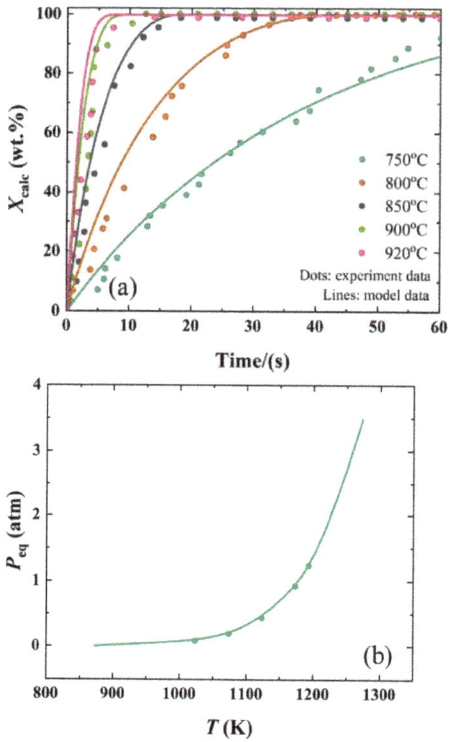

Figure 2. The effect of temperature: (**a**) comparison of model results with MFB-TGA experimental data; (**b**) the relation between CO_2 equilibrium partial pressure and temperature.

The conversion of limestone particles increases significantly with temperature. At 750, 800, 850, 900, and 920 °C, the time for the particle conversion to attain ~90% corresponds to 54 s, 25 s, 15 s, 8 s, and 6 s, respectively. This is mainly attributable to the calcium carbonate calcination reaction being an endothermic reaction with a standard enthalpy of +178 kJ/mol. In addition, as shown in Figure 2b, the thermodynamic equilibrium partial pressure of CO_2 corresponding to calcium carbonate calcination increases exponentially with temperature. Based on Equation (29), the calcination driving force (P_{eq}-P_g) increases, resulting in a gradual increase in the calcination rate. The above discussion indicates that the model can predict well the effect of temperature on calcination characteristics.

2.3. Effect of CO_2 Concentration on the Calcination Kinetics

The effect of CO_2 concentration in the gas phase on the limestone calcination is displayed in Figure 3. With the increase of CO_2 concentration in the gas phase, the conversion of limestone decreases significantly under a certain temperature. At 900 °C, when the CO_2 concentration is 0 vol.%, the particles only require ~11 s to achieve complete conversion, while under 30 vol.% atmosphere, the particle conversion is only 70 wt.% within 80 s. At 920 °C, the particles can be almost completely converted. The calcination rate of limestone under different CO_2 concentrations can be well predicted using the kinetic model established in Section 5.

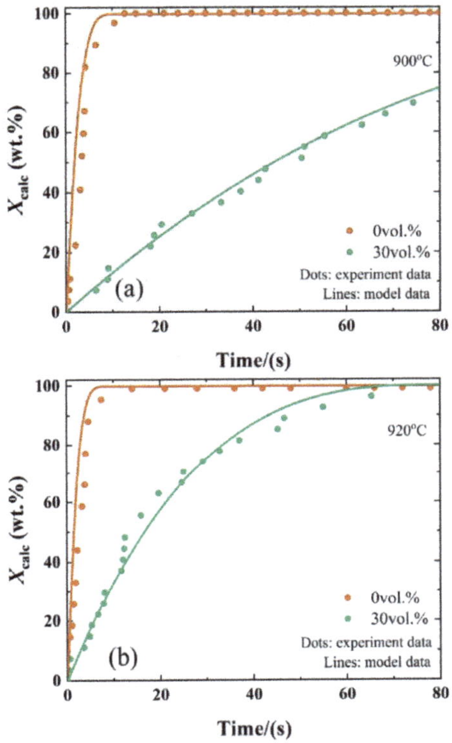

Figure 3. Comparison between the model results and the MFB-TGA experimental data at different CO_2 concentrations under (**a**) 900 °C and (**b**) 920 °C.

The MFB-TGA experiments are conducted under atmospheric pressure, and thus the volume fraction of CO_2 determines the CO_2 partial pressure in the gas phase. It can be obtained from Equation (15) that the partial pressure of CO_2 has a significant effect on the chemical reaction rate of calcium carbonate calcination. With the increase in CO_2 partial pressure P_g in the gas phase from 0 vol.% to 30 vol.% under 900 °C, the comprehensive chemical reaction constant k_c decreases dramatically by ~2 orders, which significantly reduces the calcination rate of limestone particles. With the increase in CO_2 concentration, the intraparticle diffusion rate of CO_2 generated by particle calcination from the reaction surface to the particle surface decreases, especially at the initial stage of the reaction. As the reaction proceeds, the driving force decreases (P_i-P_g), P_i is close to P_g, and the influence of diffusion is no longer significant.

2.4. Effect of Particle Size on the Calcination Kinetics

Particle size is an important factor determining calcination kinetics and generally affects external and intraparticle gas diffusion resistance. Limestone particles of different particle sizes are applied in practice. Thus, it is necessary to investigate the calcination characteristics of particles in different particle size ranges. The calcination process of 150–1250 μm limestone particles under 900 °C and 0 vol.% and 30 vol.% CO_2 concentration is explored in this work. As shown in Figure 4, the limestone particles conversion decreases with the particle size from 150 μm to 1250 μm. For the smaller particles of 150 μm, it only takes ~10 s to complete the calcination process, while for the larger particles of 1250 μm, the conversion of the particles is only ~90 wt.% within 80 s. An interesting finding in the work is that when the CO_2 concentration in the gas phase increases from 0 vol.% to 30 vol.%, the decrease in the conversion of limestone particles becomes less obvious with the increase of particle size. This indicates that the effect of CO_2 concentration in the gas phase on the reduction of the calcination rate of small particle limestone is higher than that of large particle size. Furthermore, it can be seen that the results of the prediction model are in good agreement with the experimental values.

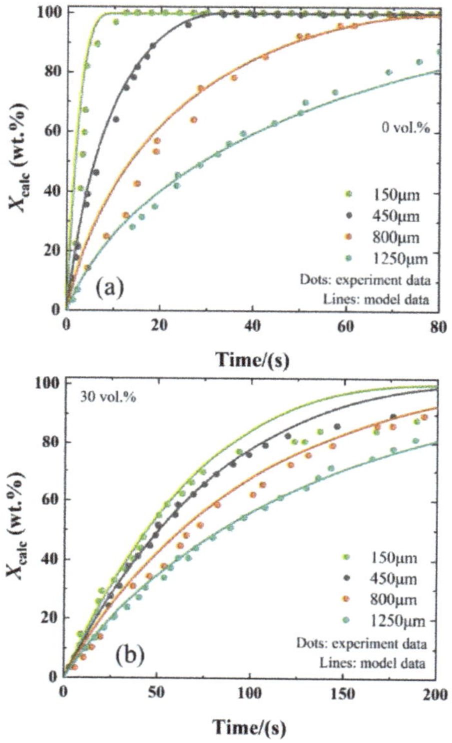

Figure 4. Comparison between the model results and the MFB-TGA experimental data at different particle size under (**a**) 0 vol.%; (**b**) 30 vol.%.

For larger particle sizes, the gas mass transfer resistance is larger both for external and intraparticle diffusion. Thus, the CO_2 partial pressure on the particle surface is higher than for smaller particles. With the continuous progress of the chemical reaction, the change of CO_2 partial pressure on the particle surface under high CO_2 concentration in the gas phase is smaller than that under lower CO_2 concentration, which results in the change of particle conversion with particle size no longer being significant at high CO_2 concentrations.

2.5. Determination of Kinetic Parameters

Based on the comparison of the experimental results with the model established in Section 5, the value of the activation energy can be obtained from the slope of the fitted line, and the value of the pre-exponential factor can be obtained from the vertical intercept of the fitted line in Figure 5. The kinetic parameter calculation is summarized in Table 1. The activation energy of the chemical reaction rate constant k_f is 182 kJ/mol, which indicates that the temperature has an extremely distinct effect on the value of the chemical reaction rate constant in the range of 750–920 °C. Activation energies have been reported to be in the range of 100–230 kJ/mol depending on the specific calcination conditions and the established model [24]. Escardino et al. [45] obtained an apparent activation energy value of 175 kJ/mol for the direct chemical reaction of thermal calcination. Garcia et al. [19] studied limestone and dolomite, with activation energies ranging from 114–166 kJ/mol. The E values measured in the work are close to those reported by other researchers [19,22,27,29]. The activation energy of the thermodynamic equilibrium constant K_1 in the reaction kinetics in this work is 150 kJ/mol, which is consistent with the value reported by Jose et al. [33]. Other parameters in the model are summarized in Table 1.

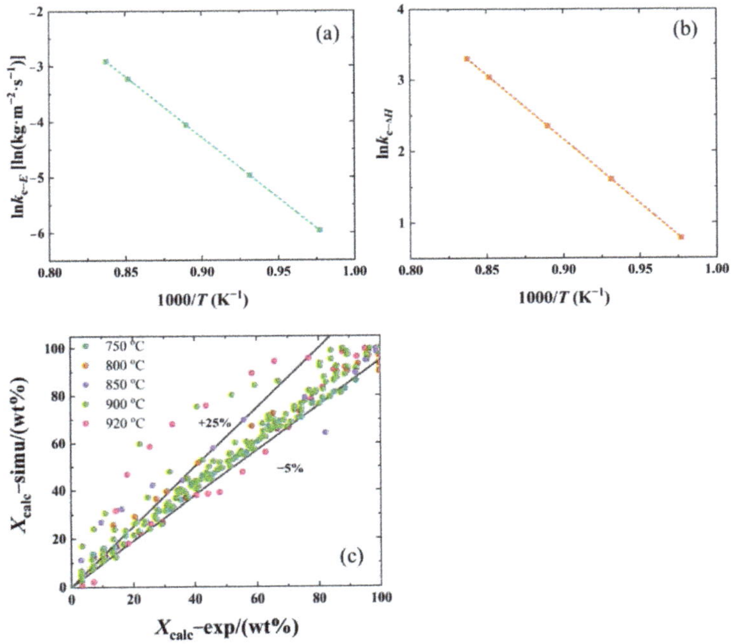

Figure 5. Arrhenius plots to calculate the kinetic parameters: (**a**) $\ln k_{c-E}$ versus 1/T; (**b**) $\ln k_{c-\Delta H}$ versus 1/T; (**c**) comparison of model results to experimental data.

Table 1. Kinetic parameters in the model calculation.

A (kg/m^2/s)	E (kJ/mol)	a (/)	ΔH (kJ/mol)	C_1 (s/K$^{0.75}$)	S_0 (m^2/kg)	ε_0 (/)	$\bar{r}_0$(μm)
5.051×10^6	182	9.99×10^7	150	5×10^{-12}	22.80	0.48	330

Figure 5c compares the model results and the experimental data. Most of the errors between the model and experimental results are between −5% and +20%, indicating that the model can reasonably predict the decomposition conversion within the experimentally investigated range.

3. Discussion

Based on Equation (10), the fraction occupied by $^*CO_2$ in the adsorbed state on the particle surface, θ, is calculated as:

$$\theta = \frac{P_i}{K_2 + P_i} = \frac{P_{eq} - \frac{K_r}{k_c} \cdot (P_{eq} - P_g)}{\frac{P_{eq}}{K_1} + P_{eq} - \frac{K_r}{k_c} \cdot (P_{eq} - P_g)} \tag{1}$$

θ represents the fraction of active sites on the surface of the limestone particles occupied by CO_2 produced by calcination. The relationship between θ and CO_2 partial pressure in the gas phase is displayed in Figure 6. It can be seen that with the increase in CO_2 partial pressure, the fraction θ occupied by CO_2 on the particle surface gradually increases. That is, the R2 desorption process of the CO_2 produced by calcium carbonate from the particle surface through the reaction R1 is limited by the increase of CO_2 in the gas phase. Especially in the low CO_2 partial pressure stage, θ increases rapidly with the increase of CO_2 partial pressure. When the partial pressure of CO_2 in the gas phase achieves ~30,000 Pa (~30 vol.% at atmospheric pressure) at 900 °C, the fraction θ occupied by active sites on the particle surface increases from 0 to 0.8. However, when the partial pressure of CO_2 in the gas phase is higher than ~60,000 Pa (~60 vol.% at atmospheric pressure), with the increase of the partial pressure of CO_2, θ increases slowly. In this case, the desorption process of CO_2 from the particle surface becomes very slow and the calcination reaction is limited.

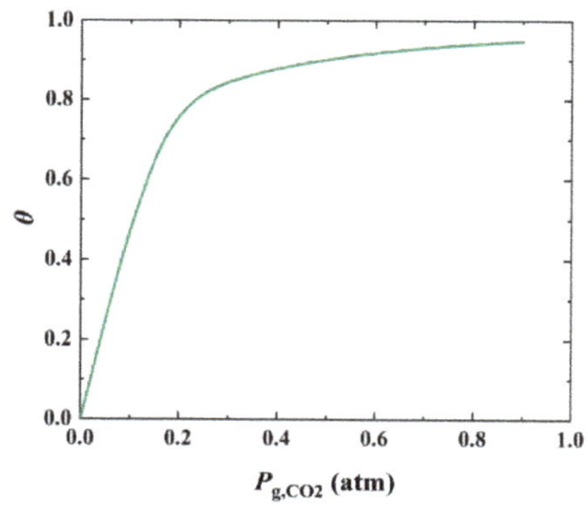

Figure 6. The fraction of surface active sites occupied θ under 900 °C.

Figure 7 shows the change of the occupied fraction θ of active sites on the particle surface under different particle sizes. It can be found that for larger particles, the fraction θ occupied by adsorbed $^*CO_2$ on the particle surface increases with the particle size, indicating that the CO_2 desorption rate becomes slower with an increase in particle size. This illustrates that there is a certain correlation between the limestone calcination reaction and particle size, which in turn verifies the limestone calcination model in Section 5. The occupied fraction θ on the particle surface rises rapidly in the initial stage and soon reaches saturation. The variation of particle surface θ with particle size under high CO_2 concentration is smaller than that under low CO_2 concentration, which also leads to the change in particle conversion with particle size no longer being significant.

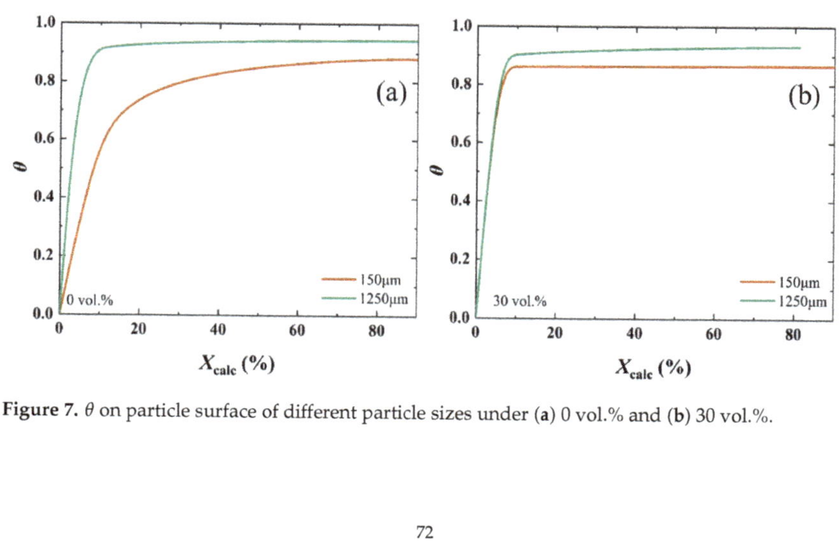

Figure 7. θ on particle surface of different particle sizes under (**a**) 0 vol.% and (**b**) 30 vol.%.

Figure 8a shows the effect of the temperature on the reaction-controlling mechanism calculated using the established model in Section 5. It can be seen that for the limestone calcination process, there is an excellent linear relationship between the consumption rate of particles and the temperature from 600–1050 °C. This reveals that the consumption of limestone particles is mainly controlled by a combination of chemical reactions and gas diffusion in the investigated range.

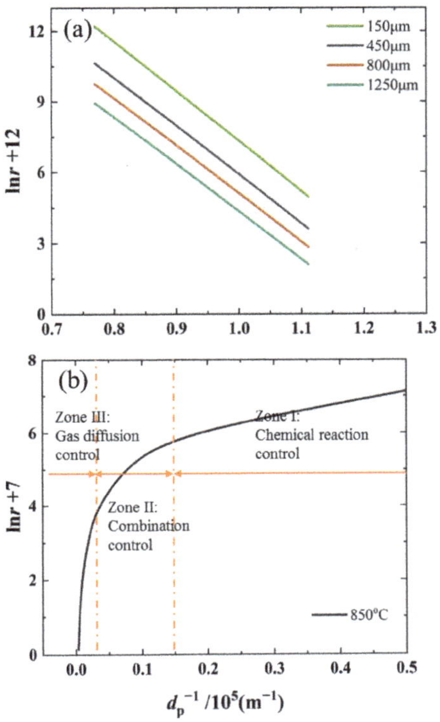

Figure 8. Calculated results of the effect on the limestone calcination reaction controlling mechanism at X_{calc} = 0.5: (**a**) temperature; (**b**) particle size. (*r* is the calcination consumption rate of particles).

Figure 8b shows the effect of the particle size on the limestone calcination reaction controlling mechanism. It can be seen that the consumption rate of limestone and particle size have clear three-zone characteristics. When the particle size is smaller than ~80 μm, the change of the particle size has less effect on the limestone particle consumption rate. In this case, the particle consumption rate is mainly controlled by chemical reactions, referred to as Zone I. While the particle size is larger than ~450 μm, the particle consumption rate varies significantly with particle size; this is referred to as Zone III. At this time, the calcination reaction is mainly controlled by gas diffusion, including the gas external diffusion step and the intraparticle diffusion step. As the particle size increases, the calcination rate decreases rapidly; that is, the gas diffusion resistance within the particle and the external gas diffusion resistance increase rapidly. In the particle size range of 80–450 μm, that is, Zone II, the gas diffusion resistance and chemical reaction resistance are in the same order of magnitude, and the calcination process is controlled by both gas diffusion and chemical reaction.

The above analysis can guide the optimization of limestone particle calcination conditions at different temperatures and CO_2 concentrations in practical applications. The chemical reaction rate, internal diffusion rate and external diffusion rate of the particles are comprehensively considered to determine the appropriate operating temperature and particle size.

4. Experiment

4.1. Microfluidized Bed Thermogravimetric Analysis (MFB-TGA)

A natural limestone from Hubei Province with calcium carbonate content exceeding 99.4% is explored in the work. After crushing and grinding, the limestone is carefully sieved to particle size ranges of 150–200 μm, 450–500 μm, 800–900 μm, and 1250–1410 μm. Each sieving process is repeated three times.

The calcination experiments of limestone are conducted in a microfluidized bed thermogravimetric analysis (MFB-TGA) apparatus, as shown in Figure 9a, which consisted of a quartz reactor, an electric furnace, a gas supply system, and a measurement system [46]. The quartz reactor is designed to be a fluidized bed with an inner diameter of 28 mm, the inner diameter of the top sample inlet of 10 mm, and a porous distribution plate with a thickness of 5 mm to make the air flow evenly into the reactor. The gas supply system consists of mass flow controllers calibrated with a soap bubble flowmeter prior to the experiment, solenoid valves, and gas cylinders. By using a sufficiently thin silicone hose to connect the reactor and the inlet pipe, hard connections can be avoided and interference with the quality signal can be eliminated. The measurement system realizes real-time online monitoring and recording of temperature, pressure and mass signals. The mass signal is the core of the apparatus and is measured with a precise weighing transducer with an accuracy of 1 mg. During the experiment, it is necessary to ensure that no part of the quartz reactor is in contact with the furnace so as not to affect the measurement of the quality signal. The temperature signal is measured with a wireless K-type thermocouple. The pressure difference between the gas inlet and outlet is monitored using a differential pressure sensor to judge the fluidization properties.

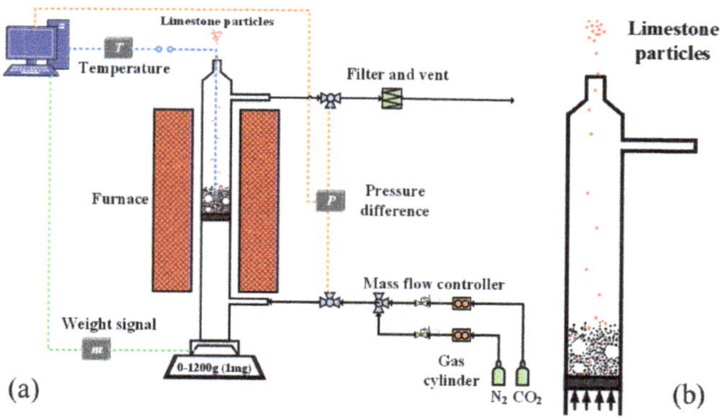

Figure 9. Schematic diagrams of (**a**) the MFB-TGA apparatus and (**b**) the limestone particles being injected into the reactor.

A novel experimental method is proposed and implemented to ensure that limestone particles are calcined under isothermal conditions in the work. As shown in Figure 9b, the limestone particles are injected into the hot sands which are fluidized in the high temperature quartz reactor, and the limestone particles are instantly heated to the reaction temperature and decomposed under the set CO_2 atmosphere. In a typical experiment, ~16 g of quartz sand with a particle size of 300–355 μm is added to the quartz reactor as the inert bed material (black particles in Figure 9b). The reaction gas is passed into the reactor, and the flow rate is ~3 U_{mf} (U_{mf} is the minimum fluidization speed). This indicates that as the temperature increases, the feed gas flow rate should decrease accordingly. The furnace temperature is raised to 850 °C and stabilized for 30 min. The fluctuation of the mass signal within 120 s is ±1 mg in the steady state, as shown in Figure 10a, indicating that the experimental system is highly stable. After that, the limestone particles (red particles in Figure 9b) are quickly injected into the fluidized bed from the top inlet of the reactor, and the mass increased rapidly by 64 mg within 0.2 s (equal to the mass of injected limestone), as shown in Figure 10b. The limestone particles are blended with the inert bed material in the reactor, heated, and calcined in a fluidizing condition. During the experiment, the mass injected of limestone is 60–70 mg to minimize the effect of mass transfer. Each condition is repeated several times to ensure reproducibility.

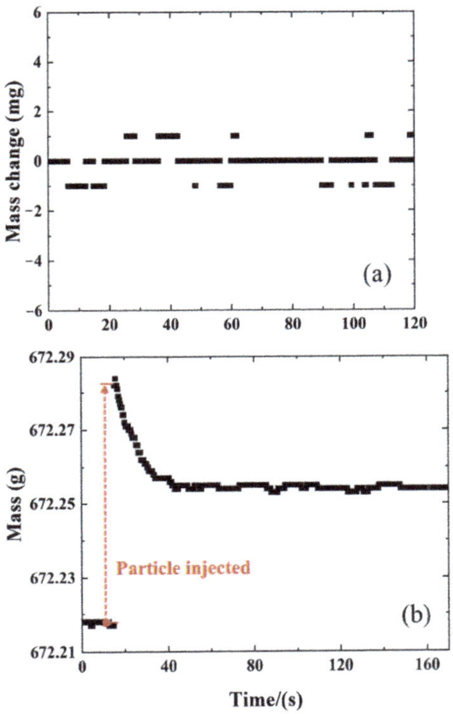

Figure 10. The mass signal of the weighing transducer: (**a**) the initial steady mass before limestone particles are injected; (**b**) real-time mass signal of limestone particles being injected.

4.2. Experimental Condition and Data Evaluation

The calcination characteristics of limestone particles are studied at 750–920 °C in the work. The CO_2 concentration is in the range of 0 and 30 vol.%, and N_2 is the balance gas. The detailed experimental conditions are displayed in Table 2.

Table 2. Experiment condition.

Sample Mass/mg	Particle Size/μm	Temperature/°C	CO_2/%	Superficial Velocity
60–70	150/450/800/1250	750–920	0/30	~3 U_{mf}

The calculation of the conversion of limestone particles depends on the mass value (limestone mass + quartz reactor mass) from the weighing transducer, and the conversion of limestone X_{Calc} is calculated as Equation (2), where $m(t)$ is the real-time mass during the calcination, m_0 is the initial steady mass before limestone particles are added and m_1 is the final steady mass after the decomposition's completion.

$$X_{Calc} = \left(1 - \frac{m(t) - m_0}{m_1 - m_0} \cdot \frac{M_{w,CaO}}{M_{w,CaCO_3}}\right) \cdot \frac{M_{w,CaCO_3}}{M_{w,CO_2}} \times 100\% \quad (2)$$

where $M_{w,CaCO_3}$, $M_{w,CaO}$, and M_{w,CO_2} are the molecular weights of $CaCO_3$, CaO, and CO_2, respectively.

5. Models

The limestone particles are considered to be dense and nonporous spherical particles with an initial radius of r_0, as displayed in Figure 11. Therefore, a shrinking core model is adopted in the work to describe the limestone calcination process. As shown in Figure 11, when the calcination reaction occurs, the calcium carbonate on the surface of the particles is first decomposed, forming a porous calcium oxide shell evenly covering the outer surface. While the reaction occurs, the calcium carbonate core with radius r_2 continues to decompose. It is worth noting that the CO_2 generated from the calcination of the limestone core must pass through the pores of the calcium oxide shell to diffuse to the outer surface, known as intraparticle diffusion. Then, the CO_2 on the outer surface of the particles is transferred into the gas stream through external diffusion. As the reaction continues, the calcium carbonate core inside the particle gradually shrinks, while the outer calcium oxide shell gradually increases until the inner calcium carbonate core is completely consumed and the reaction is completed.

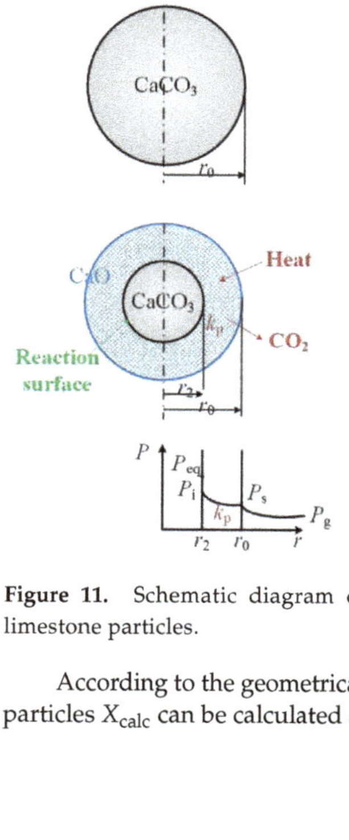

Figure 11. Schematic diagram of the shrinking core model describing the calcination of limestone particles.

According to the geometrical relation shown in Figure 11, the conversion of limestone particles X_{calc} can be calculated as:

$$X_{calc} = \frac{r_0^3 - r_2^3}{r_0^3} \quad (3)$$

Then, the conversion rate of the particles is obtained as:

$$\frac{dX_{calc}}{dt} = -\frac{3r_2^2}{r_0^3}\frac{dr_2}{dt} \quad (4)$$

Based on Equation (3), it is deduced that the correlation of the radius r_2 of the calcium carbonate core and the initial particle radius r_0 is:

$$r_2 = r_0 \cdot (1 - X_{calc})^{1/3} \quad (5)$$

The consumption rate of limestone particle volume is calculated as:

$$\frac{dV_{CaCO_3}}{dt} = 4\pi r_2^2 \frac{dr_2}{dt} = -4\pi r_2^2 \cdot \Lambda \cdot V_{CaCO_3}^{M} \cdot R_{calc} \tag{6}$$

where R_{calc} is the chemical reaction rate, 1/s; Λ is the active site density at the interface, mol/m^2; and $V_{CaCO_3}^{M}$ is the molar volume of calcium carbonate, m^3/mol.

Combining Equations (4)–(6),the conversion of the limestone particle is expressed as:

$$\frac{dX_{calc}}{dt} = \frac{3}{r_0}(1 - X_{calc})^{2/3} \Lambda \cdot V_{CaCO_3}^{M} \cdot R_{calc} \tag{7}$$

The decomposition process of calcium carbonate is a typical gas–solid surface reaction. In addition to chemical surface reactions, the diffusion of CO_2 through the porous CaO layer and around the particles should be considered in the developed model to reflect the effect of CO_2 concentration. Therefore, in the subsequent derivation, we assume that the surface reaction of spherical calcium carbonate particles occurs through the shrinking core model, and the model of intraparticle and external diffusion processes of CO_2 is established. Finally, combining the above three steps, the expression of the effective rate constant is derived to describe the detailed calcium carbonate decomposition process.

5.1. Calcination Kinetics

Assume an ideal case of the calcination of an infinite calcium carbonate plane under uniform CO_2 partial pressure and temperature conditions. The surface reaction process of calcium carbonate can be described by two steps of chemical calcination and desorption [19,32]. The first step is the chemical calcination of calcium carbonate on the surface, resulting in CO_2 molecules being adsorbed on the surface and occupying active sites. The second step is the desorption of CO_2 molecules from the active site. The two-step calcination process is characterized by reactions R1 and R2 as follows:

$$\underset{1-\theta}{CaCO_3 + *} \underset{k_{1r}}{\overset{k_{1f}}{\rightleftharpoons}} \underset{\theta}{CaO + *CO_2} \tag{R1}$$

$$\underset{\theta}{*CO_2} \underset{k_{2r}}{\overset{k_{2f}}{\rightleftharpoons}} \underset{1-\theta \;\; P_i}{* + CO_2} \tag{R2}$$

where k_{1f}, k_{1r}, k_{2f}, and k_{2r} are the forward and reverse reaction rate constants of R1 and R2, respectively. * is an active site on the particle surface; θ is defined as the fraction of the active site on the particle surface occupied by CO_2. P_i is the CO_2 partial pressure at the interface between CaO and $CaCO_3$ inside the particle.

Based on the microscopic reversibility general principle, when the reaction reaches chemical equilibrium, the forward reaction rate of the elementary reaction equal to reverse reaction rate is obtained:

$$\begin{aligned} k_{1f}(1-\theta_{eq}) &= k_{1r}\theta_{eq} \\ k_{2f}\theta_{eq} &= k_{2r}(1-\theta_{eq})P_{eq} \end{aligned} \tag{8}$$

Then, the CO_2 partial pressure at equilibrium is obtained as:

$$P_{eq} = \frac{k_{1f}k_{2f}}{k_{1r}k_{2r}} = K_1 K_2 \tag{9}$$

where the subscript eq indicates that the reaction is at equilibrium, and K_1 and K_2 are the thermodynamic equilibrium constants of R1 and R2, respectively.

It is assumed that the adsorbed $*CO_2$ is in a quasi-steady state, and the desorption rate of CO_2 is a generally extremely fast progress compared with the chemical calcination rate; that is, the control step of calcium carbonate calcination is the reaction R1, and R2 $\approx$ 0. Therefore, the fraction occupied θ is calculated as:

$$R_2 \approx 0 \Rightarrow \theta \approx \frac{k_{2r}P_{\mathrm{i}}}{k_{2f}+k_{2r}P_{\mathrm{i}}} = \frac{P_{\mathrm{i}}}{K_2+P_{\mathrm{i}}} \tag{10}$$

The chemical reaction rate of calcium carbonate calcination is also obtained as:

$$R_{\mathrm{calc}} \approx R_1 = k_{1f}\left(1-\frac{P_{\mathrm{g}}}{P_{\mathrm{eq}}}\right)\frac{1}{1+K_1\cdot\frac{P_{\mathrm{i}}}{P_{\mathrm{eq}}}} \tag{11}$$

It can be seen that the above derivation is based on the ideal case of an infinite calcium carbonate plane, while the limestone is actually spherical particles. Therefore, a modification is introduced in Equation (11) in order to choose the one that provided the best fit with the experimental results and given as:

$$R_{\mathrm{calc}} = k_{1f}\left(1-\frac{P_{\mathrm{g}}}{P_{\mathrm{eq}}}\right)\frac{1}{1+K_1\cdot\frac{P_{\mathrm{i}}}{P_{\mathrm{eq}}}\cdot\frac{\overline{r}_0}{r_0}} \tag{12}$$

where $\overline{r}_0$ is the average radius of particles in the investigated range, calculated as 330 μm, and r_0 is the particles' radius.

Combined with Equation (7), the conversion rate of the particles is obtained as:

$$\frac{\mathrm{d}X_{\mathrm{calc}}}{\mathrm{d}t} = \frac{3}{r_0\rho}\cdot k_{\mathrm{c}}\cdot\left(P_{\mathrm{eq}}-P_{\mathrm{g}}\right) \tag{13}$$

$$k_{\mathrm{c}} = k_{1f}\cdot\rho\cdot\Lambda\cdot V_{\mathrm{CaCO_3}}^{\mathrm{M}}\cdot(1-X_{\mathrm{calc}})^{2/3}\frac{1}{P_{\mathrm{eq}}+K_1P_{\mathrm{i}}\cdot\frac{\overline{r}_0}{r_0}} \tag{14}$$

where k_{c} is the comprehensive chemical reaction rate constant. Combining the constant terms in Equation (14), it is obtained that:

$$k_{\mathrm{c}} = k_f\cdot(1-X_{\mathrm{calc}})^{2/3}\frac{1}{P_{\mathrm{eq}}+K_1P_{\mathrm{i}}\cdot\frac{\overline{r}_0}{r_0}} \tag{15}$$

where the reaction rate constant k_f is calculated based on the Arrhenius form, and the thermodynamic equilibrium constant K_1 is calculated based on the van't Hoff equation in Equation (15):

$$k_f = k_{1f}\cdot\rho\cdot\Lambda\cdot V_{\mathrm{CaCO_3}}^{\mathrm{M}} = A\cdot e^{-\frac{E}{RT}} \tag{16}$$

$$K_1 = \frac{\Delta S}{R}\cdot e^{-\frac{\Delta H}{RT}} = a\cdot e^{-\frac{\Delta H}{RT}} \tag{17}$$

where A and E are the pre-exponential factor and activation energy, respectively, of k_{f}. ΔS and ΔH are the standard entropy and enthalpy change of the $CaCO_3$ calcination reaction, and a is the pre-exponential factor of K_1. The four kinetic parameters—A, E, a, and ΔH—in the reaction term are parameters to be determined.

The equilibrium concentration of CO_2, P_{eq} is obtained based on the work of Hu et al., Pa [47]:

$$P_{\mathrm{eq}} = 1.826\times10^6\times\exp\left(-19680/T_{\mathrm{p}}\right)\times10^6 \tag{18}$$

5.2. Intraparticle Diffusion

The diffusion of CO_2 through the calcium oxide layer can be described by Fick's law of diffusion, and the diffusion flow rate of CO_2 is given as:

$$J_{\mathrm{CO_2}} = -4\pi r^2\cdot D_{\mathrm{e}}\cdot\frac{dC_{\mathrm{CO_2}}}{dr} \tag{19}$$

where C_{CO_2} is the CO_2 concentration in the calcium oxide layer and D_e is the gas diffusion coefficient, which is calculated based on the empirical formula [28]:

$$D_e = 0.00881 T^{1/2} S_0^{-1} \varepsilon_0^2 \tag{20}$$

where S_0 is the initial specific surface area of particles, taken as 22.80 m^2/g, and ε_0 is the initial porosity of the CaO layer, taken as 0.48 according to the work of Wang et al. [43].

The boundary conditions are that:

$$\begin{cases} r = r_2 \; C_{CO_2} = C_i \\ r = r_0 \; C_{CO_2} = C_s \end{cases} \tag{21}$$

Integrate Equation (21) and simplify to obtain:

$$J_{CO_2} = \frac{4\pi D_e (C_i - C_s)}{\frac{1}{r_2} - \frac{1}{r_0}} = 4\pi r_0^2 \frac{D_e (C_i - C_s)}{\frac{r_0}{r_2}(r_0 - r_2)} \tag{22}$$

According to the ideal gas equation of state, combine with Equation (5) to obtain:

$$J_{CO_2} = 4\pi r_0^2 \frac{D_e (P_i - P_s)}{r_0 \frac{[1-(1-X_{calc})^{1/3}]}{(1-X_{calc})^{1/3}} RT} \tag{23}$$

Then the conversion rate of the particles is obtained as:

$$\frac{dX_{calc}}{dt} = \frac{J_{CO_2} \cdot M_{w,CaCO_3}}{m_0} = \frac{3}{r_0 \rho} \cdot k_p \cdot (P_i - P_s) \tag{24}$$

$$k_p = \frac{D_e \cdot M_{w,CaCO_3}}{r_0 \cdot \frac{[1-(1-X_{calc})^{1/3}]}{(1-X_{calc})^{1/3}} RT} \tag{25}$$

where m_0 is the initial mass of the limestone particle and k_p is the gas intraparticle diffusion rate constant.

5.3. External Diffusion

The particle conversion rate due to CO_2 external diffusion from the surface of limestone particle to the main flow is:

$$\frac{dX_{calc}}{dt} = \frac{3}{r_0 \rho} \cdot k_D (P_s - P_g) \tag{26}$$

where k_D is the diffusion coefficient of CO_2 to the bulk atmosphere, which is given based on the work of Field [48] and Graham [49] and related to temperature and particle size:

$$k_{O_2} = C_1 \frac{[(T_p + T_g)/2]^{0.75}}{d_p} \tag{27}$$

$$k_D = k_{O_2} \sqrt{\frac{M_{w,O_2}}{M_{w,CO_2}}} \tag{28}$$

where T_p and T_g are the particle temperature and gas temperature, respectively; M_{w,O_2} and M_{w,CO_2} are the molecular weights of O_2 and CO_2, respectively; and C_1 is an empirical constant with a value of 5×10^{-12} s·$K^{-0.75}$.

5.4. Calcination Kinetic Model

The effects of chemical reaction, gas intraparticle and external diffusion during limestone calcination are considered in Sections 5.1–5.3. In order to eliminate the unknown parameters P_i and P_s in Equations (13) and (24), combing Equations (13), (14), (24), (25), and (26) gives:

$$\frac{1}{\frac{3}{r_0\rho}\cdot k_c}\frac{dX_{calc}}{dt}+\frac{1}{\frac{3}{r_0\rho}\cdot k_p}\frac{dX_{calc}}{dt}+\frac{1}{\frac{3}{r_0\rho}\cdot k_D}\frac{dX_{calc}}{dt}=-(P_{eq}-P_i)-(P_i-P_s)-(P_s-P_g)=-(P_{eq}-P_g) \quad (29)$$

Thus, the conversion rate of the limestone particle is:

$$\frac{dX_{calc}}{dt}=\frac{3}{r_0\rho}\cdot K_r\cdot(P_{eq}-P_g) \quad (30)$$

$$K_r=\frac{1}{\frac{1}{k_c}+\frac{1}{k_p}+\frac{1}{k_D}} \quad (31)$$

where K_r is the comprehensive calcination rate constant of the particle. It is worth noting that in Equation (15), the partial pressure of CO_2 at the intraparticle surface of the particle, P_i, is unknown. Thus, P_g is adopted as an approximate substitute in the calculation process.

The flowchart of the calculation algorithm is displayed in Figure 12. The algorithm-solving steps are: (i) initializing the particle parameters, making X_{calc} = 0.0001 to avoid meaningless k_p calculation, and inputting the constant parameters; (ii) calculating the external diffusion rate constant k_D; (iii) calculating the comprehensive chemical reaction rate constant k_c; (iv) calculating the intraparticle diffusion rate constant k_p; (v) updating the particle conversion X_{calc} at time t and then proceeding to the next time step $t + \Delta t$.

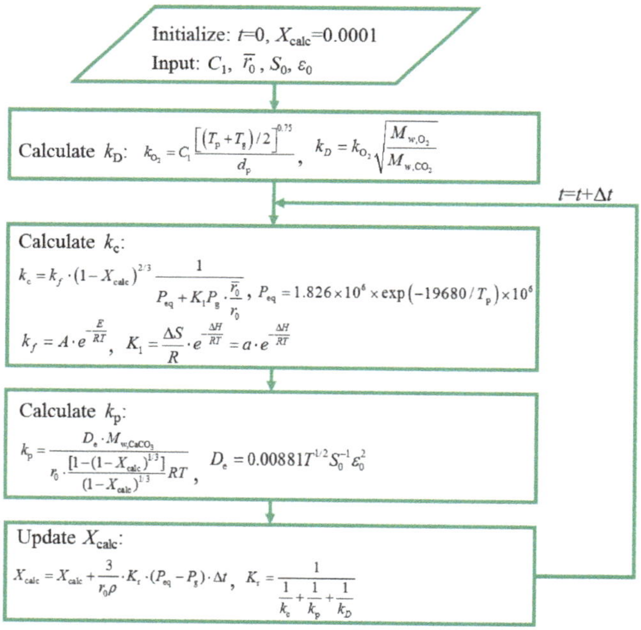

Figure 12. Flowchart of the calculation algorithm.

6. Conclusions

The calcination kinetics of the fluidized limestone particles under rapid heating and isothermal conditions are measured by injecting them into the microfluidized bed thermogravimetric analyzer (MFB-TGA). The calcination characteristics of limestone with different particle sizes are studied in the temperature range of 750–920 °C and the CO_2 concentration of 0 and 30 vol.%. A calcination kinetic model considering surface reaction, CO_2 partial pressure, particle structure change, and intraparticle and external diffusion is established, and the key parameters of the model are obtained; that is, $A = 5.051 \times 10^6$ kg·m^{-2}·s^{-1}, $E = 182$ kJ/mol, $a = 9.99 \times 10^7$, $\Delta H = 150$ kJ/mol. The results show that it only takes ~10 s to achieve complete conversion at temperatures >900 °C. Increasing the CO_2 concentration in the gas phase severely impairs the calcination rate of the particles. An interesting point is that as the partial pressure of CO_2 increases, the effect of particle size decreases. In the investigated temperature range, the particle consumption is controlled by both the chemical reaction and gas diffusion. The effect of particle size on limestone calcination kinetics presents three-zone characteristics of the reaction controlling mechanism. In Zone I, the particle size is <~80 μm, the particle size has little effect on the consumption rate, and calcination is mainly controlled by chemical reaction. In Zone III, the particle size is >~450 μm, the calcination process is mainly controlled by gas diffusion, and the effect of particle size is significant. In Zone II, in the particle size range of 80–450 μm, the calcination is controlled by both gas diffusion and chemical reaction. The model prediction results are in good agreement with the experimental data, indicating that the experimental results and the calcination model will provide valuable information for the application of CaL technology.

Author Contributions: Conceptualization, D.L. and Z.L.; methodology, D.L., Y.W. and Z.L.; software, D.L.; validation, D.L., Y.W. and Z.L.; formal analysis, D.L. and Z.L.; investigation, D.L.; resources, Z.L.; data curation, D.L. and Z.L.; writing—original draft preparation, D.L.; writing—review and editing, D.L. and Z.L.; visualization, D.L.; supervision, Z.L.; project administration, Z.L.; funding acquisition, Z.L. All authors have read and agreed to the published version of the manuscript.

Funding: This research was supported by National Natural Science Foundation of China (Grant No. 51876095) and supported by the International Joint Mission On Climate Change and Carbon Neutrality.

Data Availability Statement: The data presented in this study are available in the article.

Conflicts of Interest: The authors declare no conflict of interest.

References

1. Han, R.; Wang, Y.; Xing, S.; Pang, C.; Hao, Y.; Song, C.; Liu, Q. Progress in reducing calcination reaction temperature of Calcium-Looping CO_2 capture technology: A critical review. *Chem. Eng. J.* **2022**, *450*, 137952. [CrossRef]
2. Perejón, A.; Romeo, L.M.; Lara, Y.; Lisbona, P.; Martínez, A.; Valverde, J.M. The Calcium-Looping technology for CO_2 capture: On the important roles of energy integration and sorbent behavior. *Appl. Energy* **2016**, *162*, 787–807. [CrossRef]
3. Rivero, M.A.; Rodrigues, D.; Pinheiro, C.I.; Cardoso, J.P.; Mendes, L.F. Solid-gas reactors driven by concentrated solar energy with potential application to calcium looping: A comparative review. *Renew. Sust. Energ. Rev.* **2022**, *158*, 112048. [CrossRef]
4. Qin, C.; He, D.; Zhang, Z.; Tan, L.; Ran, J. The consecutive calcination/sulfation in calcium looping for CO_2 capture: Particle modeling and behaviour investigation. *Chem. Eng. J.* **2018**, *334*, 2238–2249. [CrossRef]
5. Di Giuliano, A.; Gallucci, K. Sorption enhanced steam methane reforming based on nickel and calcium looping: A review. *Chem. Eng. Process.* **2018**, *130*, 240–252. [CrossRef]
6. Blamey, J.; Anthony, E.; Wang, J.; Fennell, P. The calcium looping cycle for large-scale CO_2 capture. *Prog. Energy Combust. Sci.* **2010**, *36*, 260–279. [CrossRef]
7. Hashemi, S.M.; Sedghkerdar, M.H.; Mahinpey, N. Calcium looping carbon capture: Progress and prospects. *Can. J. Chem. Eng.* **2022**, *100*, 2140–2171. [CrossRef]
8. Hornberger, M.; Spörl, R.; Scheffknecht, G. Calcium looping for CO_2 capture in cement plants-pilot scale test. *Energy Procedia* **2017**, *114*, 6171–6174. [CrossRef]
9. Haaf, M.; Stroh, A.; Hilz, J.; Helbig, M.; Ströhle, J.; Epple, B. Process modelling of the calcium looping process and validation against 1 MWth pilot testing. *Energy Procedia* **2017**, *114*, 167–178. [CrossRef]
10. Bui, M.; Adjiman, C.S.; Bardow, A.; Anthony, E.J.; Boston, A.; Brown, S.; Fennell, P.S.; Fuss, S.; Galindo, A.; Hackett, L.A. Carbon capture and storage (CCS): The way forward. *Energy Environ. Sci.* **2018**, *11*, 1062–1176. [CrossRef]
11. Li, Z.; Cai, N.; Huang, Y.; Han, H. Synthesis, experimental studies, and analysis of a new calcium-based carbon dioxide absorbent. *Energy Fuels* **2005**, *19*, 1447–1452. [CrossRef]
12. Rodriguez, N.; Alonso, M.; Grasa, G.; Abanades, J.C. Heat requirements in a calciner of $CaCO_3$ integrated in a CO_2 capture system using CaO. *Chem. Eng. J.* **2008**, *138*, 148–154. [CrossRef]
13. Grasa, G.S.; Abanades, J.C. CO_2 capture capacity of CaO in long series of carbonation/calcination cycles. *Ind. Eng. Chem. Res.* **2006**, *45*, 8846–8851. [CrossRef]

14. Hyatt, E.P.; Cutler, I.B.; Wadsworth, M.E. Calcium carbonate decomposition in carbon dioxide atmosphere. *J. Am. Ceram. Soc.* **1958**, *41*, 70–74. [CrossRef]
15. Khinast, J.; Krammer, G.; Brunner, C.; Staudinger, G. Decomposition of limestone: The influence of CO_2 and particle size on the reaction rate. *Chem. Eng. Sci.* **1996**, *51*, 623–634. [CrossRef]
16. Hills, A. The mechanism of the thermal decomposition of calcium carbonate. *Chem. Eng. Sci.* **1968**, *23*, 297–320. [CrossRef]
17. Borgwardt, R.H. Calcination kinetics and surface-area of dispersed limestone particles. *AlChE J.* **1985**, *31*, 103–111. [CrossRef]
18. Rajeswara, R. Kinetics of calcium carbonate decomposition. *Chem. Eng. Res. Des.* **1989**, *67*, 38–47.
19. Garcıa-Labiano, F.; Abad, A.; De Diego, L.; Gayán, P.; Adánez, J. Calcination of calcium-based sorbents at pressure in a broad range of CO_2 concentrations. *Chem. Eng. Sci.* **2002**, *57*, 2381–2393. [CrossRef]
20. Ylätalo, J.; Parkkinen, J.; Ritvanen, J.; Tynjälä, T.; Hyppänen, T. Modeling of the oxy-combustion calciner in the post-combustion calcium looping process. *Fuel* **2013**, *113*, 770–779. [CrossRef]
21. Szekely, J.; Evans, J. A structural model for gas-solid reactions with a moving boundary. *Chem. Eng. Sci.* **1970**, *25*, 1091–1107. [CrossRef]
22. Ar, I.; Dogu, G. Calcination kinetics of high purity limestones. *Chem. Eng. J.* **2001**, *83*, 131–137. [CrossRef]
23. Escardino, A.; Garcia-Ten, J.; Feliu, C.; Saburit, A.; Cantavella, V. Kinetic study of the thermal decomposition process of calcite particles in air and CO_2 atmosphere. *Ind. Eng. Chem. Res.* **2013**, *19*, 886–897. [CrossRef]
24. Galwey, A.K.; Brown, M.E. Application of the Arrhenius equation to solid state kinetics: Can this be justified? *Thermochim. Acta* **2002**, *386*, 91–98. [CrossRef]
25. Ramezani, M.; Tremain, P.; Doroodchi, E.; Moghtaderi, B. Determination of carbonation/calcination reaction kinetics of a limestone sorbent in low CO_2 partial pressures using TGA experiments. *Energy Procedia* **2017**, *114*, 259–270. [CrossRef]
26. Guler, C.; Dollimore, D.; Heal, G. The investigation of the decomposition kinetics of calcium carbonate alone and in the presence of some clays using the rising temperature technique. *Thermochim. Acta* **1982**, *54*, 187–199. [CrossRef]
27. Dennis, J.; Hayhurst, A. The effect of CO_2 on the kinetics and extent of calcination of limestone and dolomite particles in fluidised beds. *Chem. Eng. Sci.* **1987**, *42*, 2361–2372. [CrossRef]
28. Silcox, G.D.; Kramlich, J.C.; Pershing, D.W. A mathematical model for the flash calcination of dispersed calcium carbonate and calcium hydroxide particles. *Ind. Eng. Chem. Res.* **1989**, *28*, 155–160. [CrossRef]
29. Rao, T.R. Kinetics of calcium carbonate decomposition. *Chem. Eng. Technol.* **1996**, *19*, 373–377. [CrossRef]
30. Fuertes, A.; Marban, G.; Rubiera, F. Kinetics of thermal decomposition of limestone particles in a fluidized bed reactor. *Chem. Eng. Re.s Des.* **1993**, *71*, 421–428.
31. Wang, Y.; Lin, S.; Suzuki, Y. Limestone calcination with CO_2 capture (II): Decomposition in CO_2/steam and CO_2/N_2 atmospheres. *Energy fuels* **2008**, *22*, 2326–2331. [CrossRef]
32. Valverde, J.M.; Sanchez-Jimenez, P.E.; Perez-Maqueda, L.A. Limestone calcination nearby equilibrium: Kinetics, CaO crystal structure, sintering and reactivity. *J. Phys. Chem. C* **2015**, *119*, 1623–1641. [CrossRef]
33. Ramon Fernandez, J.; Turrado, S.; Carlos Abanades, J. Calcination kinetics of cement raw meals under various CO_2 concentrations. *React. Chem. Eng.* **2019**, *4*, 2129–2140. [CrossRef]
34. Barker, R. The reversibility of the reaction $CaCO_3 \rightleftarrows CaO + CO_2$. *J. Chem. Technol. Biotechnol.* **1973**, *23*, 733–742. [CrossRef]
35. Calvo, E.G.; Arranz, M.; Leton, P. Effects of impurities in the kinetics of calcite decomposition. *Thermochim. Acta* **1990**, *170*, 7–11. [CrossRef]
36. Valverde, J.M. On the negative activation energy for limestone calcination at high temperatures nearby equilibrium. *Chem. Eng. Sci.* **2015**, *132*, 169–177. [CrossRef]
37. Wang, Y.; Lin, S.; Suzuki, Y. Study of limestone calcination with CO_2 capture: Decomposition behavior in a CO_2 atmosphere. *Energy Fuels* **2007**, *21*, 3317–3321. [CrossRef]
38. Scaltsoyiannes, A.; Lemonidou, A. $CaCO_3$ decomposition for calcium-looping applications: Kinetic modeling in a fixed-bed reactor. *Chem. Eng. Sci. X* **2020**, *8*, 100071. [CrossRef]
39. Lv, G.; Lu, J.; Cai, L.; Xie, X.; Liu, Z. Experimental study on the dynamic process of NO reduction in a precalciner. *Ind. Eng. Chem. Res.* **2011**, *50*, 4366–4372.
40. Khraisha, Y.; Dugwell, D. Coal combustion and limestone calcination in a suspension reactor. *Chem. Eng. Sci.* **1992**, *47*, 993–1006. [CrossRef]
41. Li, Y.; Li, Z.; Wang, H.; Cai, N. CaO carbonation kinetics determined using micro-fluidized bed thermogravimetric analysis. *Fuel* **2020**, *264*, 116823. [CrossRef]
42. Wang, Y.; Li, Z.; Cai, N. Redox reaction kinetics of a Fe-Cu-based oxygen carrier measured with microfluidized bed thermogravimetric analysis. *Energy Fuels* **2022**, *36*, 9672–9686. [CrossRef]
43. Wang, H.; Li, Z.; Li, Y.; Cai, N. Reduced-order model for CaO carbonation kinetics measured using micro-fluidized bed thermogravimetric analysis. *Chem. Eng. Sci.* **2021**, *229*, 116039. [CrossRef]
44. Giammaria, G.; Lefferts, L. Catalytic effect of water on calcium carbonate decomposition. *J. CO_2 Util.* **2019**, *33*, 341–356. [CrossRef]
45. Escardino, A.; Garcia-Ten, J.; Feliu, C. Kinetic study of calcite particle (powder) thermal decomposition: Part I. *J. Eur. Ceram. Soc.* **2008**, *28*, 3011–3020. [CrossRef]
46. Li, Y.; Li, Z.; Liu, L.; Cai, N. Measuring the fast oxidation kinetics of a manganese oxygen carrier using microfluidized bed thermogravimetric analysis. *Chem. Eng. J.* **2020**, *385*, 123970. [CrossRef]

47. Hu, N.; Scaroni, A.W. Calcination of pulverized limestone particles under furnace injection conditions. *Fuel* **1996**, *75*, 177–186. [CrossRef]
48. Field, M.A. Rate of combustion of size-graded fractions of char from a low-rank coal between 1200 K and 2000 K. *Combust. Flame* **1969**, *13*, 237–252. [CrossRef]
49. Graham, T. On the law of the diffusion of gases. *J. Membr. Sci.* **1995**, *100*, 17–21.sss. [CrossRef]

catalysts

MDPI

Article

Sustainable Aromatic Production from Catalytic Fast Pyrolysis of 2-Methylfuran over Metal-Modified ZSM-5

Shengpeng Xia [1,2,3,4], Chenyang Wang [1,2,3,4], Yu Chen [1,2,3,5], Shunshun Kang [1,2,3,4], Kun Zhao [1,2,3,5,*], Anqing Zheng [1,2,3,5,*], Zengli Zhao [1,2,3,5] and Haibin Li [1,2,3,4]

1 Guangzhou Institute of Energy Conversion, Chinese Academy of Sciences, Guangzhou 510640, China
2 CAS Key Laboratory of Renewable Energy, Guangzhou 510640, China
3 Guangdong Provincial Key Laboratory of New and Renewable Energy Research and Development, Guangzhou 510640, China
4 University of Chinese Academy of Sciences, Beijing 100049, China
5 School of Energy Science and Engineering, University of Science and Technology of China, Guangzhou 510640, China
* Correspondence: zhaokun@ms.giec.ac.cn (K.Z.); zhengaq@ms.giec.ac.cn (A.Z.); Tel.: +86-0208-7057-716 (K.Z. & A.Z.); Fax: +86-0208-7057-737 (K.Z. & A.Z.)

Abstract: The catalytic fast pyrolysis (CFP) of bio-derived furans offers a promising approach for sustainable aromatic production. ZSM-5 modified by different metal species (Zn, Mo, Fe, and Ga) was employed in the CFP of bio-derived furans for enhancing aromatic production. The effects of metal species, metal loadings, and the weight hourly space velocity (WHSV) on the product distributions from the CFP of 2-methylfuran (MF) were systemically investigated. It is found that the introduction of Zn, Mo, Fe, and Ga on ZSM-5 significantly increases the MF conversion and aromatic yields. The maximum MF conversions of 75.49 and 69.03% are obtained, respectively, by Fe-ZSM-5 and Ga-ZSM-5, which boost the aromatic yield by 34.5 and 42.7% compared to ZSM-5. The optimal loading of Fe on ZSM-5 is 2%. Additionally, the highest aromatic yield of 40.03% is achieved by 2%Fe-ZSM-5 at a WHSV of 2 h^{-1}. The catalyst characterization demonstrates that the synergistic effect of Brønsted and Lewis acid sites in Fe-ZSM-5 is responsible for achieving the efficient aromatization of MF. The key to designing improved zeolite catalysts for MF aromatization is the introduction of large numbers of new Lewis acid sites without a significant loss of Brønsted acid sites in ZSM-5. These findings can provide guidelines for the rational design of better zeolite catalysts used in the CFP of biomass and its derived furans.

Keywords: biomass; bio-derived furan; aromatization; ZSM-5; catalyst design

1. Introduction

Nowadays, with increasing concerns regarding the upcoming scarcity of fossil fuel resources and the environmental issues associated with their usage, unequaled endeavors have been made to produce liquid fuels and chemicals from renewable energy resources [1–3]. Biomass is the only renewable resource capable of producing chemicals and liquid fuels [4]. Pyrolysis has been recognized as one of the most promising technologies that can yield diverse phase products comprising char, gas, and bio-oil from biomass [5–7]. However, bio-oil derived from biomass pyrolysis has a complex composition, low calorific value, high oxygen content, strong acidity, and poor stability, making it unsuitable for use in the current petroleum infrastructure. Multiple techniques have been applied to bio-oil upgrading, including raw material pretreatment, catalytic hydrodeoxygenation, and catalytic fast pyrolysis (CFP). Recently, CFP of biomass has been extensively explored, seeking to selectively produce value-added chemicals from biomass, especially aromatics [8]. Monocyclic aromatics—namely, benzene, toluene, ethylbenzene, and xylene (BTEX)—are bulk organic chemicals for the production of rubber, fibers, plastics, resins, coatings, dyes, pesticides, and pharmaceuticals [9]. In addition, they are also significant blending components of high-octane clean gasoline (accounting for around 21% of the gasoline composition) [10].

Previous studies have established that zeolite is the optimum catalyst for generating aromatics by the CFP of biomass due to its unique physicochemical features with acidity and pore structure. A succession of zeolite catalysts has been applied to the CFP of biomass, including HZSM-5, HY, H-beta, SAPO-34, MCM-41, etc. It has been proven that HZSM-5 has a moderate pore size and excellent shape selectivity for aromatics, whereas HY can accommodate more intermediate products. Xiangyu Li and coworkers observed that the CFP of Kraft lignin with HZSM-5 zeolite could produce a variety of valuable chemicals such as BTX, ethene, and propene [11]. Jae-Young Kim et al. examined the influence of the acidity of HZSM-5 on the production of aromatics from the pyrolysis of lignin. They found that the aromatic yield increased from 0.76% (Si/Al mole ratio of 280) to 2.62% (Si/Al mole ratio of 30) in proportion to increasing catalyst acidity [12]. Shanmugam Thiyagarajan et al. demonstrated that renewable aromatics may be synthesized from furans using HY zeolite, which exhibited the best activity and can be readily reused after calcination [13].

It was observed that transition-metal-modified zeolite catalysts may greatly boost monocyclic aromatic yields and facilitate the transformation of polycyclic aromatics into monocyclic aromatics. Evgeny A. Uslamin and coworkers reported a mechanistic investigation of the aromatization of furans over Ga-modified HZSM-5 [14–17]. They observed that the introduction of Ga can enhance aromatic selectivity and influence the deactivation pathway, leading to the formation of PAHs [17]. It has been reported that the addition of Fe into ZSM-5 zeolite contributed to the formation of MAHs and simultaneously hindered the further polymerization reaction of MAHs and other oxygenates during the catalytic fast pyrolysis of biomass [15]. As mentioned above, the introduction of transition metals can significantly increase aromatic yields. Here, transition-metal-modified HZSM-5 was applied to enhance the yield of aromatics.

Bio-derived furans, such as 2-methylfuran (MF), furfural, and 2,5-dimethylfuran, are key platform molecules obtained from the pyrolysis or hydrolysis of cellulose and hemicellulose, which have been widely used as the model compounds of cellulose and hemicellulose [18–30]. Yu-Ting Cheng and George W. Huber investigated the cofeeding of olefins (ethylene and propylene) with a series of furans over ZSM-5 catalysts and found that aromatics can be produced directly by Diels–Alder cycloaddition reactions of furans with olefins [22]. Huiyan Zhang et al. explored the catalytic conversion of furfural to olefins and aromatics and proposed a catalytic reaction pathway that starts with furfural decarboxylation to form furan and then form olefins and aromatics with further reactions [31]. However, there are relatively few studies on the production of aromatics by the pyrolysis of MF alone and the main pathway from furans to aromatics is through the Diels–Alder reaction [19,22,28]. Here, MF, a typical representative of bio-derived furans, is used as a model compound of cellulosic biomass.

In this work, the CFP of MF over ZSM-5 modified by different metals (Mo, Zn, Ga, and Fe) was carried out in a fixed bed reactor. In addition, different metal loadings and WHSVs were examined to further improve the catalytic performance. The metal-modified ZSM-5 catalysts were characterized to establish the structure-reactivity relationship.

2. Results and Discussion

2.1. The Product Distribution from the CFP of MF over ZSM-5 Modified by Different Metal Species

Figure 1 illustrates the MF conversion and carbon yields of BTEX, olefins, and coke from the CFP of MF over ZSM-5 modified by different metal species. It is observed that the introduction of four types of metal (Zn, Mo, Fe, and Ga) can significantly improve the conversion of MF and the carbon yield of BTEX and aromatics. Compared with raw ZSM-5 zeolite, metal-modified ZSM-5 zeolites can produce more alkanes, CO, and CO_2, indicating that metal sites can efficiently promote the deoxygenation reactions of MF and the formation of alkanes, primarily C_1–C_3. However, the carbon yield of coke from the CFP of MF over ZSM-5 increases by metal modification, except for the Mo modification. The results could be due to the change in the acid amount of the ZSM-5 catalysts caused by metal modification. It is also observed that metal-modified ZSM-5 catalysts largely improve the carbon output of benzene and toluene, but the carbon yields of ethylbenzene and xylene change slightly. The maximum carbon yields of BTEX (23.6%) and aromatics (30.1%) are obtained by Ga-ZSM-5, whereas the maximum conversion of MF (75.5%) is obtained by Fe-ZSM-5, which provides the aromatic yield of 28.3%. The Fe and Ga modifications can boost the aromatic yield by 34.5 and 42.7% compared to raw ZSM-5. Interestingly, the highest carbon yield of alkylbenzene is gained over Fe-ZSM-5 and the highest carbon yield of benzene is gained over Ga-ZSM-5, demonstrating that Fe exhibits a stronger aromatization ability and Ga displays a better dealkylation ability. Meanwhile, Mo and Fe can promote the formation of olefins (primarily ethylene and propylene). It is noteworthy that when the metal loading increases from 1 to 4%, the carbon yields of BTEX, aromatics, and conversion of MF decrease. Similarly, Ga and Fe modification affords the highest MF conversion and aromatic yields. It is thus concluded that the loading of Zn, Mo, Fe, and Ga on ZSM-5 zeolite can effectively enhance the MF conversion and the yields of BTEX and aromatics. Fe- and Ga-modified ZSM-5 catalysts exhibit the maximum MF conversion and aromatic yield, respectively.

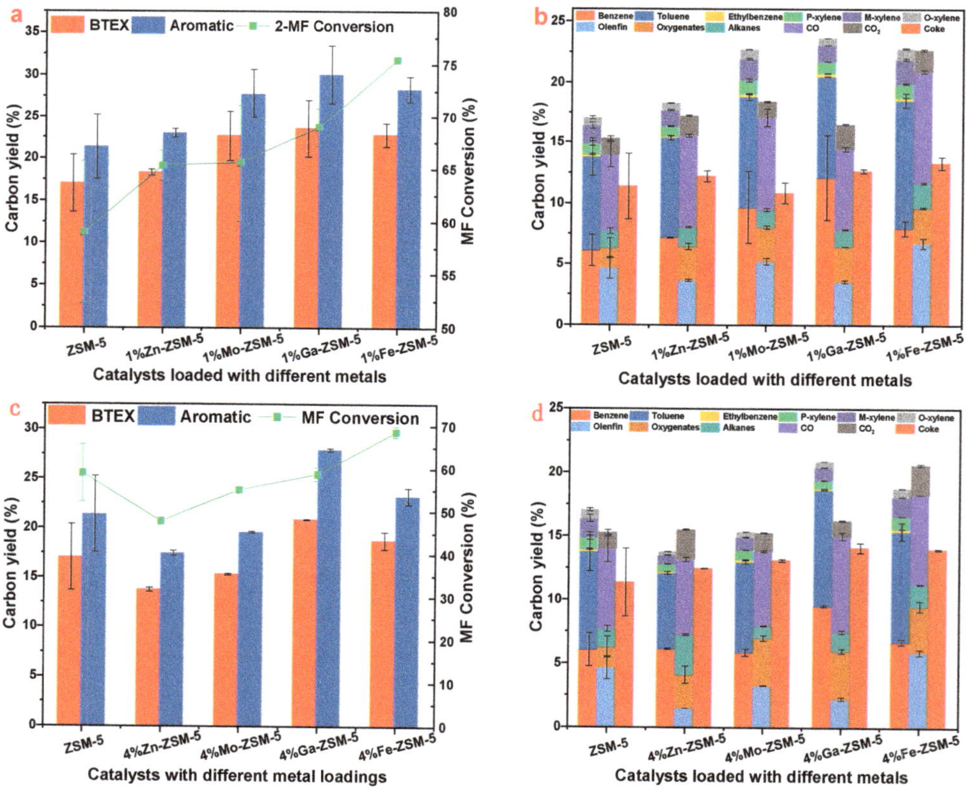

Figure 1. The effect of metal species on the product distribution from CFP of MF over metal-modified ZSM-5: (**a**,**c**) carbon yields of BTEX, aromatics, and MF conversion; (**b**,**d**) carbon yields of benzene, toluene, ethylbenzene, xylene, olefins, oxygenates, alkane, CO, CO_2, and coke. Reaction conditions: 550 °C, WHSV: 4 h^{-1}, carrier gas flow rate: 120 mL/min, reaction time: 15 min.

2.2. The Effect of Fe Loading on the Product Distribution from the CFP of MF over Fe-ZSM-5

Fe-ZSM-5 with different Fe loadings is employed in the CFP of MF since Fe-ZSM-5 exhibits the highest MF conversion among the four types of metal-modified ZSM-5. The resulting product distributions are shown in Figure 2. With the increase in Fe loading, the MF conversion and the carbon yields of BTEX, aromatics, and coke first increase and then reach their respective maximum values of 83.3%, 24.3%, 30.3%, and 15.7% at Fe loadings of 3%, 2%, 2%, and 3%. However, the MF conversion and the yield of aromatics and coke at an Fe loading of 0.5% are higher than those at an Fe loading of 1%. This may be owing to the high dispersion of Fe with a 0.5% loading forming more effective active sites. Simultaneously, the maximum yields of alkylbenzene, olefins, alkane, CO, and CO_2 are achieved at a 2% loading, indicating that 2% is optimum for the CFP of MF. As the Fe loading increases, the number of acid sites in the Fe-ZSM-5 catalyst increases, and the pores become blocked, so that the carbon yields of BTEX, aromatics, etc., begin to drop. Appropriate metal loading and metal dispersion not only improve product distribution but effectively reduce the consumption of metals.

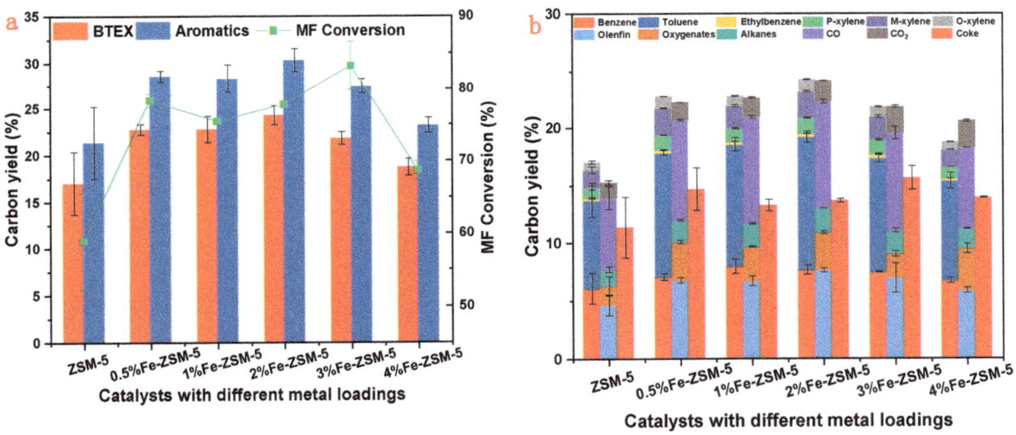

Figure 2. Effect of metal loadings on the product distribution from CFP of bio-derived MF: (**a**) carbon yields of BTEX, aromatics, and MF conversion; (**b**) carbon yields of benzene, toluene, ethylbenzene, xylene, olefins, oxygenates, alkane, CO, CO_2, and coke. Reaction conditions: 550 °C, WHSV: 4 h^{-1}, carrier gas flow rate: 120 mL/min, reaction time: 15 min.

2.3. The Effect of WHSV on the Product Distribution from the CFP of MF

Figure 3 showed the detailed product distribution from the conversion of MF at different WHSVs. The WHSV is defined as the mass flow rate of MF divided by the mass of the catalyst used in the reactor. In this experiment, the mass of the catalyst was altered from 0.5 to 2 g, while the mass flow rate of MF was maintained constant. As shown in Figure 3, the yields of BTEX, aromatics, olefins, and coke are a function of the WHSV. As the WHSV decreases from 8 h^{-1} to 2 h^{-1}, the yields of BTEX, aromatics, and coke increase from 9.23%, 11.72%, and 7.68% to 31.93%, 40.03%, and 23.69%, respectively, while the yield of olefins increases from 3.61% to 9.01%, and then remains unchanged. As indicated in Figure 3, the conversion of MF increases with decreasing WHSV and achieves 98.52% at a WHSV of 2 h^{-1}. It is inferred that the increasing amount of catalyst used can provide more active sites for the aromatization reaction of MF, leading to an increase in MF conversion and aromatic yield. These findings reveal that the appropriate WHSV for the conversion of MF was less than or equal to 2.67 h^{-1}.

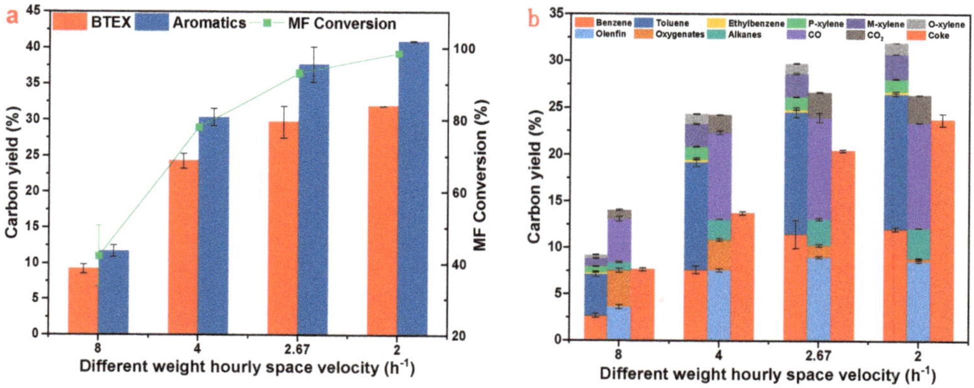

Figure 3. Effect of WHSV on the product distribution from CFP of bio-derived MF: (**a**) carbon yields of BTEX, aromatics, and MF conversion; (**b**) carbon yields of benzene, toluene, ethylbenzene, xylene, olefins, oxygenates, alkane, CO, CO_2, and coke. Reaction conditions: 550 °C, catalyst: 2%Fe-ZSM-5, carrier gas flow rate: 120 mL/min, reaction time: 15 min.

2.4. The Structure-Reactivity Relationship of Metal-Modified ZSM-5 during the CFP of MF

The acid sites and shape selectivity of the zeolite microporous structure are key factors determining the catalytic performance of metal-modified ZSM-5 during the CFP of MF. The N_2 adsorption-desorption isotherms and pore size distribution of different catalysts are illustrated in Figure 4; the resulting BET surface area, pore volume, and pore size are tabulated in Table 1. It is well-known that the molecular-sized micropore structure of ZSM-5 catalysts provides spatial restrictions around acid sites that allow for shape-selective catalysis. It can be seen from Table 1 and Figure 4 that the loading of metal species alters the specific surface area and total pore volume since the ZSM-5 was heated to 60 °C in the metal nitrate solution to evaporate the water during the catalyst preparation process. As the Fe loading increases from 0.5 to 4%, the BET surface and total pore volume area of Fe-ZSM-5 first increase and then reach their maximum value at the 2% Fe loading. Larger specific surface area and pore volume are beneficial for the aromatization reaction of MF. The results are in line with those in Figure 2, where 2%Fe-ZSM-5 exhibits the maximum yield of aromatics. It is worth noting that a large number of mesopores are generated when the Fe loading is from 1 to 3%. The mass transfer of reactant, intermediates, and products in zeolite channels can be enhanced by the introduction of a large number of mesopores, thus facilitating the aromatization of MF.

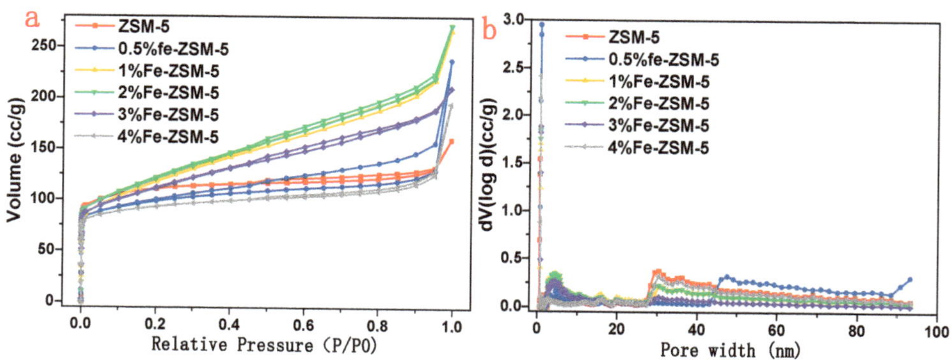

Figure 4. N_2 adsorption and desorption isotherms (**a**) and pore size distribution (**b**) diagram of different catalysts.

Table 1. The BET surface area, pore volume, and pore size of different catalysts.

Catalysts	BET Surface (m^2/g) [a]	Total Pore Volume (cm^3/g) [b]	Average Pore Size (nm)	Micropore Volume (cm^3/g) [c]	Mesopore Volume (cm^3/g)	Micropore Surface Area (m^2/g) [c]	External Surface Area (m^2/g)
ZSM-5	408.0	0.34	3.4	0.14	0.20	336.0	72.0
0.5%Fe-ZSM-5	367.4	0.37	4.0	0.12	0.25	291.5	75.8
1%Fe-ZSM-5	380.2	0.41	4.3	0.07	0.35	113.4	266.7
2%Fe-ZSM-5	423.4	0.42	4.0	0.08	0.35	160.5	262.9
3%Fe-ZSM-5	393.8	0.33	3.3	0.08	0.25	172.2	221.6
4%Fe-ZSM-5	352.3	0.30	3.4	0.13	0.17	309.0	43.3

[a] Calculated by multipoint BET method. [b] Calculated from absorbed volume of nitrogen for a relative pressure P/P_0 of 0.99. [c] Determined by t-plot method.

The acid strength and total acid amount of the different catalysts were measured by NH_3-TPD, as shown in Figure 5 and Table 2. The two peaks centered at 190 °C and 425 °C correspond to the weak and strong acid sites, respectively. It is obvious that the Fe loading increases the amount of strong acid sites, whereas it decreases the amount of weak acid sites. It is well-accepted that strong acid sites are responsible for the formation of coke, which is consistent with the results in Figure 2, where the Fe loading can promote the formation of coke [32]. The Fe loading significantly improves the total acid amount of ZSM-5, which is beneficial for the formation of aromatics during the CFP of MF. As the Fe loading increases from 0.5 to 4%, the total acid amount of Fe-ZSM-5 first increases and then decreases. The minimum total acid amount of 1149.97 μmol/g is obtained by 3%Fe-ZSM-5.

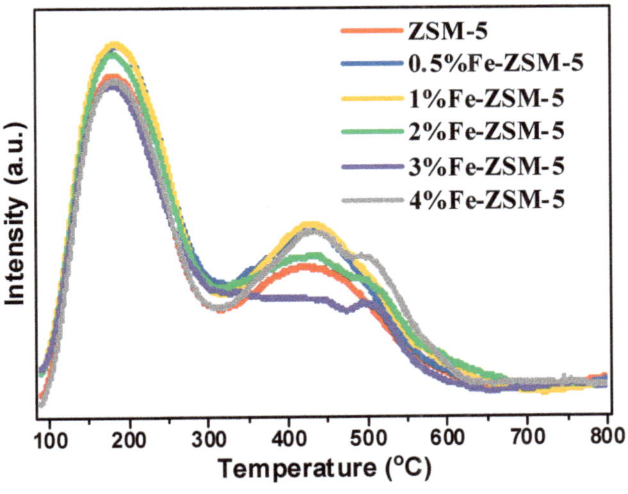

Figure 5. NH_3-TPD profiles of Fe-ZSM-5 with different Fe loadings.

Table 2. The acid strength and total acid amount of Fe-ZSM-5 with different Fe loadings.

Catalysts	The First Peak (°C)	The Second Peak (°C)	Amount of Weak Acid Sites (μmol/g)	Amount of Strong Acid Sites (μmol/g)	Total Acid Amount (μmol/g)
ZSM-5	182.50	445.00	827.64	189.08	1016.72
0.5%Fe-ZSM-5	191.277	418.22	783.26	539.58	1322.84
1%Fe-ZSM-5	190.89	426.25	763.26	536.78	1300.04
2%Fe-ZSM-5	188.01	415.39	656.50	546.78	1203.28
3%Fe-ZSM-5	184.95	389.45	614.81	535.16	1149.97
4%Fe-ZSM-5	182.9	444.7	709.90	473.41	1183.31

The Brønsted and Lewis acid sites of Fe-ZSM-5 were determined by Py-FTIR. As shown in Table 3, the Fe loading greatly alters the Brønsted to Lewis (B/L) ratio and the amount of Brønsted acid sites and Lewis acid sites compared to ZSM-5. An FE loading of 0.5% obviously improves the amount of Brønsted acid sites and Lewis acid sites of ZSM-5. The Fe loading further increases from 0.5 to 3% and the amount of Brønsted acid sites of Fe-ZSM-5 gradually drops, whereas the amount of Lewis acid sites continuously increases. The highest amount of Lewis acid sites (714.27 μmol/g) and the lowest amount of Brønsted acid sites (435.70 μmol/g) are obtained by 3%Fe-ZSM-5, which also exhibits the maximum MF conversion of 83.34%. It is speculated the MF conversion is associated with the amount of Lewis acid sites of Fe-ZSM-5. However, the significant decrease in the Brønsted acid sites could result in a reduction in the aromatic yield. Hence, 2%Fe-ZSM-5 with the proper amount of Brønsted and Lewis acid sites provides the highest aromatic yield. The synergistic effect of the Brønsted acid sites and Lewis acid sites is responsible for achieving the efficient aromatization of MF. It is thus concluded that the key to designing an improved zeolite catalyst for MF aromatization is the introduction of large numbers of new Lewis acid sites without a significant loss of Brønsted acid sites in ZSM-5.

Table 3. The amount of Brønsted and Lewis acid sites of Fe-ZSM-5.

Zeolite	B/L		Amount of Brønsted Acid Sites (μmol/g)	Amount of Lewis Acid Sites (μmol/g)
	Total (150 °C)	Strong (350 °C)		
ZSM-5	1.42	2.07	596.59	420.13
0.5%Fe-ZSM-5	1.62	1.87	817.94	504.90
1%Fe-ZSM-5	1.04	1.17	662.77	637.27
2%Fe-ZSM-5	0.85	1.07	552.86	650.42
3%Fe-ZSM-5	0.61	0.73	435.70	714.27
4%Fe-ZSM-5	1.05	1.36	606.09	577.22

3. Materials and Methods

3.1. Materials

MF (C_5H_6O, ≥98%), iron(III) nitrate nonahydrate ($Fe(NO_3)_3{\bullet}9H_2O$, ≥98.5%), and gallium nitrate hydrate ($Ga(NO_3)_3{\bullet}xH_2O$, ≥99.9%) were purchased from Shanghai Macklin Biochemical Co., Ltd. Ammonium heptamolybdate ($(NH_4)_6Mo_7O_{24}{\bullet}4H_2O$, ≥99.0%) was obtained from Tianjin Komiou Chemical Reagent Co., Ltd, Tianjin, China. Zinc nitrate hexahydrate ($Zn(NO_3)_2{\bullet}6H_2O$, ≥99.0%) was supplied by Guangzhou Chemical Reagent Factory, Guangdong, China. ZSM-5 (SiO_2/Al_2O_3 = 25) was acquired from Nankai University Catalyst Plant Co., Ltd., Tianjin, China. The ultrapure water was homemade. All reagents were used directly without further purification.

3.2. Catalyst Preparation

The metal-modified ZSM-5 catalysts were prepared by wet impregnation procedures and termed as xMe-ZSM-5 (x denotes weight ratio of metal to zeolite, x = 0.5%, 1%, 2%, 3%, 4%; Me means metal, Me = Mo, Zn, Ga, or Fe). In total, 10 g of ZSM-5 zeolite was impregnated in a 100 mL metal nitrate solution at room temperature and then rotary-evaporated at 60 °C. Subsequently, the samples were dried at 105 °C for 12 h and then calcined at 550 °C for 5 h. In the end, all of the catalysts, including raw ZSM-5, were pelleted, crushed, and sieved into 40–60 mesh for further catalytic assessment.

3.3. CFP Experiments

The conversion of bio-derived furans was conducted in a continuous fixed bed reactor, as shown in Figure 6. The fixed reactor was built from a 17 mm inner diameter quartz tube. The sieved ZSM-5 particles with a size of 0.25–0.425 mm were held in the reactor by quartz beads and quartz wool. Nitrogen (99.999%) was used as a purge gas to remove the air in the fixed bed reactor, while argon (99.999%) was used as a carrier gas for the internal standard of gas volume. The purge gas was set to 200 mL/min using a mass flow meter, and the carrier gas was set to 120 mL/min. The MF was introduced into the argon stream using a constant flow HPLC piston pump (STI 501, Hangzhou Saizhi Technology Co., Ltd, Hangzhou, China.). For a typical run, the flow rate of the mixtures was adjusted to 4.0 g/h, and 1.0 g of catalysts were used in the catalyst bed. All of the runs were performed at atmospheric pressure. A condenser at −40 °C and two-stage gas washing bottles using cyclohexane as a solvent (held at 0 °C) were used to trap the liquid products. After the reactions, the condenser and gas-washing bottles were washed with cyclohexane. The liquid products in the cyclohexane were analyzed and quantified by a TRACE 1300 gas chromatograph connected with a QD-300 mass spectrometer (GC-MS) (Thermo Fisher Scientific, Cleveland, OH, USA). The GC column was an Agilent HP-5 (30 m length, 0.25 mm internal diameter, and 0.25 μm film thickness). The gas phase products were collected with airbags and analyzed by GC (Agilent 7890A, Agilent Technologies, Inc, Santa Clara, CA, USA). The GC column was an Agilent 113-4362 GS-GASPRO (60 m length, 0.32 mm internal diameter). The coke yield was obtained by measuring the carbon content of the spent catalyst.

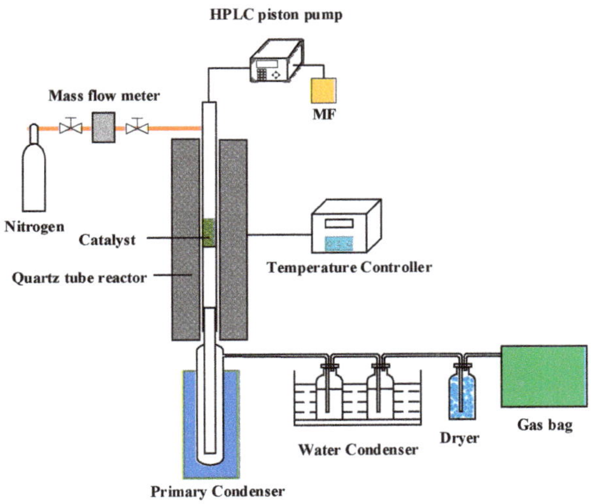

Figure 6. The schematic diagram of the fixed bed for CFP of MF.

3.4. Characterization

The textural properties of catalysts were determined by N_2 adsorption-desorption isotherms (Autosorb-iQ-2, Quantachrome). Before the test, the sample was degassed under a vacuum at 300 °C for 10 h. The NH_3-TPD analysis of different catalysts was conducted on an AutoChem II 2920 (Micromeritics) chemisorption analyzer following a TPD method. For each run, the sample was heated up to 300 °C at a rate of 10 °C/min and kept for 2 h in a He flow to remove adsorbed impurities. Then, the sample was cooled down to 90 °C for the adsorption of NH_3. After flushing with He for 1 h to remove physically adsorbed NH_3, the NH_3 desorption signal was detected using a TCD detector from 90 °C to 800 °C with a ramp of 10 °C/min. The Fourier transform infrared spectroscopy of pyridine adsorption (Py-FTIR) of different catalysts was obtained by a Thermo Nicolet Nexus 6700 FT-IR spectrometer equipped with a liquid nitrogen-cooled MCT detector. The wafers (diameter 13 mm) were purged in the IR cell at 400 °C for 2 h under a vacuum and then cooled down to room temperature for pyridine adsorption. Py-IR spectra were recorded over 32 scans at a resolution of 4 cm^{-1} in the range of 1700 to 1400 cm^{-1} at 150 and 350 °C. The Brønsted/Lewis acid (B/L) ratio was calculated as follows: B/L = 1.88IA(B)/1.42IA(L) (IA(L, B), IA = integrated absorbance of L or B band.

3.5. Methods of Data Processing

The MF conversion, the carbon yields of aromatics, oxygenates, coke, and noncondensable gases (alkanes, olefins, CO, and CO_2), and aromatic selectivity were defined as follows:

$$\text{MF conversion} = (1 - \frac{\text{Moles of carbon of MF in the product}}{\text{Moles of carbon in feedstocks}}) \times 100\%$$

$$\text{Aromatic yield} = \frac{\text{Moles of carbon in aromatic products}}{\text{Moles of carbon in feedstocks}} \times 100\%$$

$$\text{Coke yield} = \frac{\text{Moles of carbon in spent catalysts}}{\text{Moles of carbon in feedstocks}} \times 100\%$$

$$\text{Noncondensable gas yield} = \frac{\text{Moles of carbon in noncondensable gas}}{\text{Moles of carbon in feedstocks}} \times 100\%$$

$$\text{Aromatic selectivity} = \frac{\text{Moles of carbon in specific aromatic species}}{\text{Total Moles of carbon in all aromatic species}} \times 100\%$$

4. Conclusions

It is demonstrated that the introduction of Zn, Mo, Fe, and Ga on ZSM-5 evidently improves the MF conversion and the aromatic yield during the CFP of MF. Among the four types of metal-modified ZSM-5 catalysts, 1%Fe-ZSM-5 exhibits the highest MF conversion, while Ga-ZSM-5 affords higher aromatic yields. It is found that 2% is the optimal Fe loading for maximizing the aromatic yield during the CFP of MF over Fe-ZSM-5. The optimal aromatic yield of 40.03% is achieved by 2%Fe-ZSM-5 at a WHSV of 2 h^{-1}. The NH_3-TPD and Py-FTIR characterization prove that the synergistic effect of Brønsted acid sites and Lewis acid sites is the key to achieving the efficient aromatization of MF. The introduction of large numbers of new Lewis acid sites without a significant loss of Brønsted acid sites on the ZSM-5 catalyst is the guideline for the rational design of an improved zeolite catalyst for the CFP of biomass and its derived furans.

Author Contributions: Conceptualization, K.Z. and A.Z.; Methodology, Z.Z.; Validation, S.K.; Formal analysis, S.X.; Investigation, C.W., Y.C. and K.Z.; Resources, H.L.; Data curation, S.X.; Writing—original draft, S.X.; Writing—review and editing, A.Z.; Supervision, H.L.; Funding acquisition, K.Z. All authors have read and agreed to the published version of the manuscript.

Funding: This research was funded by the National Key R&D Program of China (Grant 2017YFE0124200), the National Natural Science Foundation of China (Grants 51876208 and 51776209), the Key Research and Development Program of Guangzhou (202206010122), and the Youth Innovation Promotion Association, CAS (2019341).

Data Availability Statement: Not applicable.

Acknowledgments: The authors acknowledge the National Key R&D Program of China (Grant 2017YFE0124200), the National Natural Science Foundation of China (Grants 51876208 and 51776209), the Key Research and Development Program of Guangzhou (202206010122), and the Youth Innovation Promotion Association, CAS (2019341), for their financial support of this work.

Conflicts of Interest: The authors declare no conflict of interest.

References

1. Zhao, N.; You, F.Q. Can renewable generation, energy storage and energy efficient technologies enable carbon neutral energy transition? *Appl. Energy* **2020**, *279*, 115889. [CrossRef]
2. Van Meerbeek, K.; Muys, B.; Hermy, M. Lignocellulosic biomass for bioenergy beyond intensive cropland and forests. *Renew. Sustain. Energy Rev.* **2019**, *102*, 139–149. [CrossRef]
3. Rezaei, P.S.; Shafaghat, H.; Daud, W.M.A.W. Production of green aromatics and olefins by catalytic cracking of oxygenate compounds derived from biomass pyrolysis: A review. *Appl. Catal. A: Gen.* **2014**, *469*, 490–511. [CrossRef]
4. Che, Q.; Yang, M.; Wang, X.; Chen, X.; Chen, W.; Yang, Q.; Yang, H.; Chen, H. Aromatics production with metal oxides and ZSM-5 as catalysts in catalytic pyrolysis of wood sawdust. *Fuel Process. Technol.* **2019**, *188*, 146–152. [CrossRef]

5. Butler, E.; Devlin, G.; Meier, D.; McDonnell, K. A review of recent laboratory research and commercial developments in fast pyrolysis and upgrading. *Renew. Sustain. Energy Rev.* **2011**, *15*, 4171–4186. [CrossRef]
6. Bridgwater, A.V. Review of fast pyrolysis of biomass and product upgrading. *Biomass Bioenergy* **2012**, *38*, 68–94. [CrossRef]
7. Zhang, X.; Yuan, Z.; Yao, Q.; Zhang, Y.; Fu, Y. Catalytic fast pyrolysis of corn cob in ammonia with Ga/HZSM-5 catalyst for selective production of acetonitrile. *Bioresour. Technol.* **2019**, *290*, 121800. [CrossRef]
8. Iliopoulou, E.F.; Triantafyllidis, K.S.; Lappas, A.A. Overview of catalytic upgrading of biomass pyrolysis vapors toward the production of fuels and high-value chemicals. *WIREs Energy Environ.* **2018**, *8*, e322. [CrossRef]
9. Potera, G.T. *The Effects of Benzene, Toluene and Ethylbenzene on Several Important Members of the Estuarine Ecosystem*; Lehigh University: Bethlehem, PA, USA, 1975.
10. Sun, R.; Shen, S.; Zhang, D.; Ren, Y.; Fan, J. Hydrofining of coal tar light oil to produce high octane gasoline blending components over γ-Al2O3-and η-Al2O3-supported catalysts. *Energy Fuels* **2015**, *29*, 7005–7013. [CrossRef]
11. Li, X.; Su, L.; Wang, Y.; Yu, Y.; Wang, C.; Li, X.; Wang, Z. Catalytic fast pyrolysis of Kraft lignin with HZSM-5 zeolite for producing aromatic hydrocarbons. *Front. Environ. Sci. Eng.* **2012**, *6*, 295–303. [CrossRef]
12. Kim, J.-Y.; Lee, J.H.; Park, J.; Kim, J.K.; An, D.; Song, I.K.; Choi, J.W. Catalytic pyrolysis of lignin over HZSM-5 catalysts: Effect of various parameters on the production of aromatic hydrocarbon. *J. Anal. Appl. Pyrolysis* **2015**, *114*, 273–280. [CrossRef]
13. Thiyagarajan, S.; Genuino, H.C.; van der Waal, J.C.; de Jong, E.; Weckhuysen, B.M.; van Haveren, J.; Bruijnincx, P.C.; van Es, D.S. A Facile Solid-Phase Route to Renewable Aromatic Chemicals from Biobased Furanics. *Angew. Chem. Int. Ed. Engl.* **2016**, *55*, 1368–1371. [CrossRef] [PubMed]
14. Li, G.; Yan, L.; Zhao, R.; Li, F. Improving aromatic hydrocarbons yield from coal pyrolysis volatile products over HZSM-5 and Mo-modified HZSM-5. *Fuel* **2014**, *130*, 154–159. [CrossRef]
15. Sun, L.; Zhang, X.; Chen, L.; Zhao, B.; Yang, S.; Xie, X. Comparision of catalytic fast pyrolysis of biomass to aromatic hydrocarbons over ZSM-5 and Fe/ZSM-5 catalysts. *J. Anal. Appl. Pyrolysis* **2016**, *121*, 342–346. [CrossRef]
16. Fanchiang, W.-L.; Lin, Y.-C. Catalytic fast pyrolysis of furfural over H-ZSM-5 and Zn/H-ZSM-5 catalysts. *Appl. Catal. A Gen.* **2012**, *419–420*, 102–110. [CrossRef]
17. Uslamin, E.A.; Kosinov, N.A.; Pidko, E.A.; Hensen, E.J.M. Catalytic conversion of furanic compounds over Ga-modified ZSM-5 zeolites as a route to biomass-derived aromatics. *Green Chem.* **2018**, *20*, 3818–3827. [CrossRef]
18. Li, W.; Wang, D.; Zhu, Y.; Chen, J.; Lu, Y.; Li, S.; Zheng, Y.; Zheng, Z. Efficient ex-situ catalytic upgrading of biomass pyrolysis vapors to produce methylfurans and phenol over bio-based activated carbon. *Biomass- Bioenergy* **2020**, *142*, 105794. [CrossRef]
19. Zheng, A.; Zhao, Z.; Chang, S.; Huang, Z.; Zhao, K.; Wu, H.; Wang, X.; He, F.; Li, H. Maximum synergistic effect in the coupling conversion of bio-derived furans and methanol over ZSM-5 for enhancing aromatic production. *Green Chem.* **2014**, *16*, 2580–2586. [CrossRef]
20. Chang, C.-C.; Green, S.K.; Williams, C.L.; Dauenhauer, P.J.; Fan, W. Ultra-selective cycloaddition of dimethylfuran for renewable p-xylene with H-BEA. *Green Chem.* **2014**, *16*, 585–588. [CrossRef]
21. Qi, X.; Fan, W. Selective Production of Aromatics by Catalytic Fast Pyrolysis of Furan with In Situ Dehydrogenation of Propane. *ACS Catal.* **2019**, *9*, 2626–2632. [CrossRef]
22. Cheng, Y.-T.; Huber, G.W. Production of targeted aromatics by using Diels–Alder classes of reactions with furans and olefins over ZSM-5. *Green Chem.* **2012**, *14*, 3114–3125. [CrossRef]
23. Ni, L.; Xin, J.; Jiang, K.; Chen, L.; Yan, D.; Lu, X.; Zhang, S. One-Step Conversion of Biomass-Derived Furanics into Aromatics by Brønsted Acid Ionic Liquids at Room Temperature. *ACS Sustain. Chem. Eng.* **2018**, *6*, 2541–2551. [CrossRef]
24. Do, P.T.M.; McAtee, J.R.; Watson, D.A.; Lobo, R.F. Elucidation of Diels–Alder Reaction Network of 2,5-Dimethylfuran and Ethylene on HY Zeolite Catalyst. *ACS Catal.* **2013**, *3*, 41–46. [CrossRef] [PubMed]
25. Williams, C.L.; Chang, C.-C.; Do, P.; Nikbin, N.; Caratzoulas, S.; Vlachos, D.G.; Lobo, R.F.; Fan, W.; Dauenhauer, P.J. Cycloaddition of Biomass-Derived Furans for Catalytic Production of Renewable *p*-Xylene. *ACS Catal.* **2012**, *2*, 935–939. [CrossRef]
26. Espindola, J.S.; Gilbert, C.J.; Perez-Lopez, O.; Trierweiler, J.O.; Huber, G.W. Conversion of furan over gallium and zinc promoted ZSM-5: The effect of metal and acid sites. *Fuel Process. Technol.* **2020**, *201*, 106319. [CrossRef]
27. Cheng, Y.-T.; Huber, G.W. Chemistry of Furan Conversion into Aromatics and Olefins over HZSM-5: A Model Biomass Conversion Reaction. *ACS Catal.* **2011**, *1*, 611–628. [CrossRef]
28. Cheng, Y.-T.; Jae, J.; Shi, J.; Fan, W.; Huber, G.W. Production of Renewable Aromatic Compounds by Catalytic Fast Pyrolysis of Lignocellulosic Biomass with Bifunctional Ga/ZSM-5 Catalysts. *Angew. Chem. Int. Ed.* **2012**, *51*, 1387–1390. [CrossRef]
29. Wang, C.; Si, Z.; Wu, X.; Lv, W.; Bi, K.; Zhang, X.; Chen, L.; Xu, Y.; Zhang, Q.; Ma, L. Mechanism study of aromatics production from furans with methanol over zeolite catalysts. *J. Anal. Appl. Pyrolysis* **2019**, *139*, 87–95. [CrossRef]
30. Teixeira, I.F.; Lo, B.T.W.; Kostetskyy, P.; Stamatakis, M.; Ye, L.; Tang, C.C.; Mpourmpakis, G.; Tsang, S.C.E. From Biomass-Derived Furans to Aromatics with Ethanol over Zeolite. *Angew. Chem. Int. Ed.* **2016**, *55*, 13061–13066. [CrossRef]
31. Zhang, H.; Wang, Y.; Shao, S.; Xiao, R. An experimental and kinetic modeling study including coke formation for catalytic pyrolysis of furfural. *Combust. Flame* **2016**, *173*, 258–265. [CrossRef]
32. Wang, B.; Manos, G. Role of strong zeolitic acid sites on hydrocarbon reactions. *Ind. Eng. Chem. Res.* **2008**, *47*, 2948–2955. [CrossRef]

catalysts

MDPI

Article

Migration Mechanism of Lattice Oxygen: Conversion of CO_2 to CO Using $NiFe_2O_4$ Spinel Oxygen Carrier in Chemical Looping Reactions

Da Song [1,2], Yan Lin [2], Kun Zhao [2], Zhen Huang [2,*], Fang He [1,*] and Ya Xiong [3]

1 College of Chemistry and Bioengineering, Guilin University of Technology, Guilin 541004, China
2 Guangzhou Institute of Energy Conversion, Chinese Academy of Sciences, Guangzhou 510640, China
3 Guangdong Provincial Key Laboratory of Environmental Pollution Control & School of Environmental Science & Engineering, Sun Yat-sen University, 135 Xingang Xi Road, Guangzhou 510275, China
* Correspondence: huangzhen@ms.giec.ac.cn (Z.H.); hefang@glut.edu.cn (F.H.)

Abstract: CO_2 resourceful utilization contributes to the goal of carbon neutrality. Chemical Looping Dry Reforming (CLDR) has attracted significant attention as a method for converting CO_2 to CO. $NiFe_2O_4$ oxygen carrier (OC) is found to be a potential material for CLDR. However, the migration process of lattice oxygen, which are critical for the conversion of CO_2 to CO, was not extensively investigated. In this study, the reduction and oxidation degrees of the $NiFe_2O_4$ were finely modulated in a thermogravimetric analyzer. The lattice oxygen migration mechanism of the $NiFe_2O_4$ in redox cycles was characterized by means of X-ray diffraction (XRD), transmission electron microscopy (TEM), X-ray photoelectron spectroscopy (XPS), and in-situ Raman. The novelty of this paper is clarifying the release-uptake paths of lattice oxygen during CO_2 resourceful utilization. The result indicates that the concentration gradient between the surface and the bulk drives the diffusion of lattice oxygen. The stabilization of surface lattice oxygen content is attributed to the rapid migration of O anion, which is closely associated with the movement process of Ni particles inward and outward through the spinel bulk. In addition, a highly reactive chemical reaction interface consisting of lattice oxygen and the corresponding metal atoms is always present on the surface of the oxygen carrier and is confirmed by an in-situ Raman and XPS during the whole process of CLDR. The results of this paper offer reference and basis for further development and design of CLDR using spinel OC.

Keywords: chemical looping (CL); oxygen carrier (OC); CO_2 splitting; lattice oxygen; $NiFe_2O_4$

1. Introduction

Excessive emission of greenhouse gases (mainly CO_2) by human beings is considered the primary factor leading to global warming. The CO_2 resource utilization, an effective pathway to reduce greenhouse gas emissions, contributes to the goal of global carbon neutrality [1,2]. The chemical looping dry reforming (CLDR) has attracted much attention for converting CO_2 to CO with the aid of oxygen carriers (OC) [3,4]. The CLDR separates a reforming process into three stages: (1) oxygen carrier reduction stage with a hydrocarbon fuel; (2) CO_2 oxidation stage; (3) air oxidation stage. At the reduction stage, the OC is reduced to a metallic state (Me) and meanwhile partially oxidizes the hydrocarbon fuel into high-quality syngas; At the CO_2 oxidation stage, the reduced oxygen carrier in the metallic state (Me) is oxidized by CO_2 to replenish a part of its lattice oxygen to form a metal oxide with lower oxygen content (MeO_{1-x}) and meanwhile CO_2 is converted to CO; At the air oxidation stage, the partially oxygen-recovered OC (MeO_{1-x}) is fully oxidized into its initial state (MeO), as shown in Figure 1. In the reduction stage, many fuels including but not limited to methane can be used as long as the oxygen carrier exhibits enough reactivity with the fuels in the CLDR process such as H_2, CO, pyrolysis gas, etc. Due to the separation of the redox process, CLDR has a few of potential advantages such as inherent

CO_2 separation, highly efficient activation of CO_2 with a minimal energy penalty, and low operating cost [5–8].

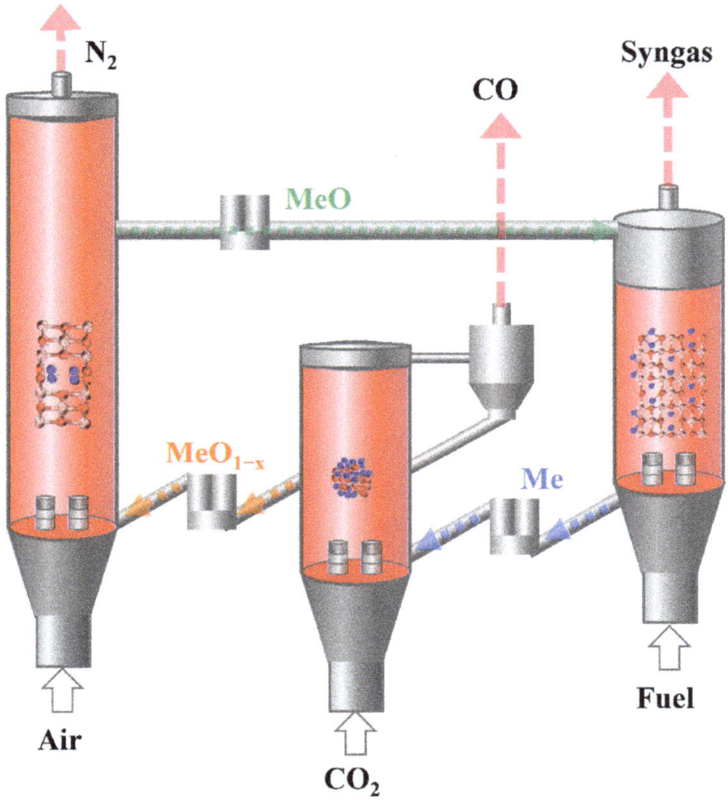

Figure 1. Conceptual illustration of CLDR.

The OC is viewed as the cornerstone of the chemical looping dry reforming [9]. Due to the considerably steady molecule structure and low free energy characteristic of CO_2, the capacity of oxygen carriers to maintain enough high reactivity toward CO_2 reforming is regarded as the most critical issue [10]. Single metal (Co, Ce, Fe, etc.) oxides have been evaluated as oxygen carriers for the CLDR process [11,12]. Among them, Fe-based OC showed the highest CO_2 reducibility. Tt is also abundant, low-cost, has more oxygen capacity, and is environmentally benign [10]. Thus, Fe-based OCs are considered excellent OC candidates for CLDR [13,14]. In addition, in order to further improve the reactivity of Fe-based oxygen carriers, the use of compound metal oxides (e.g., Ni/Fe, Mn/Fe, Co/Fe, Cu/Fe, and Ca/Fe) as oxygen carrier candidates has been proposed with promising prospect [15–17]. $CoFeO_y$ OC formed from Co-doped Fe-based oxygen carriers exhibited a high syngas yield and maintained stability for 10 cycles in a chemical looping reaction [18]. The synergistic effect of Co doping in ferrite materials to promote Fe reduction. Cu-doped Fe_2O_3 led to the formation of oxygen vacancies, which reduced the reaction activation energy [19], and its reaction rate remained high after 15 cycles. For the methane-carbon dioxide reaction system in CLDR, $MnFe_2O_4$ OC exhibits the maximum conversion of CH_4 and CO_2 are 63% and 99% [20]. Sun et al. designed a $Ca_2Fe_2O_5$ OC [21], and the cyclic performance of reduced OC is much better than that of Fe_2O_3. Furthermore, several literatures indicate that the different metals form some unique crystalline structures, such as spinel (AB_2O_4) [22,23] and perovskites (ABO_3) [24–26], can also obtain some excellent reactivity in CLDR.

In our group, we found that $NiFe_2O_4$ spinel OC is a potential OC for the CLDR of methane [27]. The capability of CO_2 reduction of reduced $NiFe_2O_4$ OC is up to 238 mL/g, which is 2.14 times that of reduced Fe_2O_3 + NiO OC, which gives an average of 185 mL of CO was generated in a single cycle in the 10 successive redox cycles, indicating that the $NiFe_2O_4$ spinel OC shows a relatively stable and good recyclability [28]. As well, the $NiFe_2O_4$ OCs evidently improve the reactivity of Fe-based OCs.

It is known that all the reactions in the three stages of the CLDR involve the migration of lattice oxygen, therefore it is considered the key factor determining the reactivity of OC [29]. However, a lack of fundamental understanding of lattice oxygen migration paths in oxygen carriers limits high-performance oxygen carriers design and development. To date, some exploratory investigations on oxygen ion migration in oxygen carriers have been performed. Chen et al. [30] studied the role of oxygen species in cerium-based oxygen carriers at 450 °C using CH_3SH as a model compound by XPS. they found that the improvement in catalytic stability is closely related to faster migration of bulk phase lattice oxygen. Liu et al. [31] studied the evolution of the active components of copper-iron-based oxygen carriers in the chemical looping combustion of CO, and the results of XPS analysis combined with DFT theory calculation indicated that the bulk phase lattice oxygen tends to migrate outward rather than CO diffusion inward into the particle. Zhao et al. [32] investigated the reaction mechanism of $La_{1.6}Sr_{0.4}FeCoO_6$ oxygen carriers in chemical looping methane steam reforming using XRD and XPS. As well, the results showed that the reaction boundary was fixed on the surface of the oxygen carriers. The oxygen vacancies formed on the surface promoted the migration of lattice oxygen from the bulk phase to the surface. In addition, there are several literatures to discuss the oxygen transfer mechanisms, especially using the density functional theory (DFT) method [33,34]. But there is still a lack of in-depth and systematic understanding of lattice oxygen migration in the bulk of oxygen carriers. As well, experimental investigations on the lattice oxygen transfer in spinel OCs were not extensively investigated. It's rarely found the studies on migration mechanism of lattice oxygen in CLDR to achieve CO_2 resource utilization. Thus, the investigations on release-recovery paths of lattice oxygen are of great significance for understanding the oxidation–reduction reaction mechanisms of OCs.

In order to clarify the release-uptake paths of lattice oxygen during the CLDR process, a thermogravimetric analyzer (TGA) is used to control the reduction degree and oxidation degree of OCs. The migration of lattice oxygen and the evolution of active components during chemical looping reactions were studied by means of XPS, XRD, TEM-EDS, and in-situ Raman. The results of this paper offer reference and basis for further development and design of CLDR using spinel OC.

2. Results and Discussion

2.1. XRD Analysis

2.1.1. Crystalline Phase Evolution of the OC in Reduction

The changes in XRD patterns of OC in the reduction stage are shown in Figure 2. The XRD patterns of the as-synthesized OC showed 12 characteristic peaks, which were found to be in perfect agreement with the standard $NiFe_2O_4$ card 00-054-0964 indicating that the as-synthesized OC by the sol-gel method is pure $NiFe_2O_4$ spinel. When the oxygen carrier was reduced by CO for 5 min, two small Ni peaks (card: 03-065-0380, marked as C) are observed. It indicates that NiO in the $NiFe_2O_4$ is firstly reduced by releasing lattice oxygen and separated from the spinel structure to form free metallic Ni. Another phenomenon was observed in that the rest of the characteristic peaks were slightly broadened and weakened due to a part of $NiFe_2O_4$ converted to $Fe_{3-x}O_4$ (card: 01-086-1355, marked as B). $NiFe_2O_4$ and $Fe_{3-x}O_4$ are both spinel structures, and the XRD spectra of them are very consistent.

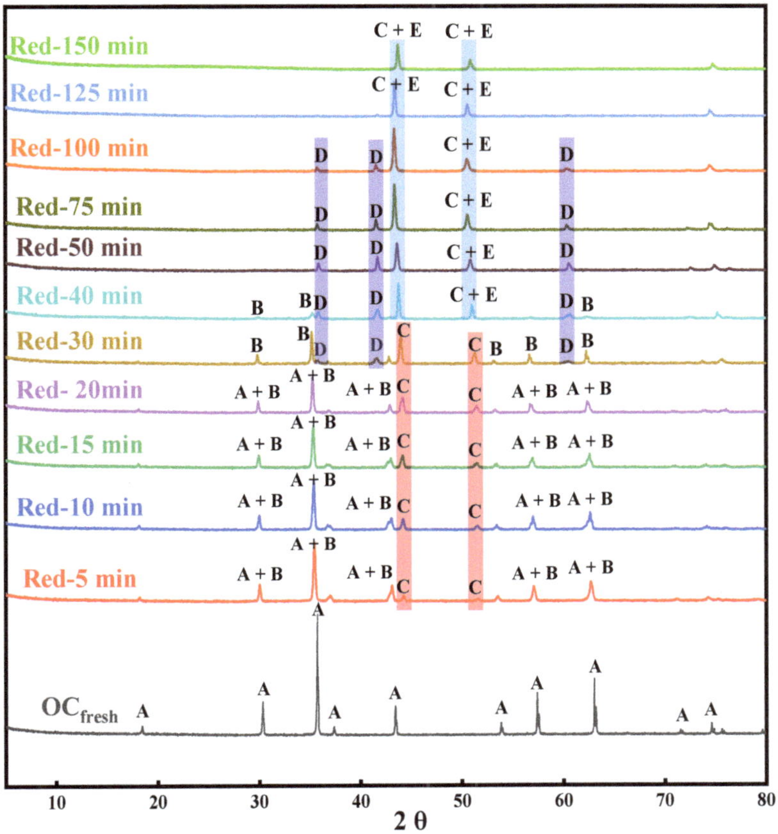

Figure 2. XRD patterns of $NiFe_2O_4$ OC at different reduction times. (A: $NiFe_2O_4$, B: $Fe_{3-x}O_4$, C: Ni, D: $Fe_{1-x}O$, E: Fe).

As the reduction degree deepens, three characteristic peaks of $Fe_{1-x}O$ (card: 01-089-0687) are observed (marked as D). It means that the spinel structure of $NiFe_2O_4$ and $Fe_{3-x}O_4$ gradually collapsed due to Fe^{3+} ions being continuously reduced. Finally, when reduction time is above 100 min, almost Ni and Fe cations are completely reduced into metallic Ni and Fe (card: 96-901-4712).

2.1.2. Crystalline Phase Evolution of the Reduced OC in Oxidation

The metallic Fe and Ni, as the completely reduced product of $NiFe_2O_4$ OC, show a stepwise oxidation behavior under a CO_2 atmosphere. As well, no apparent carbon deposits were found during the CO_2 oxidation process. As shown in Figure 3.

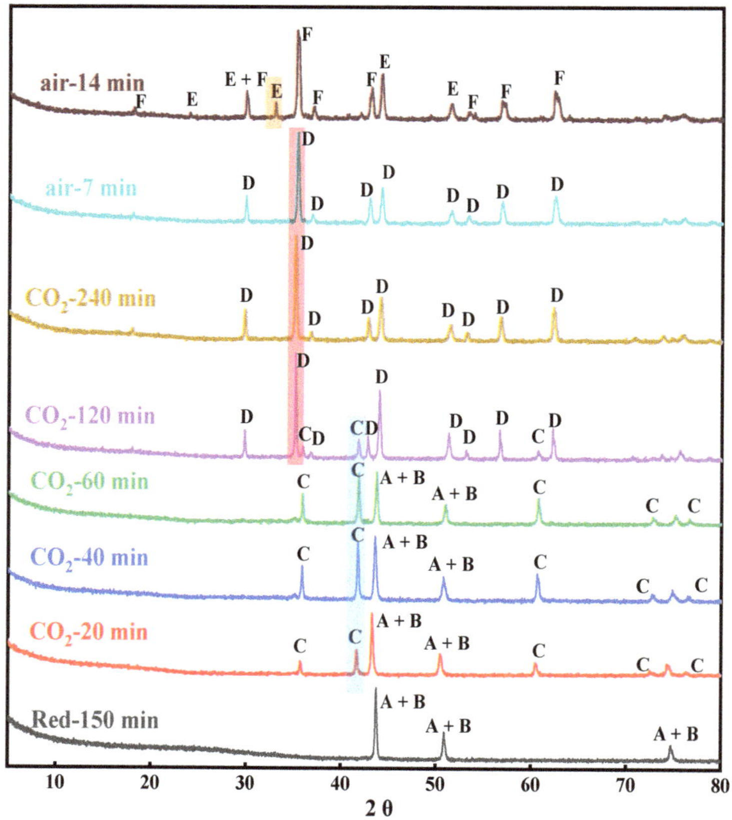

Figure 3. XRD patterns of reduced OC at different oxidation times. (A: Ni, B: Fe, C: $Fe_{1-x}O$, D: $Fe_{3-x}O_4$, E: Fe_2O_3, F: $NiFe_2O_4$).

When the oxygen carrier was oxidized by CO_2 for 5 min, a part of metallic Fe is oxidized recovering a fraction of lattice oxygen to form the $Fe_{1-x}O$ phase (marked as C). As the oxidation degree deepens, the characteristic peaks of $Fe_{3-\delta}O_4$ are observed (marked as D). It is worth noting that the Ni atoms still exist in the form of metallic Ni due to thermodynamic limitations. It is difficult to replenish all the lattice oxygen under the CO_2 atmosphere owing to the weak oxidation performance of CO_2. Thus, in order to recover all the lattice oxygen, an air oxidation step is necessary to achieve a whole chemical looping cycle. The OC after CO_2 oxidation is further oxidized by air to recover all the lattice oxygen to form the $NiFe_2O_4$ phase (marked as F) with spinel structure and a small amount of Fe_2O_3 phase (card: 00-004-0755, marked as E). Although it is difficult to re-oxidize the reduced OC back into its original state, it still exhibits good performance after multi-cycles during the CLDR [27,35].

2.2. XPS Analysis

2.2.1. The Release of Lattice Oxygen

The nature of the oxygen species and the transformation of metal species (Fe and Ni) on the near-surface of OC with reaction time were explored in detail using XPS combined with XRD to further clarify the mechanism of the migration of lattice oxygen. The experimental results are shown in Figures 4 and 5 (the samples analyzed by XPS and XRD are consistent). And the XPS analysis of the O element and the Fe, Ni elements will form a corroboration relation, which strongly supports the migration mechanisms of lattice oxygen during the reduction of $NiFe_2O_4$ OC.

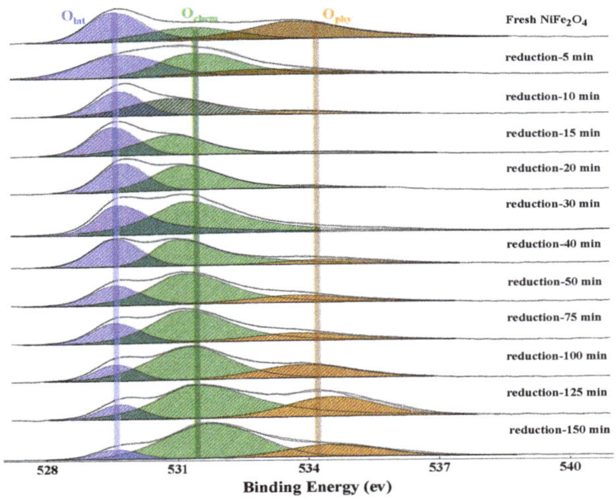

Figure 4. XPS of O1s at different reduction times. (Blue: lattice oxygen, green: chemisorbed oxygen, yellow: physiosorbed oxygen).

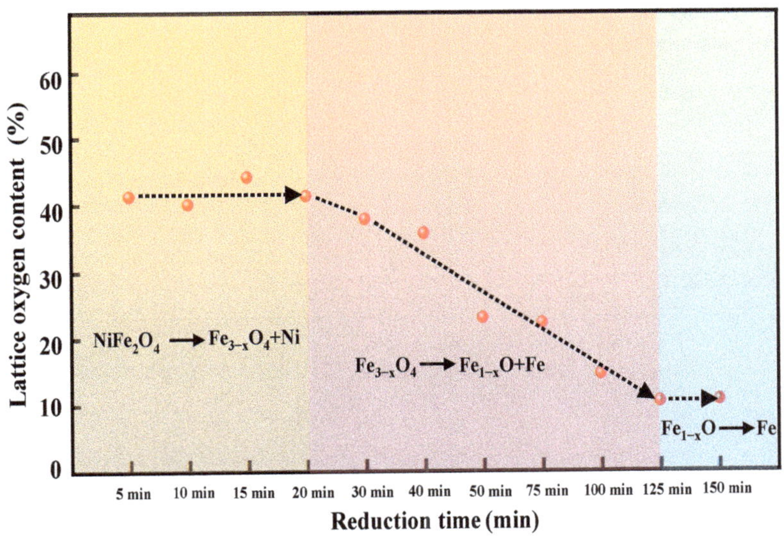

Figure 5. The relative content of lattice oxygen on the surface of OC at different reduction times combined with the evolution of the crystal phase.

The oxygen species on the oxygen carrier surface mainly include lattice oxygen, chemisorbed oxygen, and physiosorbed oxygen [30]. The binding energy of lattice oxygen, chemisorbed oxygen, and physiosorbed oxygen are 529.0–530.0 eV, 530.0–532.0 eV, and >532.0 eV, respectively [36–40]. Previous literature reported that the different oxygen species can be transformed into each other as follows: $O_2(g) \leftrightarrow O^{2-}(ads) \leftrightarrow O_2^{2-}(ads) \leftrightarrow 2O^{-}(ads) \leftrightarrow 2O^{2-}(ads) \leftrightarrow 2O^{2-}(lat)$ [41].

As shown in Figure 4, the XPS spectra shows that three oxygen species (lattice oxygen; chemisorbed oxygen; and physiosorbed oxygen) can be found on the near-surface of the fresh $NiFe_2O_4$ oxygen carrier particles. At the initial reduction stage (5–20 min), the relative content of physiosorbed oxygen gradually decreases with the increase of reduction time (between 5–20 min), which decreases from ~15% at 5 min to ~0% at 20 min. On the contrary, the relative content of chemisorbed oxygen gradually increases with reduction time, which increases from ~40% at 5 min to ~60% at 30 min. It is worth noting that the relative content of lattice oxygen tends to stabilize at ~40% for the first 20 min at the reduction stage (Figure 5). Meanwhile, the result of XRD shows that the OC maintains the spinel structure and the main phase is $Ni_{0.6}Fe_{2.4}O_4$ and $Fe_{3-x}O_4$ during the first 20 min (Figure 5). As well, the oxygen carrier exhibits a fast reduction rate at the initial stage. The results of XRD combined with XPS indicate that the physiosorbed oxygen was first consumed from the surface of the oxygen carrier due to the weak adsorption capacity for CO_2/CO. As well, the surface lattice oxygen is easily converted into chemically adsorbed oxygen due to being highly active in the chemical reaction interface [41]. Thus, the relative content of physiosorbed oxygen decreased whereas chemisorbed oxygen increased. It also indicates that the conversion rate of the surface lattice oxygen to the chemisorbed oxygen is less than the rate of bulk lattice oxygen migration to the surface, which causes the relative content of surface lattice oxygen to be stable. In other words, three reaction rates determine the changing trend of different oxygen species, namely the consumption rate of chemisorbed oxygen bound to fuel, the conversion rate of surface lattice oxygen to chemisorbed oxygen, and the migration rate of bulk lattice oxygen to the surface, respectively. In the initial stage of the reduction reaction, the migration rate of bulk phase lattice oxygen to the surface is fairly rapid due to the presence of a large amount of lattice oxygen in the bulk of OC and the concentration gradient between the surface and the bulk. This process is accompanied by a part of the Ni element in OC being reduced into free metallic nickel and divorces from the spinel lattice of the $NiFe_2O_4$. As well, it also caused the gas-solid reaction interface of was fixed at the surface of the oxygen carrier. Hence, the reaction pathway can be described as follows: CO is firstly adsorbed at the active sites on the surface of the OC particles and then it combines with chemically adsorbed oxygen at the reaction interface to form CO_2 or CO_3^{2-}. The surface lattice oxygen is rapidly and continuously converted into the chemically adsorbed oxygen, meanwhile, the bulk lattice oxygen rapidly migrates to the surface of the spinel structure to replace the consumed lattice oxygen at the initial stage of the reduction reaction. Thus, we inferred that as Ni migrates outward to the particle surface, promoting oxygen vacancies formation and the diffusion of lattice oxygen. It finally leads to a fast rate and the stabilization of the surface lattice oxygen content at the initial reduction stage.

In the intermediate stage of the reduction reaction (20–125 min), the relative content of lattice oxygen gradually decreases with the increase of reduction time, which decreases from ~40% at 20 min (Main phase: $Fe_{3-x}O_4$) to ~10% at 125 min (Main phase: metallic Fe and metallic Ni), and then it gradually tends to stabilize (Figure 5). Whereas, the content of physiosorbed oxygen gradually increases from ~0% at 20 min (Main phase: $Fe_{3-x}O_4$) to ~25% at 125 min (Main phase: metallic Fe and metallic Ni), and then it is gradually stabilized. It indicates that with the collapse of the spinel structure, the bulk lattice oxygen of OC is continuously consumed. In other words, the rate of migration of bulk-phase lattice oxygen to the surface slows down due to the concentration of the lattice oxygen decreasing gradually.

Interestingly, the XRD and XPS analysis show that even if the $NiFe_2O_4$ OC is completely reduced to the metallic Ni and Fe phases, which means the bulk lattice oxygen is almost fully consumed, and a fraction of lattice oxygen always appears at the surface of the reduced OC (Figure 5). At the same time, the physiosorbed oxygen content of the fully reduced OC (125–150 min) was significantly higher than those of the OC in the initial reduction stage (0–20 min). It indicates that the completely reduced OC, namely the metallic Fe and Ni phase, easily adsorbs CO_2. As CO_2 is adsorbed on the surface of the reduced OC, a CO_2-splitting reaction occurs on the active metallic Fe. The O anions extracted from CO_2 transform into surface adsorbed oxygen. Subsequently, the adsorbed oxygen further transforms into lattice oxygen induced by the oxygen vacancies on the surface. These two reactions are represented as R (1) [35] and R (2) [41]. Eventually, the relative concentrations of each oxygen species remain unchanged.

$$Fe + CO_2 \rightarrow Fe^{\delta+} + [O^-_{ads}] + CO \text{ R} \quad (1)$$

$$[O^-_{ads}] \leftrightarrow [O^-_{lat}]\ R \tag{2}$$

Therefore, an active interface consisting of lattice oxygen and the corresponding metallic ions is maintained during the whole reduction process.

The valence evolution of metal cations has a corroborative relationship with the surface oxygen species, thereby shedding light on the migration mechanism of lattice oxygen. It was reported that the peaks with binding energy at around ~856.2 eV, ~854 eV, and ~852.6 eV are assigned to Ni^{2+}, Ni^0 (metallic Ni), and Ni^0 (Fe-Ni), respectively [42,43]. As shown in Figure 6, only Ni^{2+} peaks can be observed in the XPS spectra of the fresh $NiFe_2O_4$ OC. After reduction for 5 min, a peak for Ni^0 appears in the Ni 2p spectra, indicating that the metallic Ni phase is formed on the surface at the initial reduction stage. This is consistent with the XRD observations. In the reduction time from 5 to 15 min, the relative content of Ni^{2+} ions decreased significantly reaching zero till 20 min, whereas the Ni^0 content increased rapidly. Combined with the O1s XPS spectra (see Figure 4), it can be observed that physiosorbed oxygen decrease is accompanied by the initial reduction stage. Thus, the presence of metallic Ni^0 on the surface probably has no effect on the adsorption capacity of OC for CO_2/CO. It was probably due to thermodynamic limitations that metallic Ni^0 cannot react with CO_2 [22]. After 15 min, the peak of Ni^{2+} disappears. All Ni^{2+} is converted to Ni^0 on the surface.

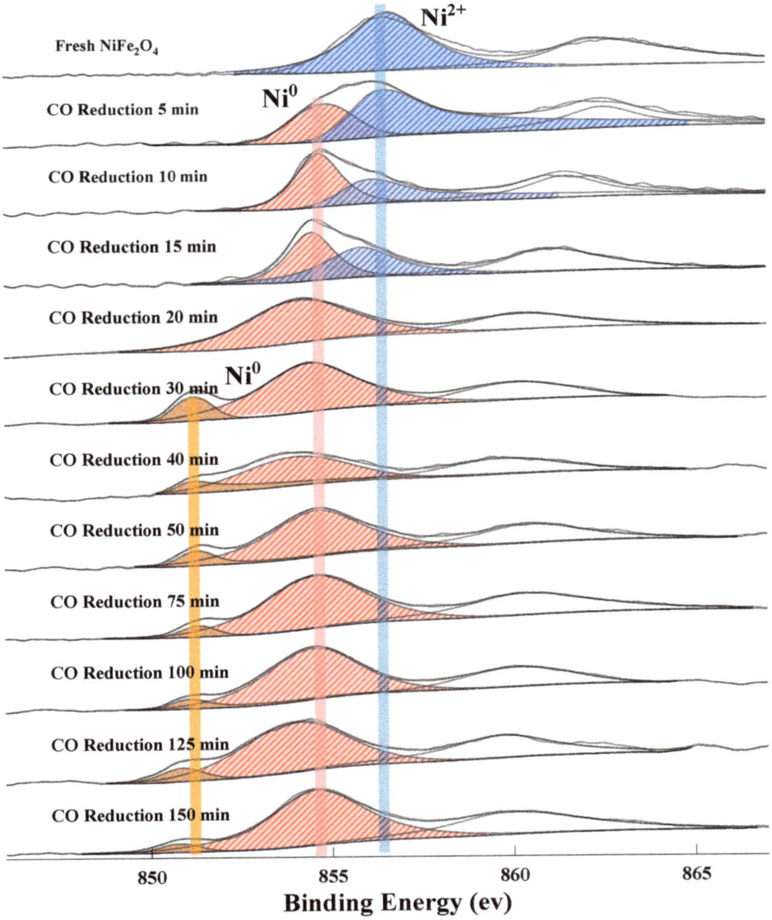

Figure 6. XPS spectrum of Ni *2p 3/2* at different reduction times.

At 30 min, a peak for Fe-Ni alloy appears, corresponding to the formation of metallic Fe with the decomposition of $Fe_{3-x}O_4$ to $Fe_{1-x}O$ and Fe. In the reduced $NiFe_2O_4$ OC, the presence of metallic Fe may create an effect on the electronic structure of metallic Ni. It further indicates that the complete reduction product of OC, is not a simple mixture of Ni and Fe, but a composite with complex space structure, which performance is better than that of the complete reduction product of NiO + Fe_2O_3 OC in CLDR [35].

The XPS spectrum of Fe appears in the form of double peaks. As shown in Figure 7. The fresh OC has two Fe^{3+} peaks located at the binding energy of 713.2 eV and 715.5 eV assigned to the spinel $NiFe_2O_4$ with the structure of the tetrahedron and octahedron structure, respectively. With the reaction proceeding, Fe^{3+} is continuously reduced to Fe^{2+} or Fe^0. Thus, the Fe^0, Fe^{2+} and Fe^{3+} peaks appear at ~706.7 eV, ~710 eV and ~713–715 eV, respectively [44]. It is worth noting that the Fe^0 peak cannot be observed on the surface of OC at an initial reduction of 30min. However, the XRD analysis of the crystal phase and the XPS analysis of Ni is similarly observed in the presence of metallic Fe^0 at 30 min. It indicated that Fe^0 is firstly formed inside of the OC particles. In other words, the lattice oxygen tends to migrate to the surface at high temperatures, therefore metallic phase begins to be formed from inside the OC bulk. Similar results have been obtained in other literatures [31,45]. Hence, the metal ions migration pathway can be described as follow: With the release of lattice oxygen, Ni^{2+} was firstly migrated to the surface and reduced to Ni^0 due to the higher reactivity of Ni. Then the lower active Fe^{2+} and Fe^{3+} are subsequently reduced thereby providing lattice oxygen to the fuel on the surface. As the reduction degree deepens, Fe cations begin to be reduced into metallic Fe^0. The presence of metallic Fe in the bulk creates an effect on the electronic structure of metallic Ni. The surface Fe/Ni ratio of the OC will be analyzed via EDS combined with HRTEM, to further clarify the phenomenon of the migration and enrichment of Fe and Ni atoms during the reduction process. In addition, it was observed that even if OC is completely reduced to the metal phase, the Fe element on the surface is still dominated by Fe^{2+} and Fe^{3+} and only a small amount of Fe^0. It is due to the strong surface adsorption of the metallic Fe in complete reduced OC, the reaction R (1) is dominated in the later stage of the reaction, thus the relative concentration of surface Fe species remains stable. As well, compared to the initial stage of the reaction, the relative content of physiosorbed oxygen gradually increases. The analysis of different elements forms a good corroboration relation, which strongly supports the migration mechanisms of lattice oxygen during the reduction of $NiFe_2O_4$ OC.

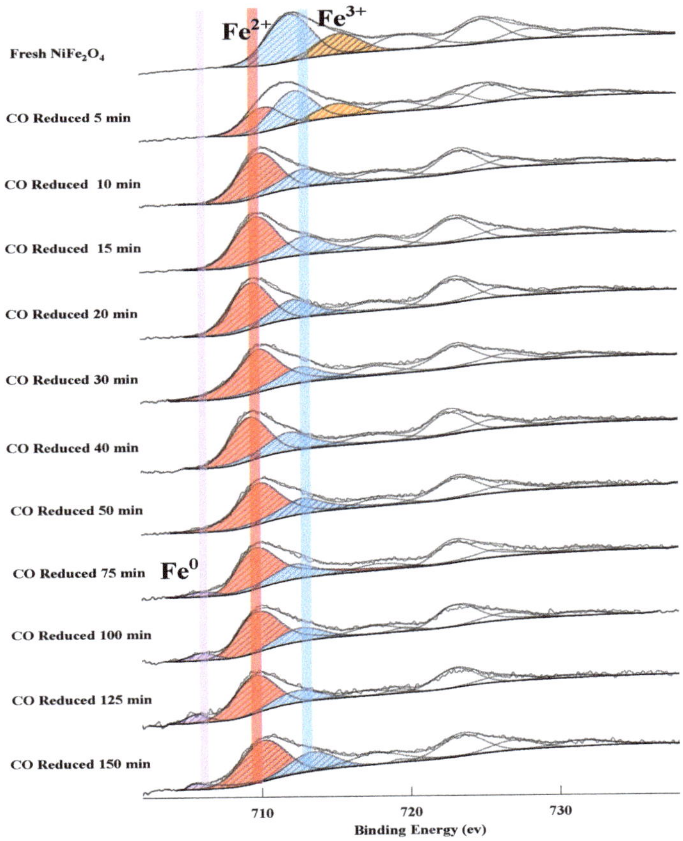

Figure 7. XPS spectrum of Fe *2p 3/2* at different reduction times.

2.2.2. The Uptake of Lattice Oxygen

The O 1s XPS spectra of the oxygen species of the reduced OC re-oxidized in the CO_2-air atmosphere at different oxidation times are shown in Figure 8. It was observed that the lattice oxygen content gradually increases from ~10% at 0 min to ~37% at 240 min of CO_2-oxidization (see Figure 9). However, the physiosorbed oxygen content gradually decreases from ~23% at 0 min to ~0% at 120 min. It is noted that the physiosorbed oxygen reappeared at oxidization for 240 min. These results should be attributed to a fact that the chemical reaction interface is fixed on the surface of the OC particle during the CO_2 oxidation. If the reaction interface migrates from the surface to the inside. The surface lattice oxygen will reach a maximum at the initial stage of the reaction and then tends to stabilize, rather than gradually increases. In other words, if the surface metal restores all lattice oxygen at the initial stage of the reaction, CO_2 molecules will penetrate through the oxide layer to the interior of the particle to react with metallic Fe. Thus, it indicates that the rate of conversion of the chemisorbed oxygen to the surface lattice oxygen is slightly quicker than the rate of surface lattice oxygen migration to bulk, and then the surface lattice oxygen gradually accumulates. The oxygen concentration gradient between the surface and the bulk continuously drives the lattice oxygen inward transportation to the bulk. In addition, with the recovery of lattice oxygen, the adsorption capacity of oxygen carriers for physisorbed oxygen gradually weakens, which is consistent with the analysis of the reduction process. However, at the end of CO_2 oxidation, a large amount of chemisorbed oxygen can be still observed due to the slower rate of CO_2 splitting and the prolonged contact between CO_2 and oxygen carriers. Because of the weak oxidizability of CO_2, it is very difficult to recover the entire lattice oxygen for the reduced OC. Thus, an air oxidation process is required to recover the full lattice oxygen of the OC.

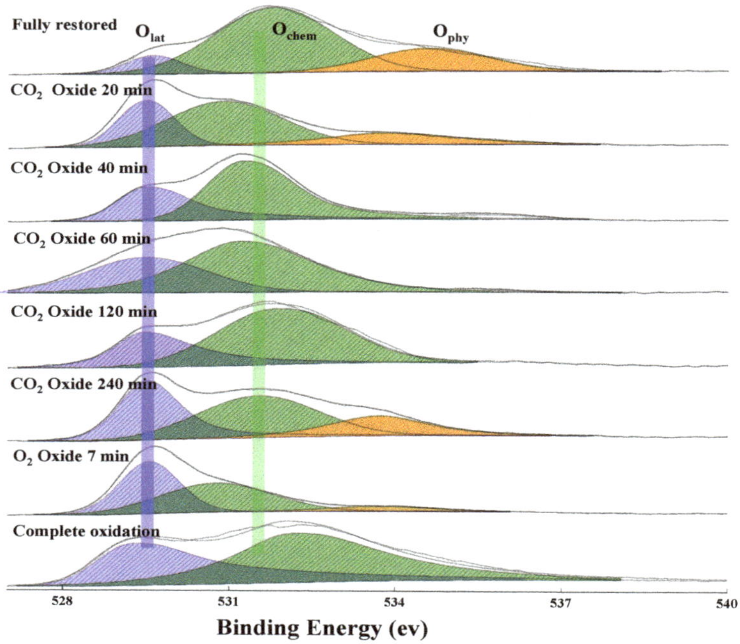

Figure 8. XPS of O1s at different oxidation times. (Blue: lattice oxygen, green: chemisorbed oxygen, yellow: physiosorbed oxygen).

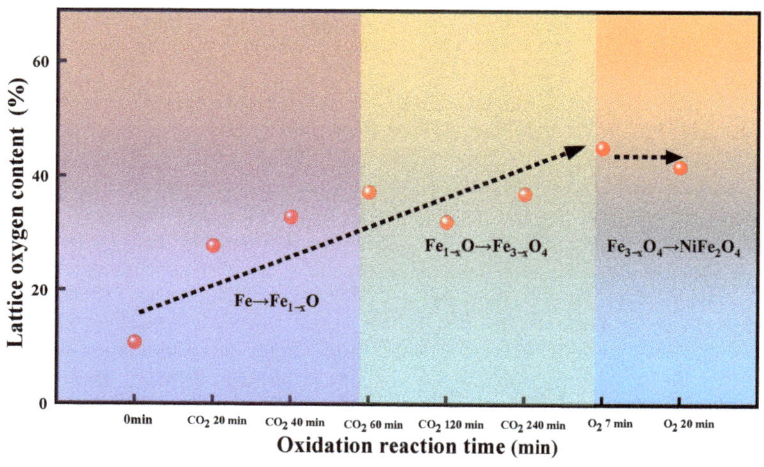

Figure 9. The relative content of lattice oxygen on the surface of OC at different oxidation times combined with the evolution of the crystal phase.

Figure 9 illustrates the lattice contents as a function of oxidation reaction time in air oxidation of the reduced OC. It was observed that the relative content of lattice oxygen increases, which increases from ~37% (at 240 min of CO_2-oxidization) to ~45%, and it tends to stabilize. The increase in the relative content of lattice oxygen is due to metallic Ni^0 recovery lattice oxygen on the surface, namely the formation of $NiFe_2O_4$ species. The stability of the relative content of lattice oxygen is due to the rate of conversion of the chemisorbed oxygen to the surface lattice oxygen is much quicker than the rate of surface lattice oxygen migration to bulk. There is a concentration gradient of lattice oxygen between the surface and the bulk, thus the lattice oxygen is continuously diffused from the surface to the bulk to fill the oxygen vacancies and a relatively stable equilibrium is attained. Thus, the concentrations of lattice oxygen remain stable. Through the above analysis, we infer that the concentration gradient between the surface and the bulk drives the transmission of lattice oxygen to achieve the reduction or oxidation of OC. The migration rate of lattice oxygen is associated with the movement process of Ni particles into and out of spinel bulk.

The Ni 2p XPS spectra (Figure 10) show that the nickel element on the surface of the OC is kept in form of Ni^0 in the course of CO_2 oxidation due to metallic Ni cannot be oxidized in thermodynamics. The peak of two kinds of Ni^0 was observed, namely metallic Ni (Ni-Ni) and other kinds of Ni^0 that the presence of Fe creates an effect on the electronic structure of metallic Ni. The relative content of the two kinds of Ni^0 fluctuates due to elemental Fe gradually restoring lattice oxygen. During the air oxidation stages, the peak of Ni^{2+} was observed. The result is very consistent with the analysis of oxygen species. In addition, the signal of the Ni element on the surface disappears at 60 min due to the Fe^0 quickly recovering the lattice oxygen to form Fe_xO and the generated oxide (Fe_xO) can form an oxide film on the surface. The surface Ni atoms are covered by the Fe atoms. As well, the signal of the Ni element is related to the thickness of the surface oxide (Fe_xO). Two peaks of Ni^{2+} were observed for fully oxidized oxygen carriers, which is probably due to the presence of different spatial structures of Fe_2O_3 and $NiFe_2O_4$.

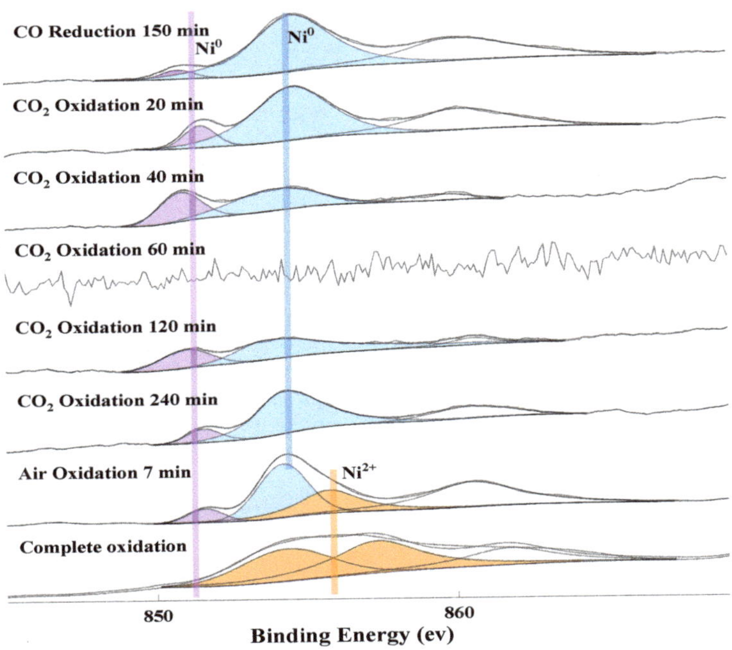

Figure 10. XPS spectrum of Ni *2p 3/2* at different oxidation times.

As shown in Figures 7 and 11, the Fe element in the completely reduced OC is mainly in form of Fe^{2+}, Fe^{3+}, and a minor amount of Fe^0. As the reduced OC is oxidized, the Fe^0 and Fe^{2+} are continuously oxidized into Fe^{3+}. When the OC is oxidized by CO_2 for above 20 min, the XPS peak of Fe^0 disappears. After CO_2-oxidation for 240 min, the Fe element in the reduced $NiFe_2O_4$ OC can be reoxidized into Fe_3O_4 by such a weak oxidant, CO_2. In the completely oxidized OC, the surface Fe element of OC has two Fe^{3+} peaks due to the complex spatial structure of the $NiFe_2O_4$ and Fe_2O_3 particles.

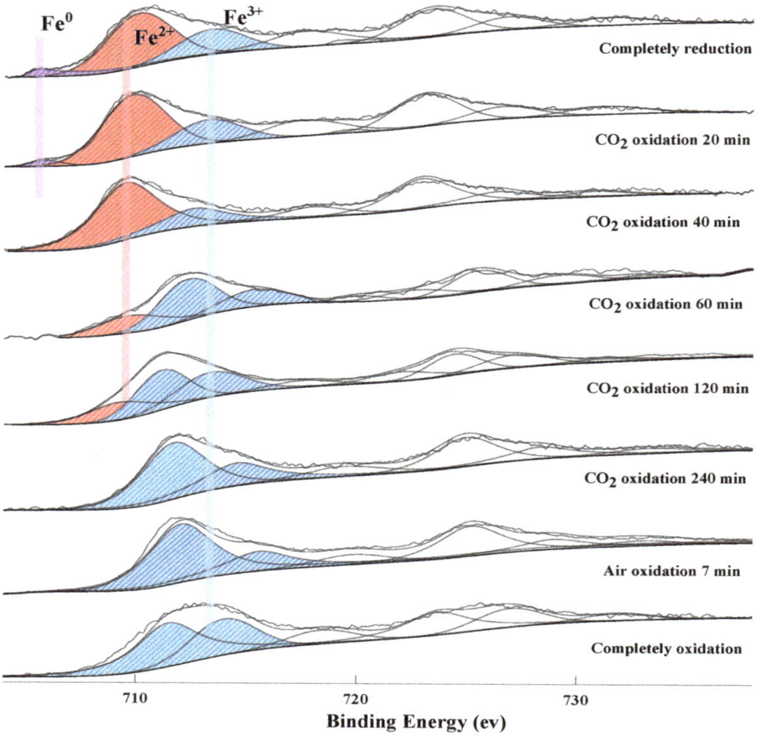

Figure 11. XPS spectrum of Fe *2p 3/2* at different oxidation times.

2.3. *In Situ Raman Analysis*

The vibrations changes of the chemical bonding in the course of redox reactions on the surface of the OC were studied via in-situ Raman. Compared to non-in situ characterization, in-situ characterization makes it easier to obtain the real results of the materials. As shown in Figure 12, Raman characterizations show five peaks attributed to $NiFe_2O_4$ [46]. The strongest peak at ~$695 cm^{-1}$ is attributed to the symmetric stretching vibration of the O atom along the Fe-O bond direction. The peak at $481 cm^{-1}$ is attributed to the vibration of the Ni-O bond [47]. The peak at $330 cm^{-1}$ is attributed to the symmetric bending vibration of the Fe-O bond [48,49].

At the initial stage of the CO-reduction reaction, the peak of Ni-O ($481 cm^{-1}$) and Fe-O ($695 cm^{-1}$) display a red shift and gradually becomes weaker, which is due to the formation of oxygen vacancy during lattice oxygen release. Then a very important process is found that the peak of Fe-O ($695 cm^{-1}$) changes from weak to strong. Meanwhile, the peak of Ni-O ($481 cm^{-1}$) disappeared. It indicates that the element Ni underwent Ni^{2+} to Ni^0 transition on the surface and Fe atoms combined with remaining lattice oxygen to form a new structure. With the reaction proceeding, the peak of Fe-O gradually becomes weaker from the new structure. However, no matter how long the reduction reaction time lasts, even if the XRD analysis show OC completely reduced to the metal phase, there will always be a peak of Fe-O. The in-situ Raman analysis and the XPS analysis are consistent.

For the OC oxidized by CO_2, the peak of Fe-O changes from weak to strong. As well, no Ni-O vibration peaks were found. The Raman spectra of the oxygen carrier after air oxidation are different from the fresh samples. But it is very similar to the Raman bands of the pure Fe_2O_3 [49]. It is due to the formation of Fe_2O_3 on the surface of oxygen carriers [50].

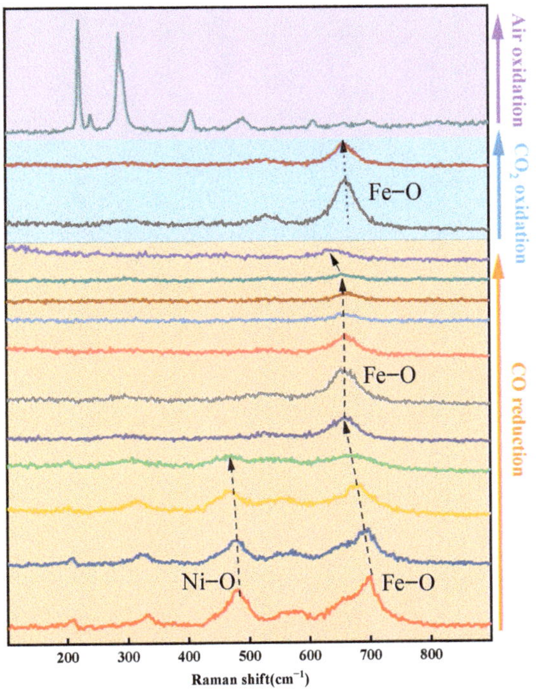

Figure 12. Raman spectra of the redox process of OC.

2.4. TEM Analysis

The morphology of the fresh $NiFe_2O_4$ OC, completely reduced OC, reoxidized OC by CO_2, and completely reoxidized OC by air was investigated by HRTEM, as shown in Figure 13. It is observed that the Ni, Fe, and O elements distribute uniformly throughout the particle in the fresh oxygen carrier. The lattice spacing of the TEM under high magnification was also performed. The crystal forms of the fresh oxygen carrier, ~2.49 Å of the lattice fringe spacing ascribed to (311) plane of $NiFe_2O_4$, can be observed, which are located in the exogenous parts of the oxygen carrier particle. The ratio of surface Fe/Ni~1.83 is very consistent with the theoretical value (Table 1).

Table 1. The relative content of Fe and Ni from EDS mapping.

Element (Atom %)	Fresh OC	Completely Reduced OC	Completely Oxidized OC by CO_2	Completely Oxidized OC by Air
Fe	47.71%	59.90%	38.19%	37.49%
Ni	25.98%	9.53%	3.67%	3.43%

From Figure 13b, it can be found that the particle size of samples shows an increase due to the collapse of the spinel structure and the enrichment of metallic phases. Small circular particles were observed in distribution inside of the oxygen carrier, which is presumed to be enriched Ni^0. Under high magnification, a fuzzy boundary located in the exogenous parts of the oxygen carrier particle, and ~2.02 Å of the lattice fringe spacing ascribed to (111) plane of metallic Fe^0, can be observed. A fuzzy boundary probably has a relationship with the highly reactive chemical reaction interface, which consists of lattice oxygen and the corresponding metallic ions. The EDS mapping shows the ratio of surface Fe/Ni~6.28, which indicates that the surface Ni atoms are covered by the Fe atoms.

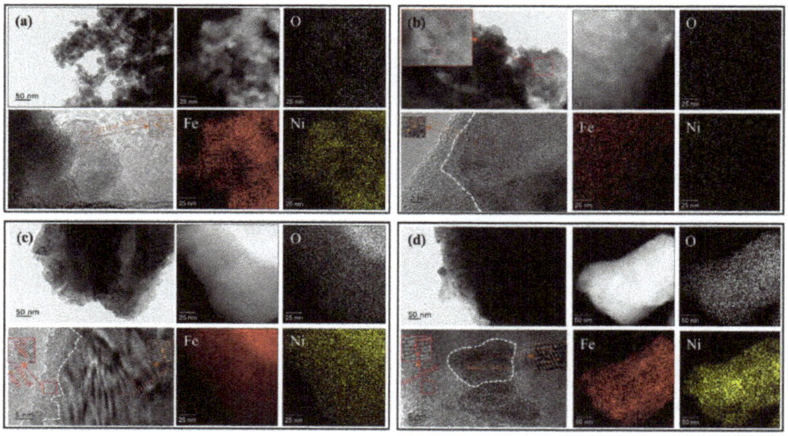

Figure 13. HRTEM and EDS mapping: (**a**) (Fresh $NiFe_2O_4$ OC); (**b**) (Completely reduced OC); (**c**) (Completely oxidized OC by CO_2); (**d**) (Completely oxidized OC by Air).

As shown in Figure 13c, many small circular particles were still observed distributed inside of the oxygen carrier. It is probably due to Ni^0 cannot be oxidized by CO_2. The lattice spacing of Fe_3O_4(311) and a boundary are observed. From Figure 13d, in the middle of the Fe_2O_3(311), there is a layer with the lattice fringe spacing of ~1.47 Å related to the (440) plane of $NiFe_2O_4$. The EDS mapping shows that the ratio of surface Fe/Ni did not return to the original state. It indicates that Fe_2O_3 exists on the surface of the completely oxidized oxygen carrier [50].

According to the above analysis, the potential reaction paths and mechanism of $NiFe_2O_4$ in the CLDR process can be proposed as shown in Figure 14. It can observe that the process of CLDR of methane consists of a successive three-step chemical looping scheme, where $NiFe_2O_4$ OC is supposed to: (i) convert various fuels (e.g., CH_4, biomass) to value-added products (e.g., heat, syngas), (ii) reduce CO_2 to produce CO, and (iii) consume oxygen from the air to replenish full lattice oxygen meanwhile sustaining the thermal balance of the system. The chemical reaction interface is fixed on the surface of OC particles (Figure 14). During the oxygen carrier reduction process, the bulk lattice oxygen migrated to the particle surface and converted to adsorbed oxygen, which is consumed by fuel. During the oxygen carrier oxidation process, CO_2 is adsorbed on the particle surface and dissociates into O ions, which are converted from adsorbed oxygen to lattice oxygen and migrate to the bulk. The diffusion of oxygen anions maintains the overall charge neutrality due to the oxygen chemical potential gradient. It is worth noting that the phase segregation caused by the outward diffusion of metal cations is not conducive to the cyclic stability of oxygen carriers. Adding an inert component to control the movement of active metal atoms is one of the directions worthy of our efforts in the future to achieve better performance. Through the above analysis, we infer that utilizing or controlling the reversible dynamic changes behavior of particles will be an effective means to enhance the reactivity and stability of oxygen carriers in future chemical looping applications.

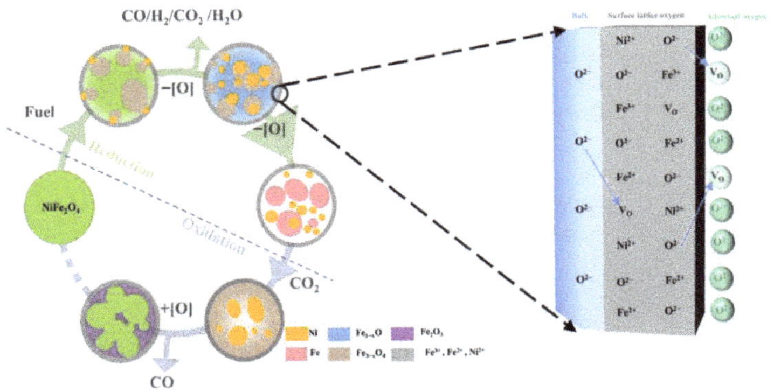

Figure 14. Redox mechanism description in CLDR process.

3. Materials and Methods

3.1. Preparation of Oxygen Carrier

The sol−gel methods were applied to prepare $NiFe_2O_4$ nanoparticles. Nickel nitrate hexahydrate [$Ni(NO_3)_2{\cdot}6H_2O$] (Aladdin, ≥99.99%) and iron nitrate nonahydrate [$Fe(NO_3)_3{\cdot}9H_2O$] (Aladdin, ≥99.99%) were used as raw materials and dissolved in deionized water, forming 1 M of $Ni(NO_3)_2$ solution and 2 M of $Fe(NO_3)_3$ solution, respectively. The citric acid (Macklin, ≥99.5%) with the equivalent molar $Ni(NO_3)_2$ was added to the original mixed solution. Subsequently, obtain a pH of 7.5. Finally, the solution was heated at 65–75 °C to evaporate 2/3 of the water, and the gel formed when the solution was cooled to room temperature. The precursors were dried in a vacuum evaporator at 120 °C and calcined for the formation of the ferrite phase in a muffle furnace under an air atmosphere according to the same temperature program. The dried materials were calcined by increasing the temperature from 25 °C to 400 °C with a ramp of 8 °C/min and keeping at 400 °C for 2 h. Then the sample was increased to 1200 °C at a heating rate of 4 °C/min, and kept at 1200 °C for 2 h [46].

3.2. Thermo-Gravimetric (TG) Analysis

The three stages of the CLDR experiment are carried out by the STA-409/PC thermogravimetric analyzer, which is produced by NETZSCH, Selb, Germany. The three different gases (CO, CO_2, and air) flow rates are all set to 100 mL/min. The experimental procedure is as follows: $NiFe_2O_4$ OC (m = 150 mg ± 0.1 mg) was added to the thermogravimetric reactor. the reaction temperature is heated up to 950 °C at a heating rate of 10 °C/min in an inert atmosphere. Then, CO is introduced for the reduction reaction. After the reduction experiment, CO_2 and air is introduced for oxidation reaction, respectively. Obtain 19 samples with different reaction times for subsequent analysis, which are recorded as OC_{fresh}, reduction-5 min, reduction-10 min, reduction-15 min, reduction-20 min, reduction-30 min, reduction-40 min, reduction-50 min, reduction-75 min, reduction-100 min, reduction-125 min, reduction-150 min, CO_2-oxidation-20 min, CO_2-oxidation-40 min, CO_2-oxidation-60 min, CO_2-oxidation-120 min, CO_2-oxidation-240 min, air-oxidation-7 min, air-oxidation-14 min, respectively.

3.3. Characterization

3.3.1. X-ray Photoelectron Spectroscopy (XPS)

An X-ray photoelectron spectroscopy technology (XPS, Thermo Fisher Scientific ESCALAB 250Xi, Waltham, MA, USA) was carried out to investigate the valence state of variable metal ions and oxygen species of OCs, with an Al Ka X-ray source (1486.6 eV photons) at an operating voltage of 20 kV and a current of 10 mA, respectively.

3.3.2. X-ray Diffraction (XRD)

An X-ray diffractometer (XRD) using 40 kV, 40 mA Cu Kα radiation was used to analyze the crystal structure of fresh and reacted samples. XRD measurement scanning rate is 2 °/min from 2θ = 5–80° with a step size of 0.02°.

3.3.3. In Situ Raman

A Laser Confocal Micro-Raman Spectral System (LabRAM HR800, HORIBA Scientific, Piscataway, NJ, USA) was used to analyze the surface structure of the oxygen carrier. Laser full range of 50 mw, selected a laser probe with a wavelength of 432.521 cm^{-1}. Added $NiFe_2O_4$ oxygen carrier mass of 100 mg in the in-situ stage, under 20 mL/min inert atmosphere, warmed up at a rate of 50 K/min to 950 °C. When the temperature rises to 950 °C, a gas (the CO reduction and CO_2-air oxidation) is continuously introduced into the reactor, respectively. A total of 20 mL/min CO gas (10 vol%CO, N_2 equilibrium) for reduction. Then switch to CO_2 (10 vol%CO_2, N_2 equilibrium) with a flow rate of 20 mL/min for oxidation reaction after the end of reduction. Finally, pass air (20 mL/min) for strong oxidation.

3.3.4. Transmission Electron Microscopy (TEM)

The microstructure of the sample was analyzed using a high-resolution field emission transmission electron microscope (JEOL 125 JEM-2100F, Tokyo, Japan). EDS energy spectrometer manufactured by American Thermoelectric Noran (Waltham, MA, USA).

4. Conclusions

In this study, we elaborated on the reversible phase transition law of $NiFe_2O_4$ using the sol-gel method for the preparation of the chemical looping CO_2 splitting process. Then valence evolution of active components of the oxygen carriers was analyzed in detail by XPS. As well, we explained the chemical mechanism of the change in the relative content of different oxygen species. A large number of evidence about bulk lattice oxygen migration was obtained by different characterizations.

A fact was confirmed by an in-situ Raman and XPS that a highly reactive chemical reaction interface consisting of lattice oxygen and the corresponding metal atoms is always present on the surface of the oxygen carrier during the CLDR. In summary, the diffusion of oxygen anions maintains the overall charge neutrality due to the oxygen chemical potential gradient. Oxygen anions move from a high concentration area to a low concentration area, namely the surface lattice oxygen is converted to chemisorbed oxygen and forms oxygen vacancies. The bulk lattice oxygen to the surface and healed the surface vacancy via outward diffusion. The diffusion of metallic Fe and metallic Ni changes the morphology and crystalline structure of OC. The extraordinary performance of $NiFe_2O_4$ for CLDR is due to the movement of Ni particles into and out of spinel bulk, which promote oxygen migration and diffusion. In the reduction period under CO, a part of the Ni element in OC is first separated from the spinel structure to become the free metallic Ni. It was confirmed that exsolved Ni nanoparticles were not redissolved into the spinel in the oxidation under CO_2, and the active phase (Ni) was dissolved into the matched spinel under an air atmosphere. This understanding of the migration of active metals and lattice oxygen is of great significance for material design for the applications of CLDR, especially through structure design and utilizing the dynamic phase change to stabilize the metal particles. As well, establishing the relationship between the dynamic behavior and performance of the metal is one of the directions worthy of our efforts in the future to achieve better performance of OC in CLDR; it also can promote the process of CO_2 resource utilization and carbon neutrality goals.

Author Contributions: Conceptualization, D.S. and Y.L.; methodology, Z.H.; software, D.S.; validation, D.S., Z.H. and F.H.; formal analysis, K.Z. and F.H.; investigation, D.S.; resources, Y.X.; data curation, D.S.; writing—original draft preparation, D.S.; writing—review and editing, F.H.; visualization, Z.H.; supervision, Z.H.; project administration, Y.L.; funding acquisition, Z.H. All authors have read and agreed to the published version of the manuscript.

Funding: This research was funded by the National Natural Science Foundation of China (52006224, 52076209, 22179027), the Foundation and Applied Foundation Research of Guangdong Province (2019B1515120022, 2019A1515110828, 2020A1515110138, 2021A1515010459, 2022B1515020045), the Joint Fund of the Yulin University and the Dalian National Laboratory for Clean Energy (2021021), Guangxi Natural Science Foundation (2018GXNSFDA281005), Youth Innovation Promotion Association, CAS (2019341).

Conflicts of Interest: The authors declare no conflict of interest.

References

1. Donat, F.; Müller, C.R. CO_2-free conversion of CH_4 to syngas using chemical looping. *Appl. Catal. B Environ.* **2020**, *278*. [CrossRef]
2. Wang, J.; Huang, L.; Yang, R.; Zhang, Z.; Wu, J.; Gao, Y.; Wang, Q.; O'Hare, D.; Zhong, Z. Recent advances in solid sorbents for CO_2 capture and new development trends. *Energy Environ. Sci.* **2014**, *7*, 3478–3518. [CrossRef]
3. Kortlever, R.; Shen, J.; Schouten, K.J.P.; Calle-Vallejo, F.; Koper, M.T.M. Catalysts and Reaction Pathways for the Electrochemical Reduction of Carbon Dioxide. *J. Phys. Chem. Lett.* **2015**, *6*, 4073–4082. [CrossRef]
4. Adanez, J.; Abad, A.; Garcia-Labiano, F.; Gayan, P.; de Diego, L.F. Progress in Chemical-Looping Combustion and Reforming technologies. *Prog. Energy Combust. Sci.* **2012**, *38*, 215–282. [CrossRef]
5. Kim, J.; Johnson, T.A.; Miller, J.E.; Stechel, E.B.; Maravelias, C.T. Fuel production from CO_2 using solar-thermal energy: System level analysis. *Energy Environ. Sci.* **2012**, *5*, 8417–8429. [CrossRef]
6. Ekström, C.; Schwendig, F.; Biede, O.; Franco, F.; Haupt, G.; de Koeijer, G.; Papapavlou, C.; Røkke, P.E. Techno-Economic Evaluations and Benchmarking of Pre-combustion CO_2 Capture and Oxy-fuel Processes Developed in the European ENCAP Project. *Energy Procedia* **2009**, *1*, 4233–4240. [CrossRef]
7. Meng, W.X.; Banerjee, S.; Zhang, X.; Agarwal, R.K. Process simulation of multi-stage chemical-looping combustion using Aspen Plus. *Energy* **2015**, *90*, 1869–1877. [CrossRef]
8. Ksepko, E.; Lysowski, R. Reactivity Study of Bimetallic Fe-Mn Oxides with Addition of TiO_2 for Chemical Looping Combustion Purposes. *Catalysts* **2021**, *11*, 1437. [CrossRef]
9. Hossain, M.M.; de Lasa, H.I. Chemical-looping combustion (CLC) for inherent CO_2 separations—A review. *Chem. Eng. Sci.* **2008**, *63*, 4433–4451. [CrossRef]
10. Tang, M.C.; Xu, L.; Fan, M.H. Progress in oxygen carrier development of methane-based chemical-looping reforming: A review. *Appl. Energy* **2015**, *151*, 143–156. [CrossRef]
11. Galvita, V.V.; Poelman, H.; Bliznuk, V.; Detavernier, C.; Marin, G.B. CeO_2-Modified Fe_2O_3 for CO_2 Utilization via Chemical Looping. *Ind. Eng. Chem. Res.* **2013**, *52*, 8416–8426. [CrossRef]
12. Galvita, V.V.; Poelman, H.; Detavernier, C.; Marin, G.B. Catalyst-assisted chemical looping for CO_2 conversion to CO. *Appl. Catal. B Environ.* **2015**, *164*, 184–191. [CrossRef]
13. Song, T.; Shen, T.; Shen, L.; Xiao, J.; Gu, H.; Zhang, S. Evaluation of hematite oxygen carrier in chemical-looping combustion of coal. *Fuel* **2013**, *104*, 244–252. [CrossRef]
14. Linderholm, C.; Knutsson, P.; Schmitz, M.; Markström, P.; Lyngfelt, A. Material balances of carbon, sulfur, nitrogen and ilmenite in a 100 kW CLC reactor system. *Int. J. Greenh. Gas Control* **2014**, *27*, 188–202. [CrossRef]
15. Cocchi, S.; Mari, M.; Cavani, F.; Millet, J.-M.M. Chemical and physical behavior of CoFe2O4 in steam–iron process with methanol. *Appl. Catal. B Environ.* **2014**, *152*, 250–261. [CrossRef]
16. Siriwardane, R.; Tian, H.; Simonyi, T.; Poston, J. Synergetic effects of mixed copper–iron oxides oxygen carriers in chemical looping combustion. *Fuel* **2013**, *108*, 319–333. [CrossRef]
17. Tseng, Y.-H.; Ma, J.-L.; Chin, C.-P.; Kuo, Y.-L.; Ku, Y. Preparation of composite nickel–iron oxide as highly reactive oxygen carrier for chemical-looping combustion process. *J. Taiwan Inst. Chem. Eng.* **2014**, *45*, 174–179. [CrossRef]
18. Qiu, Y.; Zhang, S.; Cui, D.; Li, M.; Zeng, J.; Zeng, D.; Xiao, R. Enhanced hydrogen production performance at intermediate temperatures through the synergistic effects of binary oxygen carriers. *Appl. Energy* **2019**, *252*, 113454. [CrossRef]
19. Qin, L.; Guo, M.; Liu, Y.; Cheng, Z.; Fan, J.A.; Fan, L.-S. Enhanced methane conversion in chemical looping partial oxidation systems using a copper doping modification. *Appl. Catal. B Environ.* **2018**, *235*, 143–149. [CrossRef]
20. Zhou, H.; Yi, Q.; Wei, G.; Zhang, Y.; Hou, Y.; Huang, Z.; Zheng, A.; Zhao, Z.; Li, H. Reaction performance and lattice oxygen migration of $MnFe_2O_4$ oxygen carrier in methane-carbon dioxide reaction system. *Int. J. Hydrogen Energy* **2020**, *45*, 30254–30266. [CrossRef]
21. Sun, Z.; Wu, X.; Russell, C.K.; Dyar, M.D.; Sklute, E.C.; Toan, S.; Fan, M.; Duan, L.; Xiang, W. Synergistic enhancement of chemical looping-based CO_2 splitting with biomass cascade utilization using cyclic stabilized $Ca_2Fe_2O_5$ aerogel. *J. Mater. Chem. A* **2018**, *7*, 1216–1226. [CrossRef]
22. Huang, Z.; Deng, Z.; Chen, D.; Wei, G.; He, F.; Zhao, K.; Zheng, A.; Zhao, Z.; Li, H. Exploration of Reaction Mechanisms on Hydrogen Production through Chemical Looping Steam Reforming Using $NiFe_2O_4$ Oxygen Carrier. *ACS Sustain. Chem. Eng.* **2019**, *7*, 11621–11632. [CrossRef]

23. Chen, J.; Zhao, K.; Zhao, Z.; He, F.; Huang, Z.; Wei, G. Identifying the roles of MFe_2O_4 (M=Cu, Ba, Ni, and Co) in the chemical looping reforming of char, pyrolysis gas and tar resulting from biomass pyrolysis. *Int. J. Hydrogen Energy* **2019**, *44*, 4674–4687. [CrossRef]
24. Daza, Y.A.; Kent, R.A.; Yung, M.M.; Kuhn, J.N. Carbon Dioxide Conversion by Reverse Water–Gas Shift Chemical Looping on Perovskite-Type Oxides. *Ind. Eng. Chem. Res.* **2014**, *53*, 5828–5837. [CrossRef]
25. Haeussler, A.; Abanades, S.; Jouannaux, J.; Julbe, A. Non-Stoichiometric Redox Active Perovskite Materials for Solar Thermochemical Fuel Production: A Review. *Catalysts* **2018**, *8*, 611. [CrossRef]
26. Carrillo, A.J.; Bork, A.H.; Moser, T.; Sediva, E.; Hood, Z.D.; Rupp, J.L.M. Modifying $La_{0.6}Sr_{0.4}MnO_3$ Perovskites with Cr Incorporation for Fast Isothermal CO_2-Splitting Kinetics in Solar-Driven Thermochemical Cycles. *Adv. Energy Mater.* **2019**, *9*, 13. [CrossRef]
27. Huang, Z.; Jiang, H.; He, F.; Chen, D.; Wei, G.; Zhao, K.; Zheng, A.; Feng, Y.; Zhao, Z.; Li, H. Evaluation of multi-cycle performance of chemical looping dry reforming using CO_2 as an oxidant with Fe–Ni bimetallic oxides. *J. Energy Chem.* **2016**, *25*, 62–70. [CrossRef]
28. Lin, Y.; Wang, H.; Huang, Z.; Liu, M.; Wei, G.; Zhao, Z.; Li, H.; Fang, Y. Chemical looping gasification coupled with steam reforming of biomass using $NiFe_2O_4$: Kinetic analysis of DAEM-TI, thermodynamic simulation of OC redox, and a loop test. *Chem. Eng. J.* **2020**, *395*, 125046. [CrossRef]
29. Liu, W. Controlling lattice oxygen activity of oxygen carrier materials by design: A review and perspective. *React. Chem. Eng.* **2021**, *6*, 1527–1537. [CrossRef]
30. Chen, D.; He, D.; Lu, J.; Zhong, L.; Liu, F.; Liu, J.; Yu, J.; Wan, G.; He, S.; Luo, Y. Investigation of the role of surface lattice oxygen and bulk lattice oxygen migration of cerium-based oxygen carriers: XPS and designed H2-TPR characterization. *Appl. Catal. B Environ.* **2017**, *218*, 249–259. [CrossRef]
31. Liu, F.; Zhao, J.; Xuan, G.; Zhang, F.; Yang, L. Spatial evolution characteristics of active components of copper-iron based oxygen carrier in chemical looping combustion. *Fuel* **2021**, *306*, 121650. [CrossRef]
32. Zhao, K.; Zheng, A.; Li, H.; He, F.; Huang, Z.; Wei, G.; Shen, Y.; Zhao, Z. Exploration of the mechanism of chemical looping steam methane reforming using double perovskite-type oxides $La_{1.6}Sr_{0.4}FeCoO_6$. *Appl. Catal. B Environ.* **2017**, *219*, 672–682. [CrossRef]
33. Wang, M.J.; Zhang, S.X.; Xia, M.; Wang, M.K. A Theoretical Study of the Oxygen Release Mechanisms of a Cu-Based Oxygen Carrier during Chemical Looping with Oxygen Uncoupling. *Catalysts* **2022**, *12*, 332. [CrossRef]
34. Cheng, Z.; Baser, D.S.; Nadgouda, S.G.; Qin, L.; Fan, J.A.; Fan, L.-S. C_2 Selectivity Enhancement in Chemical Looping Oxidative Coupling of Methane over a Mg–Mn Composite Oxygen Carrier by Li-Doping-Induced Oxygen Vacancies. *ACS Energy Lett.* **2018**, *3*, 1730–1736. [CrossRef]
35. Huang, Z.; He, F.; Chen, D.; Zhao, K.; Wei, G.; Zheng, A.; Zhao, Z.; Li, H. Investigation on reactivity of iron nickel oxides in chemical looping dry reforming. *Energy* **2016**, *116*, 53–63. [CrossRef]
36. Cao, J.-L.; Wang, Y.; Yu, X.-L.; Wang, S.-R.; Wu, S.-H.; Yuan, Z.-Y. Mesoporous CuO–Fe_2O_3 composite catalysts for low-temperature carbon monoxide oxidation. *Appl. Catal. B Environ.* **2008**, *79*, 26–34. [CrossRef]
37. Cao, W.; Tan, O.; Pan, J.; Zhu, W.; Reddy, C.V.G. XPS characterization of $x\alpha$-Fe_2O_3–$(1-x)ZrO_2$ for oxygen gas sensing application. *Mater. Chem. Phys.* **2002**, *75*, 67–70. [CrossRef]
38. Dupin, J.-C.; Gonbeau, D.; Vinatier, P.; Levasseur, A. Systematic XPS studies of metal oxides, hydroxides and peroxides. *Phys. Chem. Chem. Phys.* **2000**, *2*, 1319–1324. [CrossRef]
39. Wang, D.; Jin, L.; Li, Y.; Hu, H. Partial oxidation of vacuum residue over Al and Zr-doped α-Fe_2O_3 catalysts. *Fuel* **2017**, *210*, 803–810. [CrossRef]
40. Aronniemi, M.; Sainio, J.; Lahtinen, J. XPS study on the correlation between chemical state and oxygen-sensing properties of an iron oxide thin film. *Appl. Surf. Sci.* **2007**, *253*, 9476–9482. [CrossRef]
41. Wei, Y.; Wang, H.; Li, K. Ce-Fe-O mixed oxide as oxygen carrier for the direct partial oxidation of methane to syngas. *J. Rare Earths* **2010**, *28*, 560–565. [CrossRef]
42. Biesinger, M.C.; Payne, B.P.; Lau, L.W.M.; Gerson, A.; Smart, R.S.C. X-ray photoelectron spectroscopic chemical state quantification of mixed nickel metal, oxide and hydroxide systems. *Surf. Interface Anal.* **2009**, *41*, 324–332. [CrossRef]
43. Kuhn, J.N.; Zhao, Z.; Felix, L.G.; Slimane, R.B.; Choi, C.W.; Ozkan, U.S. Olivine catalysts for methane- and tar-steam reforming. *Appl. Catal. B Environ.* **2008**, *81*, 14–26. [CrossRef]
44. Weckhuysen, B.M.; Wang, D.; Rosynek, M.P.; Lunsford, J.H. Conversion of Methane to Benzene over Transition Metal Ion ZSM-5 Zeolites: II. Catalyst Characterization by X-ray Photoelectron Spectroscopy. *J. Catal.* **1998**, *175*, 347–351. [CrossRef]
45. Zeng, L.; Cheng, Z.; Fan, J.A.; Fan, L.-S.; Gong, J. Metal oxide redox chemistry for chemical looping processes. *Nat. Rev. Chem.* **2018**, *2*, 349–364. [CrossRef]
46. Liu, S.; He, F.; Huang, Z.; Zheng, A.; Feng, Y.; Shen, Y.; Li, H.; Wu, H.; Glarborg, P. Screening of $NiFe_2O_4$ Nanoparticles as Oxygen Carrier in Chemical Looping Hydrogen Production. *Energy Fuels* **2016**, *30*, 4251–4262. [CrossRef]
47. Yang, H.; Gong, L.; Wang, H.; Dong, C.; Wang, J.; Qi, K.; Liu, H.; Guo, X.; Xia, B.Y. Preparation of nickel-iron hydroxides by microorganism corrosion for efficient oxygen evolution. *Nat. Commun.* **2020**, *11*, 5075. [CrossRef]
48. Louie, M.W.; Bell, A.T. An Investigation of Thin-Film Ni–Fe Oxide Catalysts for the Electrochemical Evolution of Oxygen. *J. Am. Chem. Soc.* **2013**, *135*, 12329–12337. [CrossRef]

49. Testa-Anta, M.; Ramos-Docampo, M.A.; Comesaña-Hermo, M.; Rivas-Murias, B.; Salgueiriño, V. Raman spectroscopy to unravel the magnetic properties of iron oxide nanocrystals for bio-related applications. *Nanoscale Adv.* **2019**, *1*, 2086–2103. [CrossRef]
50. Ma, Z.; Zeng, D.; Zhang, S.; Xiao, R. Effect of Supports on the Redox Performance of $NiFe_2O_4$ in a Chemical Looping Process. *Energy Technol.* **2019**, *7*, 1900374. [CrossRef]

catalysts

MDPI

Article

In Situ Removal of Benzene as a Biomass Tar Model Compound Employing Hematite Oxygen Carrier

Zhen Huang [1], Yonghao Wang [1,2], Nanhang Dong [2], Da Song [1,3], Yan Lin [1,*], Lisheng Deng [1,4] and Hongyu Huang [1,4]

1 Guangzhou Institute of Energy Conversion, Chinese Academy of Sciences, Guangzhou 510640, China
2 School of Energy and Power Engineering, Northeast Electric Power University, No. 169 Changchun Road, Chuanying District, Jilin City 132012, China
3 College of Chemistry and Bioengineering, Guilin University of Technology, Guilin 541004, China
4 Southern Marine Science and Engineering Guangdong Laboratory (Guangzhou), No. 1119 Haibin Road, Nansha Street, Nansha District, Guangzhou 511458, China
* Correspondence: linyan@ms.giec.ac.cn

Abstract: Tar is an unavoidable biomass gasification byproduct. Tar formation reduces gasification efficiency and limits the further application of biomass gasification technology. Hence, efficient tar removal is a major problem to be solved in the formation and application of biomass gasification technology. Chemical looping gasification (CLG), a novel and promising gasification technology has attracted extensive attention owing to its low tar generation. Active oxygen carriers (OCs), the reduced OC in CLG, are considered to be excellent catalysts for tar cracking. In this study, the use of benzene as a typical tar model compound for tar removal using the iron ore OC is investigated. In the blank experiment, where an inert material (SiO_2) is used as the carrier, the benzene cracking is relatively low, and the benzene conversion, H_2 yield, and carbon conversion are 53.65%, 6.33%, and 1.24%, respectively. The addition of hematite promotes benzene cracking. A large amount of oxygen-containing gases (CO and CO_2) are generated. Additionally, the conversion degrees for benzene, H_2 and carbon are about 67.75%, 21.55%, and 38.39%, respectively. These results indicate that hematite performs both oxidation and catalysis during benzene cracking. The extension of the residence time facilitates benzene removal, owing to the good interaction between the gas phase and solid phase. The addition of water vapor inhibits the benzene conversion and promotes the conversion of carbon deposition. The lattice oxygen reactivity of hematite OC shows an uptrend as the cycle number is increased during the benzene conversion cycle. The experimental results confirm that CLG has a low-tar advantage and that hematite is an effective OC for benzene removal.

Keywords: benzene reforming; chemical looping; oxygen carrier; hematite; biomass tar; model compound

1. Introduction

China's renewable energy source is equivalent to 2.15 billion standard coal, of which over 50% is biomass energy, which is two times as powerful as hydro energy and three and a half times as powerful as wind energy [1]. Biomass offers a significant advantage over fossil fuels in terms of CO_2 emission reduction. For instance, the CO_2 emissions of a biomass-fired plant with a capacity of 25,000 kW are 220 thousand tons less than that of a coal-fired plant. Additionally, biomass is a low-sulfur, low-nitrogen, and low-ash energy source with reduced environmental pollution levels during utilization [2,3]. Owing to the many similarities between biomass and fossil fuel, including molecular structure and utilization route, the industrial system of fossil fuel can be used for biomass conversion with minor adjustments, generating solid, liquid, and gas fuel or industrial raw materials. With the goal of carbon neutrality, it is crucial to advance biomass energy technology.

One of the practical technologies for biomass utilization is gasification, which converts biomass to syngas via thermo-chemical conversion [4]. Syngas mainly includes CO and

H_2, which can be used to generate heat and power and as a platform compound for the synthesis of liquid fuels and chemicals [5–7]. Biomass gasification has several advantages including high efficiency, low emissions, and simplicity of integration [8]. However, a significant drawback of the conventional gasification process is the high cost of oxygen sources. During gasification, solid fuel is partially oxidized, and air, oxygen, and H_2O steam are commonly used as gasified agents [9]. Using air as a gasification medium will result in the low heating value of the gas products because of nitrogen dilution [10]; using oxygen is beneficial for obtaining high heating-value gas products, but the high cost of oxygen separation cannot be ignored [11]; producing H_2O steam requires an external boiler, which consumes a lot of energy [12,13]. Additionally, tar, which accounts for 5%~15% of total energy in raw biomass, is a byproduct of gasification [14]. Therefore, it is crucial to design a biomass gasification method that has a low-cost oxygen source, produces less tar, and produces high-heating-value gas products.

Chemical looping gasification (CLG) is an effective solution for the abovementioned biomass gasification technology demand [15–17]. As shown in Figure 1, typical gasification is decomposed into two steps in CLG. In the fuel reactor (FR), the metal oxide is used as the oxygen carrier (OC) [18–25], which serves the dual purpose of oxidation and catalysis in the gasification [26]. The OC catalyzes to facilitate biomass gasification as well as a lattice oxygen source to replace molecular oxygen. After reacting with biomass, the OC is reduced, and it is then sent to the air reactor for air calcining to regenerate. Thereafter, the OC reenters the FR and AR cycle, as well as creating a reduction and oxidation loop (chemical looping). CLG does not require high-cost production of pure oxygen or water steam, and the regulation of syngas components can be achieved by introducing CO_2/H_2O. Additionally, the OC effectively promotes tar conversion, which is a great advantage for the thermo-chemical conversion of highly volatile solid fuels such as biomass.

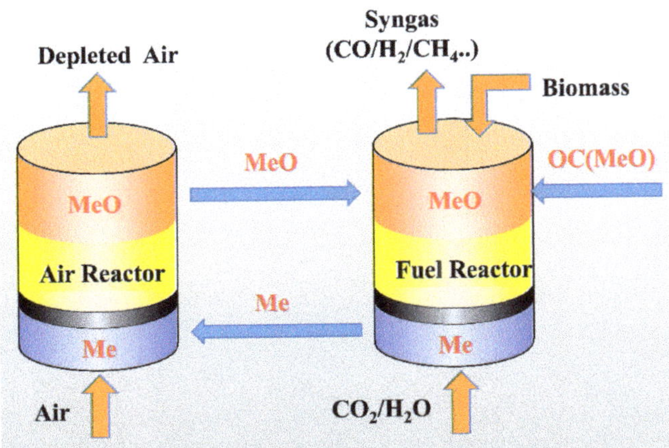

Figure 1. The schematic of biomass chemical looping gasification.

Biomass tar is a complex mixture, which consists of several polycyclic organic compounds. The major compositions of biomass tar are presented in Table 1. Benzene accounts for the highest proportion of tar compositions, and the majority of other compositions are benzene derivatives [27]. The benzene ring is common to see in the molecular structure of tar compounds, and this phenomenon might be attributed to the extreme stability of the benzene ring [28,29]. Both benzene and naphthalene barely decompose below 900 °C, and their decompositions are not significant until 1000 °C. Moreover, the thermal stability of benzene is higher than that of naphthalene [30]. According to certain studies, naphthalene must undergo a complete breakdown at a reaction temperature higher than 1300 °C, while benzene must undergo a complete decomposition at a reaction temperature higher than 1400 °C [31]. It can be inferred that the decomposition of benzene is a key barrier to the complete conversion of biomass tar because benzene is present in large amounts and has nearly the highest thermal stability. Currently, several methods have been suggested for benzene elimination in gasification. These methods include catalytic steam reforming [32,33], photothermal steam reforming [34], steam-oxygen gasification [35,36], microwave-based treatment [37], and plasma technology [38,39]. Only a few studies in the literature report the in situ removal of benzene in CLG.

Table 1. Typical compositions of biomass tar [40].

Species	Benzene	Naphthalene	Toluene	Xylene	Indene	Phenols	2~3 Rings Compounds
Composition (wt.%)	28	15	14	7	7	7	21

To investigate the effect of CLG operation on the tar conversion, benzene was used as the biomass tar model compound in the CLG experiments to simulate the in situ removal of biomass tar decomposition using a fixed bed reactor, and hematite was employed as the OC. The effects of residence time, steam/carbon ratio, and cycle number on benzene conversion were investigated. The results provide some important references for biomass tar elimination in CLG operation.

2. Results and Discussion

2.1. Benzene Conversion Characteristics Using Hematite

The characteristics of benzene cracking between inert SiO_2 and hematite were compared to highlight the role of iron ore OC, as shown in Figure 2 and Tables 2–5. The hematite mass, benzene flow, and N_2 flow were set as 0.5 g, 40 mL/min, and 0.05 mL/min, respectively. The total reaction time was 60 min, and each point was collected every 3 min using a sampling bag.

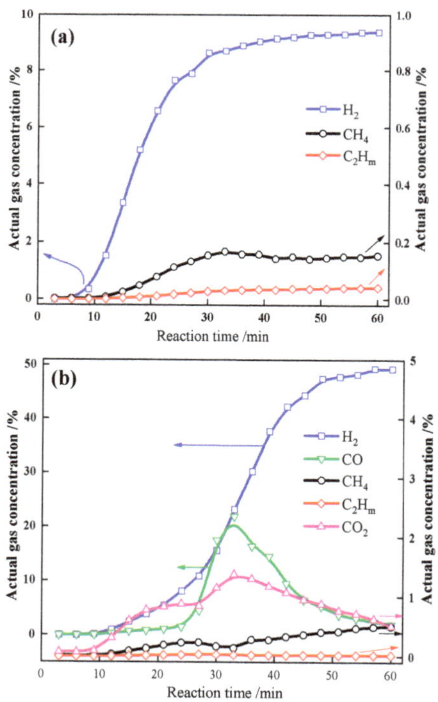

Figure 2. The behavioral comparisons of benzene cracked by: (**a**) SiO_2 and (**b**) Hematite.

Table 2. The role of hematite in benzene cracking.

Carrier	Relative Gas Concentration (%)					η_t (%)	Y_{H2} (%)	Y_c (%)
	H_2	CO	CH_4	C_2H_m	CO_2			
SiO_2	95.40	-	3.91	0.69	-	55.37	7.22	1.98
Hematite	72.25	24.42	0.37	0.09	2.87	69.05	21.96	39.13

Table 3. Distribution of liquid products during benzene cracked by two carriers.

Carrier	Mass of Liquid Products (g)				
	Biphenyl $C_{12}H_{10}$	Meta-Terphenyl $C_{18}H_{14}$	Para-Terphenyl $C_{18}H_{14}$	Benzanthracene $C_{18}H_{12}$	Sum
SiO_2	0.91	0.15	0.14	-	1.20
Hematite	0.24	0.07	0.06	0.18	0.55

Table 4. Distribution of carbon-containing species in the products.

Carrier	The Mass of Carbon (C) in the Products (g)					Carbon Balance (%)
	C-Gas	C-Liquid	C-Residual Benzene	C-Solid	C-Total	
SiO_2	0.01	1.13	1.09	0.06	2.29	93.85
Hematite	0.05	0.51	0.75	1.00	2.31	94.67

Table 5. XRF analysis of hematite.

Element	Fe	O	Si	Al	K	Ca	P	Ti	Mn
Composition (wt/%)	61.9	32.4	3.3	1.7	0.3	0.1	0.1	0.1	0.1

In the blank experiment (using SiO_2), the gas concentrations were low, the main gas component was H_2 (95.40%) and the CH_4 (3.91%) and C_2H_m (0.69%) contents were low. The oxygenated gases (CO and CO_2) were not observed. H_2 concentration increased gradually in the first 30 min before stabilizing. Additionally, the benzene conversion, H_2 yield, and carbon deposition were 55.37%, 7.22%, and 1.98%, respectively. These results indicate that the self-cracking performance of benzene at high temperatures was low, since the inert SiO_2 carrier had no oxidation and catalysis ability. During this process, the reaction (R1) played a major role [41]. Although the benzene conversion was 55.37%, the majority of benzene only partially cracked and only a small portion of its C-H/C-C bonds was completely broken. Thus, the H_2 yield (gas product) and carbon deposition (solid product) were very low, and a large amount of benzene series (liquid product) was generated, as shown in (R2). Benzene dehydrogenation mostly generates benzene radicals, which later combine with other radicals to form a new, larger benzene series. The mass of the generated liquid products was 1.20 g and the main component (biphenyl) accounted for 75.83% of that mass.

However, hematite addition accelerated the benzene cracking. Along with the H_2, CH_4, and C_2H_m, a large amount of oxygen-containing gases (CO and CO_2) was generated. In the initial stage, a small amount of absorbed oxygen reacted with the benzene on the surface of the hematite, which leads to the formation of CO and CO_2. After 25 min, the CO concentration increased and reached its peak at 33 min, indicating that lattice oxygen was involved in the benzene oxidation. In the final stage, lattice oxygen was absent, making it impossible to meet the oxygen demand of benzene oxidation, but part of the OC was reduced to metal Fe, which catalyzed the benzene cracking. Therefore, CO concentration declined gradually, and the H_2 concentration increased continuously.

Furthermore, compared to the blank experiment, the benzene conversion, H_2 yield and carbon deposition increased to 69.05%, 21.96%, and 39.13%, respectively. Hematite has the dual function of oxidation and catalysis. First, benzene underwent an oxidation-reduction reaction with hematite to generate a large amount of oxygen-containing gas, while the hematite was reduced to metallic Fe or Fe_3C, as shown in (R3). Subsequently, the generated Fe or Fe_3C acted as a catalyst to catalytically crack the benzene, and large amounts of hydrogen and carbon deposits were generated, as shown in (R4). Additionally, when compared to the blank experiment, 0.55 g of liquid products were generated during the process. This was attributed to the relatively low oxidation and catalytic ability of hematite. Thus, the secondary reaction R5 occurred in the process.

$$C_6H_6 \rightarrow \text{Gas } (H_2/CH_4/C_2H_m) + \text{Liquid } (C_xH_y, x > 4) + \text{Solid (C)} \quad \Delta H > 0 \qquad \text{(R1)}$$

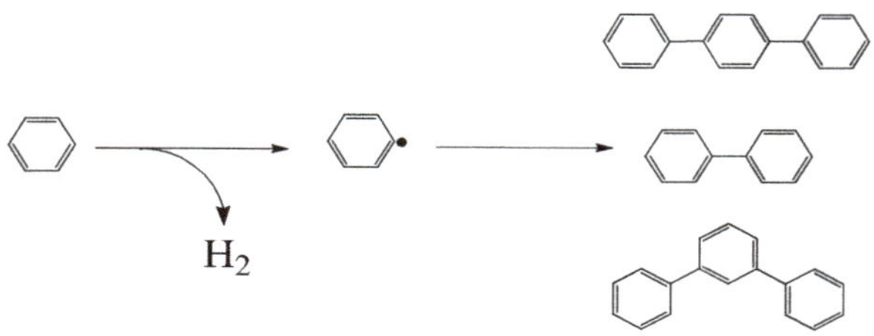

$$\Delta H > 0 \qquad \text{(R2)}$$

$$C_6H_6 + Fe_2O_3 \rightarrow H_2/CH_4/CO/CO_2/C_2H_m + Fe/Fe_3C \quad \Delta H > 0 \qquad \text{(R3)}$$

$$C6H6 \xrightarrow{Fe/Fe_3C} \text{Gas } (H_2/CH_4/C_2H_m) + \text{Liquid } (C_xH_{y(x>4)}) + \text{Solid (C)} \ \Delta H > 0 \qquad \text{(R4)}$$

2.2. The Residence Time

This section investigates the effect of residence time on benzene cracking, and the result is shown in Figure 3. The hematite mass and benzene flow were set as 0.5 g and 0.05 mL/min, respectively. The total reaction time was 60 min. The residence time was adjusted by controlling the N_2 flow(40, 60, 80, 100 mL/min).

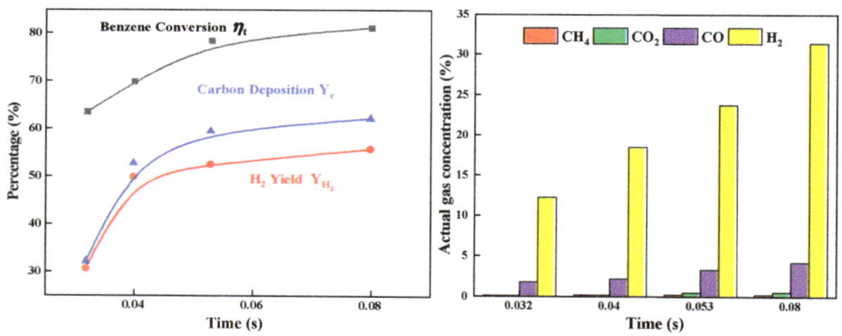

Figure 3. Effect of residence time on benzene cracking.

The benzene conversion degree increased from 63.5% to 78.3% when the residence time increased from 0.032 to 0.08 s. Over 0.053 s, the benzene conversion increased from 78.3% to 81.1%, but at a slower rate. There was a significant positive correlation between the benzene conversion degree and the residence time. Long residence time allows for increased interaction between benzene and hematite, which is beneficial for benzene cracking. The increased conversion degree reduced slightly after 0.053 s, because the benzene-cracking capacity of the OC was close to the saturation at this condition. Carbon deposition and H_2 are the typical byproducts of benzene cracking (R1), and their percentage–time curves were similar to that of the benzene conversion degree. The difference was in the period of 0.032 to 0.053 s, which showed a rapid increase in the carbon deposition and H_2 yield curves. The restricted amount of lattice oxygen capacity made it increasingly difficult to completely oxidize input benzene. Therefore, there was a rapid increase in both carbon deposition and H_2 yield.

The main gas components were H_2 and CO. A small amount of CH_4 and CO_2 was observed when the residence time was above 0.04 s. Over the residence time, the H_2 concentration increased from 12.5% to 31.9%, and the CO concentration increased from 1.9% to 4.4%. As the residence time increased, the lattice oxygen was depleted, and the reducing gas content increased. For the benzene cracking process, OC plays a dual role in catalysis and oxidation in the range of 0.04 to 0.053 s, but the oxidation function declines after 0.053 s because lattice oxygen was exhausted, and the reducing gas content subsequently increased.

2.3. H_2O Steam Content

H_2O steam is usually used as a gasifying agent to promote biomass gasification. The effect of H_2O steam on benzene conversion degree, H_2 yield, and carbon deposition was investigated, and the Steam/Carbon ratio (S/C) was set in a range of 0.49–1.14. The results are presented in Figure 4. The hematite mass, benzene flow, and N_2 flow were set as 0.5 g, 60 mL/min, and 0.03 mL/min, respectively. The total reaction time was 60 min. The residual solid products were analyzed by X-ray diffraction (XRD, X'Pert Pro MPD) and Cu Kα radiation (λ = 0.1504 nm) at a voltage of 40 kV and a current of 40 mA. The scanning rate was 2°/min from 2θ = 5° to 80° at a step of 0.02°.

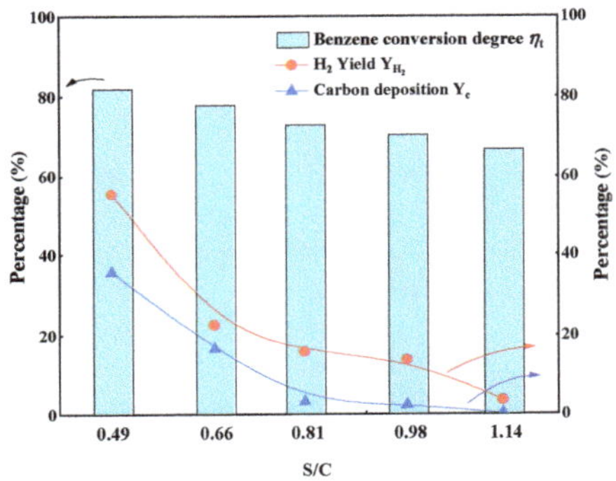

Figure 4. Effect of mass ratio on benzene cracking.

H_2O steam inhibited the benzene decomposition, and the benzene conversion degree reduced from 81.8% to 66.8% when the S/C value increased from 0.49 to 1.14. According to (R1)~(R4), benzene was decomposed with the release of H_2, as well as carbon deposition. As the benzene conversion degree decreased, H_2 yield reduced from 55.5% to 3.7%, while carbon deposition reduced from 35.7% to 0.33%.

There were two possible reasons for the H_2O inhabitation of the benzene decomposition. First, in our experimental devices, liquid H_2O was pumped into the reactor where it was gasified to H_2O steam because of the high internal temperature. The H_2O gasification had a significant impact on the gas flow in the reactor, resulting in a higher flow through the quartz tube. Consequently, the residence time of benzene and its decomposition products was shortened leading to a decrease in the benzene conversion degree. Second, the introduction of H_2O steam enhanced the oxidation performance of the atmosphere, leading to OC reduction. Thereafter, the catalysis effect, which was dependent on the reduced OC, diminished as a result of the reduced OC formation.

$$C/Fe_3C + H_2O \rightarrow CO/CO_2 + H_2 + Fe \quad \Delta H > 0 \tag{R5}$$

Figure 5 shows the variation in the XRD pattern of the reduced OC when the S/C value increased from 0.49 to 0.81. When the S/C value was set at 0.49, the reduced OC primarily contained metal Fe, carbon, and Fe_3C. Carbon and Fe_3C vanished as the S/C value was increased to 0.66. Metal Fe was the dominant phase, and its signal intensity was reduced relative to that of the S/C value of 0.49. When the S/C value was increased to 0.81, the signal intensity of metal Fe decreased further, and $FeO_{1-\delta}$ was observed in the spectrum. The above phenomenon indicated that H_2O first reacted with carbon deposition and Fe_3C before oxidizing into $FeO_{1-\delta}$ when the carbon was used up. Upon the metal Fe oxidation, the catalysis effect of the OC decreased, resulting in a lower benzene conversion degree. Therefore, the S/C should be set at a level that allows for sufficient elimination for carbon deposition, and little oxidation to the reduced OC, at the same time. When the S/C value reached 0.66, the Fe_3C pattern was too weak to be observed, and the OC maintained an almost completed reduced state. Therefore, the S/C value was set at 0.66 for the cyclic experiments.

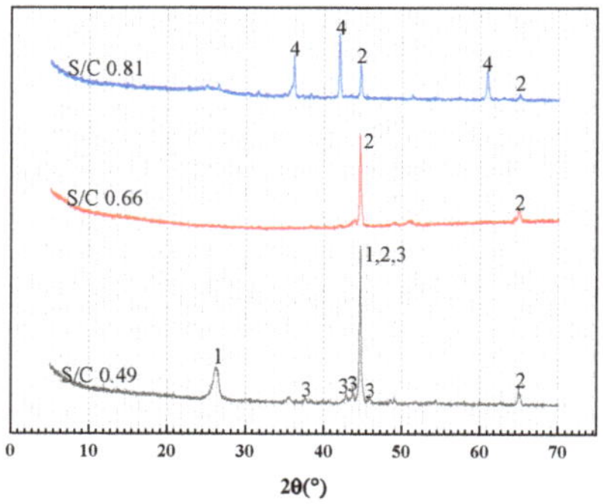

Figure 5. The effect of S/C on the XRD pattern of reduced OC. 1-C; 2-Fe; 3-Fe_3C; 4-$FeO_{1-\delta}$.

Figure 6a–c shows the morphology of the OC at different stages. Figure 6a depicts the morphology of the original OC. There were massive blocks distributed in the picture with a few small blocks adhering to the surface. These blocks were made up specifically of the reduced OC particle. After reacting with benzene, the OC particle agglomerated into blocks with a smooth edge, and some tubes, overlapping on the surface of the blocks, crossed over with each other, as shown in Figure 6b. These microtubes had distinct structures from the OC blocks. According to the fixed bed experiments, a large amount of carbon deposition was found in the reduced OC, and these microtubes could be formed from the carbon nanotube polymer, which was generated during the carbon deposition ((R1) and (R4)). Similar results were found in the literature [42]. The carbon microtubes disappeared when H_2O steam was introduced into the reaction (R5). Carbon deposition was converted to CO/CO_2 when it reacted with H_2O. Figure 6c depicted the morphology of the microtube-free OC. After reacting to H_2O steam, the OC structure became loose and irregular, and the carbon microtubes were not observed in the image.

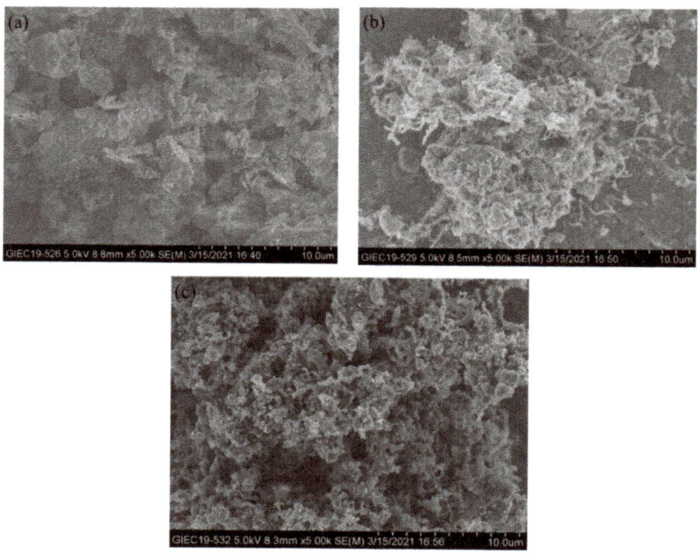

Figure 6. SEM image of OC: (**a**) original, (**b**) reduced, (**c**) reduced with H_2O steam.

2.4. The Cycle Numbers

Figure 7 shows the variations in benzene conversion during 5 chemical looping redox cycles. In the benzene conversion stage, the hematite mass, S/C, benzene flow, and N_2 flow were set as 0.5 g, 0.66, 60 mL/min, and 0.03 mL/min, respectively. The total reaction time was 60 min. In the oxidation stage, the air flow was set as 300 mL/min for 30 min.

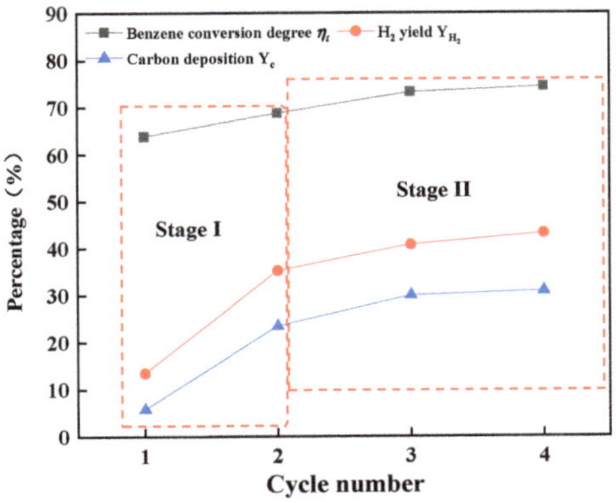

Figure 7. Influence of cycle number on the benzene conversion degree, H_2 yield and carbon deposition.

The benzene conversion degree increased from 63.9% to 74.4% as the cycle number increased. The H_2 yield increased from 13.4% to 43.2%, and the carbon deposition increased from 5.8% to 31.0%. This improvement was attributed to the increase in the catalytic effect of OC, which could be observed in the H_2 yield and carbon deposition figures.

The multiple cycles were divided into two stages. In stage I, rapid increases in H_2 yield and carbon deposition were observed in the 2nd cycle. The H_2 yield and carbon deposition of the 2nd cycle were more than twice those of the 1st cycle. However, compared to that of the 2nd cycle, the benzene conversion degree of the 1st cycle only increased by about 10%. The benzene conversion could not have increased at the same time that H_2 yield and carbon deposition were increasing so rapidly. It ought to be related to the further decomposition of medium products generated from the benzene cracking. According to Section 3.3, it could be inferred that the OC was almost reduced to metal Fe. A serious agglomeration would happen to the OC after reduction, resulting in the deactivation of the OC. Lattice oxygen could not be recovered as completely as in the original, and a part of low-valence Fe oxides was exposed to benzene, improving the catalytic effect of the OC. Consequently, the medium products generated from the benzene cracking could be further decomposed by such an OC, leading to a higher H_2 yield and a higher carbon deposition. In stage II, the primary effect was eliminated during stage I, and the OC performance was stable. The catalytic effect of the OC gradually increased as the cycle number increased, and the benzene conversion degree also increased.

Figure 8 shows the influence of cycle number on relative concentration of gas products and H_2/CO ratio. As the cycle number increased from 1 to 4, the H_2 concentration increased from 56.6% to 61.2%, and the CO and CO_2 concentration decreased from 38.8% to 35.5% and from 4.4% to 3%, respectively. This variation in gas-product relative concentration indicated that the lattice oxygen gradually deactivated after multiple redox cycles. The oxidative performance of the OC was declined, while the catalytic performance was improved. Therefore, the H_2 yield, carbon deposition, and H_2/CO ratio all increased after multiple redox cycles.

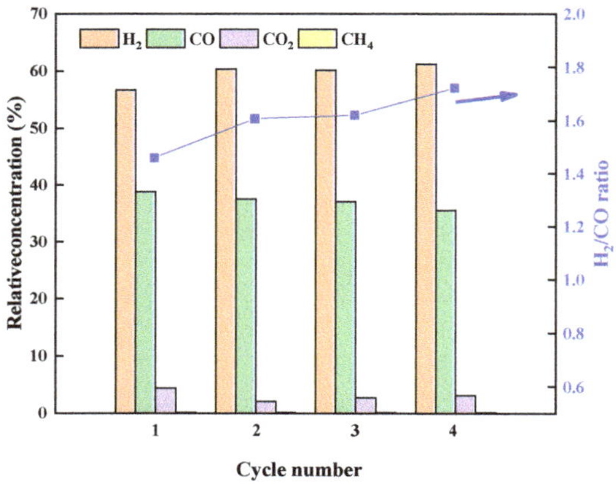

Figure 8. The influence of cycle number on relative concentration of gas products and H_2/CO.

3. Materials and Methods

3.1. Hematite

In the experiments, Australian hematite was used as the OC, and its compositions are listed in Table 5. The hematite was heated from room temperature to 1100 °C at a heating rate of 10 °C/min before being maintained at 1100 °C for 3 h. Thereafter, the hematite was naturally cooled to room temperature before being ground into a powder and sieved to a diameter less than 0.15 mm.

3.2. Fixed Bed Experiment

The fixed bed reactor is shown in Figure 9. Benzene or deionized water was injected into the valve by using a high-pressure infusion pump at a certain flow rate and a pressure of 1 atm, and the N_2 or air flow was controlled by using a mass flowmeter. The quartz tube had an outer diameter of 20 mm, and inner diameter of 15 mm, and a length of 800 mm. In the middle of the quartz tube, there was a porous bezel, which carried a certain mass of quartz cotton (about 0.02 g), and the OC was placed on top of it. To avoid condensation of liquid products, a heater (150 °C) was wrapped around the line connecting the quartz tube to the gas collector and the bottom of the quartz tube, which was placed at the center of the reactor. The reaction temperature was set as 850 °C. The exit of the reactor was connected to a gas collector, which contained isopropanol. After the collection, the solution in the collector was stored at 5 °C. The gas product was collected in the sample bag for subsequent analysis. After the benzene conversion experiment, air was used to oxidize the carrier and to eliminate the carbon deposit, and the gas product was collected by using the sample bag. The amount of carbon deposit from the benzene conversion was calculated by measuring the CO and CO_2 content in the gas product. The solution in the collector was transferred to a 200 mL volumetric flask. The gas and liquid products were analyzed using the Agilent 7890A gas chromatograph; the split flow was set as 10:1, the FID detector was set at 280 °C, the DB-5HT nonpolar capillary column was equipped with gas chromatograph, and the column was heated from 40 °C to 280 °C at a heating rate of 40 °C/min.

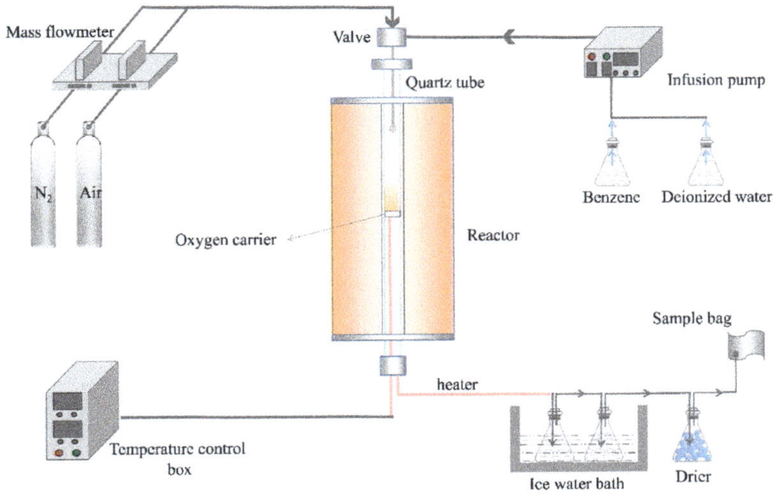

Figure 9. A schematic of the fixed bed experiment.

3.3. Data Processing

During the experiment, isopropanol was used to absorb the content of benzene (g/L) as part of the non-participating reaction, and the conversion degree of benzene was:

$$\eta_t = \frac{m_t - m_0}{m_t} \times 100\% \quad (1)$$

In the formula:

m_t——Amount of benzene fed during reaction time (g);

m_0——Quality of unreacted benzene (g).

The total accumulation of gas exports was calculated using the nitrogen balance. The formula was shown in (2).

$$V_{out} = \frac{V_{in}}{V_n} \quad (2)$$

In the formula:

V_{in}——Total volume of imported nitrogen (L);

V_{out}——Total volume of gas outlet (L);

V_n——The volume fraction of the total volume of the outlet occupied by N_2 in the dry-base state.

Carbon deposition Y_c(%):

$$Y_c(\%) = \frac{12m_c}{13(m_i - m_0)} \times 100\% \quad (3)$$

In the formula:

m_i——Amount of benzene fed during reaction time (g);

m_0——Quality of unreacted benzene (g);

m_c——Carbon mass on the surface of the OC (g), which was calculated by measuring the CO and CO_2 content during the OC oxidation.

H_2 yield is calculated as follows:

$$Y_{H_2}(\%) = \frac{2[H_2]}{6[n]_{in} + 2[m]_{in}} \times 100\% \quad (4)$$

In the formula:

$[n]$-------- The number of moles of imported benzene;
$[m]$-------- The number of moles in imported H_2O.
Steam/carbon (S/C) is the mass flow ratio of H_2O steam to benzene carbon:

$$S/C = \frac{q_{m,\ H_2O}}{q_{m,\ benezene\ carbon}} \times 100\% \quad (5)$$

In the formula:
$q_{m,\ H_2O}$-------- The mass flow of H_2O steam (g/min);
$q_{m,\ benezene\ carbon}$------- The mass flow of carbon of benzene (g/min).
The relative concentrations of gas products are obtained as follows:

$$Relative\ concentration = \frac{y_k}{y_{CO_2} + y_{CO} + y_{CH_4} + y_{H_2}} \times 100\% \quad (6)$$

where y_{gas} is the actual concentration of gas products (CO_2, CO, CH_4, and H_2) detected by GC, and y_k is the specific gas product (CO_2, CO, CH_4, C_2H_x, or H_2).

The residence times are calculated as below:

$$The\ residence\ time = \frac{V_{OC}}{N_2\ flow} \quad (7)$$

In the formula: V_{OC}-------- The volume of the OC is estimated as 0.053 cm^3 (0.5 g hematite in the quartz tube with an inner diameter of 15 mm);

N_2 *flow*------ The volume flow of N_2 (40, 60, 80, and 100 mL/min).

4. Conclusions

In this study, fixed-bed reactor tests for benzene conversion using hematite as an OC were performed. In summary, hematite exhibited better benzene conversion than an inert carrier (SiO_2). The conclusions are as follows:

(1) During benzene conversion, hematite functioned as both an oxidative and a catalytic OC. Hematite also improved the benzene conversion degree by about 20%, going from 55.37% to 69.05%.
(2) As the residence time increased from 0.032 s to 0.08 s, the benzene conversion increased from 63.5% to 81.1%. However, the positive influence of the increased residence time declined when the residence time was over 0.053 s.
(3) The introduction of H_2O steam was advantageous for the removal of carbon deposition. As the S/C ratio increased from 0.49 to 1.14, the carbon deposition decreased from 35.7% to 0.33%, and H_2O inhibited the benzene conversion, resulting in a decrease in H_2 yield.
(4) A significant primary effect was observed in the multiple redox cycles. The benzene conversion increased in the 2nd cycle. The agglomeration of hematite deactivated a part of lattice oxygen vacancy, resulting in a decline in the oxidative performance of the hematite after the redox cycle. As the cycle number increased, the benzene conversion improved.

Author Contributions: Conceptualization, Z.H. and Y.L.; methodology, Y.W.; software, Y.L.; validation, Y.W.; formal analysis, D.S.; investigation, Y.W.; resources, D.S.; data curation, Y.W. and N.D.; writing—original draft preparation, Y.W.; writing—review and editing, Y.L.; visualization, Y.L.; supervision, Y.L. and L.D.; project administration, Y.L. and L.D.; funding acquisition, Z.H., Y.L. and H.H. All authors have read and agreed to the published version of the manuscript.

Funding: This research was funded by the financial support from the National Natural Science Foundation of China (52006224, 52076209), the Key Special Project for Introduced Talents Team of Southern Marine Science and Engineering Guangdong Laboratory (Guangzhou) (GML2019ZD0108), the Foundation and Applied Foundation Research of Guangdong Province (2019B1515120022, 2019A1515110828, 2020A1515110138, 2021A1515010459, 2022B1515020045), the Joint Fund of the Yulin University and the Dalian National Laboratory for Clean Energy (2021021).

Acknowledgments: The authors gratefully acknowledge the financial support from the National Natural Science Foundation of China (52006224, 52076209), the Key Special Project for Introduced Talents Team of Southern Marine Science and Engineering Guangdong Laboratory (Guangzhou) (GML2019ZD0108), the Foundation and Applied Foundation Research of Guangdong Province (2019B1515120022, 2019A1515110828, 2020A1515110138, 2021A1515010459, 2022B1515020045), the Joint Fund of the Yulin University and the Dalian National Laboratory for Clean Energy (2021021).

Conflicts of Interest: The authors declare no conflict of interest.

References

1. C.R.E.D.S.R. Group. *China Renewable Energy Development Strategy Research Series*; China Electric Power Press: Beijing, China, 2008.
2. Chen, H.; Lu, D.; An, J.; Qiao, S.; Dong, Y.; Jiang, X.; Xu, G.; Liu, T. Thermo-Economic analysis of a novel biomass Gasification-Based power system integrated with a supercritical CO_2 cycle and a Coal-Fired power plant. *Energy Convers. Manag.* **2022**, *266*, 115860. [CrossRef]
3. Chen, C.; Bi, Y.; Huang, Y.; Huang, H. Review on slagging evaluation methods of biomass fuel combustion. *J. Anal. Appl. Pyrolysis* **2021**, *155*, 105082. [CrossRef]
4. Hanchate, N.; Ramani, S.; Mathpati, C.S.; Dalvi, V.H. Biomass gasification using dual fluidized bed gasification systems: A review. *J. Clean. Prod.* **2021**, *280*, 123148. [CrossRef]
5. Liu, R.; Tsiava, R.; Xu, S.; Chen, D. Experimental study of char gasification characteristics with high temperature flue gas. *J. Energy Inst.* **2021**, *97*, 187–193. [CrossRef]
6. Dos Santos, R.G.; Alencar, A.C. Biomass-derived syngas production via gasification process and its catalytic conversion into fuels by Fischer Tropsch synthesis: A review. *Int. J. Hydrogen Energy* **2020**, *45*, 18114–18132. [CrossRef]
7. Cheng, K.; Zhang, L.; Kang, J.; Peng, X.; Zhang, Q.; Wang, Y. Selective Transformation of Syngas into Gasoline-Range Hydrocarbons over Mesoporous H-ZSM-5-Supported Cobalt Nanoparticles. *Chem.—A Eur. J.* **2015**, *21*, 1928–1937. [CrossRef]
8. Wei, G. Chemical Looping Gasification of Biomass Coupled with CO_2 Splitting with Fe-Based Mixed Oxygen Carriers. Ph.D. Thesis, Taiyuan University of Technology, Taiyuan, China, 2020. [CrossRef]
9. Jiang, C.; Jin, X.; Xu, T.; Xiao, B.; Hu, Z.; Wang, X. Biomass chemical looping gasification for syngas production using modified hematite as oxygen carriers. *J. Environ. Sci.* **2023**, *125*, 171–184. [CrossRef]
10. González, C.A.D.; Sandoval, L.P. Sustainability aspects of biomass gasification systems for small power generation. *Renew. Sustain. Energy Rev.* **2020**, *134*, 110180. [CrossRef]
11. Yan, X.; Hu, J.; Zhang, Q.; Zhao, S.; Dang, J.; Wang, W. Chemical-looping gasification of corn straw with Fe-based oxygen carrier: Thermogravimetric analysis. *Bioresour. Technol.* **2020**, *303*, 122904. [CrossRef]
12. Sepe, A.M.; Li, J.; Paul, M.C. Assessing biomass steam gasification technologies using a multi-purpose model. *Energy Convers. Manag.* **2016**, *129*, 216–226. [CrossRef]
13. Pacioni, T.R.; Soares, D.; Domenico, M.D.; Rosa, M.F.; Moreira, R.D.F.P.M.; José, H.J. Bio-syngas production from agro-industrial biomass residues by steam gasification. *Waste Manag.* **2016**, *58*, 221–229. [CrossRef] [PubMed]
14. Li, C.; Suzuki, K. Tar property, analysis, reforming mechanism and model for biomass gasification—An overview. *Renew. Sustain. Energy Rev.* **2009**, *13*, 594–604. [CrossRef]
15. Yan, B.; Liu, Z.; Wang, J.; Ge, Y.; Tao, J.; Cheng, Z.; Chen, G. Mn-doped $Ca_2Fe_2O_5$ oxygen carrier for chemical looping gasification of biogas residue: Effect of oxygen uncoupling. *Chem. Eng. J.* **2022**, *446*, 137086. [CrossRef]
16. Wang, S.; Yin, X.; Jarolin, K.; Dymala, T.; Xu, J.; Yin, S.; Dosta, M.; Song, T.; Heinrich, S.; Shen, L. Mechanical strength evolution of biomass pellet during chemical looping gasification in fluidized bed. *Fuel Process. Technol.* **2021**, *221*, 106951. [CrossRef]
17. Liu, Q.; Hu, C.; Peng, B.; Liu, C.; Li, Z.; Wu, K.; Zhang, H.; Xiao, R. High H_2/CO ratio syngas production from chemical looping co-gasification of biomass and polyethylene with CaO/Fe_2O_3 oxygen carrier. *Energy Convers. Manag.* **2019**, *199*, 111951. [CrossRef]
18. Wen, Y.-Y.; Li, Z.S.; Xu, L.; Cai, N.S. Experimental Study of Natural Cu Ore Particles as Oxygen Carriers in Chemical Looping with Oxygen Uncoupling (CLOU). *Energy Fuels* **2012**, *26*, 3919–3927. [CrossRef]
19. Wang, L.; Lin, Y.; Huang, Z.; Zeng, K.; Huang, H. Conversion of carbon dioxide to carbon monoxide: Two-step chemical looping dry reforming using $Ca_2Fe_2O_5$–$Zr_{0.5}Ce_{0.5}O_2$ composite oxygen carriers. *Fuel* **2022**, *322*, 124182. [CrossRef]
20. Lin, Y.; Wang, H.; Huang, Z.; Liu, M.; Wei, G.; Zhao, Z.; Li, H.; Fang, Y. Chemical looping gasification coupled with steam reforming of biomass using $NiFe_2O_4$: Kinetic analysis of DAEM-TI, thermodynamic simulation of OC redox, and a loop test. *Chem. Eng. J.* **2020**, *395*, 125046. [CrossRef]
21. Sun, R.; Shen, L.; Wang, S.; Bai, H. CO conversion over $LaFeO_3$ perovskite during chemical looping processes: Influences of Ca-doping and oxygen species. *Appl. Catal. B Environ.* **2022**, *316*, 121598. [CrossRef]
22. Tian, X.; Zheng, C.; Zhao, H. Ce-modified $SrFeO_{3-\delta}$ for ethane oxidative dehydrogenation coupled with CO_2 splitting via a chemical looping scheme. *Appl. Catal. B Environ.* **2022**, *303*, 120894. [CrossRef]
23. Kang, Y.; Han, Y.; Tian, M.; Huang, C.; Wang, C.; Lin, J.; Hou, B.; Su, Y.; Li, L.; Wang, J.; et al. Promoted methane conversion to syngas over Fe-based garnets via chemical looping. *Appl. Catal. B Environ.* **2020**, *278*, 119305. [CrossRef]

24. Zeng, D.; Qiu, Y.; Li, M.; Ma, L.; Cui, D.; Zhang, S.; Xiao, R. Spatially controlled oxygen storage materials improved the syngas selectivity on chemical looping methane conversion. *Appl. Catal. B Environ.* **2021**, *281*, 119472. [CrossRef]
25. Liu, G.; Wang, H.; Deplazes, S.; Veksha, A.; Wirz-Töndury, C.; Giannis, A.; Lim, T.T.; Lisak, G. Ba–Al-decorated iron ore as bifunctional oxygen carrier and HCl sorbent for chemical looping combustion of syngas. *Combust. Flame* **2021**, *223*, 230–242. [CrossRef]
26. Nam, H.; Wang, Z.; Shanmugam, S.R.; Adhikari, S.; Abdoulmoumine, N. Chemical looping dry reforming of benzene as a gasification tar model compound with Ni- and Fe-based oxygen carriers in a fluidized bed reactor. *Int. J. Hydrogen Energy* **2018**, *43*, 18790–18800. [CrossRef]
27. Lin, Y.; Liao, Y.; Yu, Z.; Fang, S.; Ma, X. A study on co-pyrolysis of bagasse and sewage sludge using TG-FTIR and Py-GC/MS. *Energy Convers. Manag.* **2017**, *151*, 190–198. [CrossRef]
28. Bangala, D.N.; Abatzoglou, N.; Martin, J.-P.; Chornet, E. Catalytic Gas Conditioning: Application to Biomass and Waste Gasification. *Ind. Eng. Chem. Res.* **1997**, *36*, 4184–4192. [CrossRef]
29. Coll, R.; Salvadó, J.; Farriol, X.; Montané, D. Steam reforming model compounds of biomass gasification tars: Conversion at different operating conditions and tendency towards coke formation. *Fuel Process. Technol.* **2001**, *74*, 19–31. [CrossRef]
30. Claude, V.; Courson, C.; Köhler, M.; Lambert, S.D. Overview and Essentials of Biomass Gasification Technologies and Their Catalytic Cleaning Methods. *Energy Fuels* **2016**, *30*, 8791–8814. [CrossRef]
31. Ahmad, A.A.; Zawawi, N.A.; Kasim, F.H.; Inayat, A.; Khasri, A. Assessing the gasification performance of biomass: A review on biomass gasification process conditions, optimization and economic evaluation. *Renew. Sustain. Energy Rev.* **2016**, *53*, 1333–1347. [CrossRef]
32. Furusawa, T.; Miura, Y.; Kori, Y.; Sato, M.; Suzuki, N. The cycle usage test of Ni/MgO catalyst for the steam reforming of naphthalene/benzene as model tar compounds of biomass gasification. *Catal. Commun.* **2009**, *10*, 552–556. [CrossRef]
33. Park, H.J.; Park, S.H.; Sohn, J.M.; Park, J.; Jeon, J.-K.; Kim, S.-S.; Park, Y.-K. Steam reforming of biomass gasification tar using benzene as a model compound over various Ni supported metal oxide catalysts. *Bioresour. Technol.* **2010**, *101*, S101–S103. [CrossRef] [PubMed]
34. Chen, G.; Dong, X.; Yan, B.; Li, J.; Yoshikawa, K.; Jiao, L. Photothermal steam reforming: A novel method for tar elimination in biomass gasification. *Appl. Energy* **2022**, *305*, 117917. [CrossRef]
35. Chen, Y.-H.; Parvez, A.M.; Schmid, M.; Scheffknecht, G.; Chen, T.-L. 20 kW Pilot scale steam-oxygen gasification of solid recovered fuel with a focus on newly developed off-line and on-line tar measurement methods. *Fuel Process. Technol.* **2022**, *227*, 107096. [CrossRef]
36. Schmid, M.; Hafner, S.; Biollaz, S.; Schneebeli, J.; Waizmann, G.; Scheffknecht, G. Steam-oxygen gasification of sewage sludge: Reduction of tar, H_2S and COS with limestone as bed additive. *Biomass Bioenergy* **2021**, *150*, 106100. [CrossRef]
37. Li, J.; Tao, J.; Yan, B.; Jiao, L.; Chen, G.; Hu, J. Review of microwave-based treatments of biomass gasification tar. *Renew. Sustain. Energy Rev.* **2021**, *150*, 111510. [CrossRef]
38. Liu, L.; Zhang, Z.; Das, S.; Kawi, S. Reforming of tar from biomass gasification in a hybrid catalysis-plasma system: A review. *Appl. Catal. B Environ.* **2019**, *250*, 250–272. [CrossRef]
39. Delagrange, S.; Pinard, L.; Tatibouët, J.-M. Combination of a non-thermal plasma and a catalyst for toluene removal from air: Manganese based oxide catalysts. *Appl. Catal. B Environ.* **2006**, *68*, 92–98. [CrossRef]
40. Caballero, M.A.; Corella, J.; Aznar, M.-P.; Gil, J. Biomass Gasification with Air in Fluidized Bed. Hot Gas Cleanup with Selected Commercial and Full-Size Nickel-Based Catalysts. *Ind. Eng. Chem. Res.* **2000**, *39*, 1143–1154. [CrossRef]
41. Kaisalo, N.; Simell, P.; Lehtonen, J. Benzene steam reforming kinetics in biomass gasification gas cleaning. *Fuel* **2016**, *182*, 696–703. [CrossRef]
42. He, L.; Hu, S.; Jiang, L.; Liao, G.; Zhang, L.; Han, H.; Chen, X.; Wang, Y.; Xu, K.; Su, S.; et al. Co-production of hydrogen and carbon nanotubes from the decomposition/reforming of biomass-derived organics over Ni/α-Al_2O_3 catalyst: Performance of different compounds. *Fuel* **2017**, *210*, 307–314. [CrossRef]

catalysts

MDPI

Article

A Theoretical Study of the Oxygen Release Mechanisms of a Cu-Based Oxygen Carrier during Chemical Looping with Oxygen Uncoupling

Minjun Wang [1,*], Shixiong Zhang [1], Ming Xia [2] and Mengke Wang [1]

[1] School of Energy and Power, Jiangsu University of Science and Technology, Zhenjiang 212003, China; zsx4920816@163.com (S.Z.); wangmk0607@163.com (M.W.)

[2] Chuzhou Cigarette Factory, China Tobacco Anhui Industrial Co., Ltd., Chuzhou 239000, China; xiaming198708@126.com

* Correspondence: wmj403@just.edu.cn

Abstract: The Cu-based oxygen carrier is a promising material in the chemical looping with oxygen uncoupling (CLOU) process, while its performance in the CLOU is significantly dependent on the oxygen release properties. However, the study of oxygen release mechanisms in CLOU is not comprehensive enough. In this work, the detailed oxygen release mechanisms of CuO(110) and CuO(111) are researched at an atomic level using the density functional theory (DFT) method, including the formation of O_2, the desorption of O_2 and the diffusion of O anion, as well as the analysis of the density of states. The results show that (1) the most favorable pathway for O_2 formation and desorption occurs on the CuO(110) surface of O-terminated with energy barriers of 1.89 eV and 3.22 eV, respectively; (2) the most favorable pathway for O anion diffusion occurs in the CuO(110) slab with the lowest energy barrier of 0.24 eV; and (3) the total density of states for the O atoms in the CuO(110) slab shifts to a lower energy after an O vacancy formation. All of the above results clearly demonstrate that the CuO(110) surface plays a significantly important role in the oxygen release reaction, and the oxygen vacancy defect should be conducive to the reactivity of oxygen release in a Cu-based oxygen carrier.

Keywords: chemical looping with oxygen uncoupling; Cu-based oxygen carrier; oxygen release mechanisms; density functional theory

1. Introduction

Carbon emissions from fossil fuel utilization are important sources of greenhouse gases (GHGs) that cause the greenhouse effect [1]. To minimize global CO_2 emissions, it is critical to develop effective CO_2 capture technologies [2]. Chemical looping combustion (CLC) is a potential approach for capturing 100% carbon from fuel at a cheap cost and with low energy penalties [3]. A two-step reaction can be characterized as the CLC process: (i) the fuel is oxidized in the fuel reactor (FR) by an oxygen carrier, usually a metal oxide, to produce a concentrated stream of H_2O and CO_2. (ii) The reduced oxygen carrier is then re-oxidized by air in an air reactor (AR) in the next step. The oxidized carrier is then separated and transferred to the fuel reactor in preparation for the next cycle [4]. Below are the simplified reaction equations for the two-step reaction.

$$\text{Fuel} + \text{MeO}_x \rightarrow \text{MeO}_{x-1} + \text{CO}_2 + \text{H}_2\text{O} \tag{1}$$

$$\text{O}_2 + 2\text{MeO}_{x-1} \rightarrow 2\text{MeO}_x \tag{2}$$

Instead of air, the oxygen carrier is employed in CLC to carry oxygen to the fuel. As a result, the CLC has a CO_2 separation advantage over traditional combustion, where CO_2 is diluted with primary flue gas components, such as nitrogen [5]. In addition, CLC also reduces NO_x generation because the fuel burns in the FR without air, and the oxidation of the oxygen carrier occurs at a lower temperature in the absence of fuel [6]. However, solid fuels are difficult to be applied in the traditional CLC process due to the rate-limiting gasification stage of solid fuels. To address this problem, an innovative CLC technology has been suggested, named chemical looping with oxygen uncoupling (CLOU) [7]. The CLOU technology makes use of the characteristics of particular metal oxides, which, at high temperatures, produce gaseous O_2. The CLOU process has three reaction steps: (i) the decomposition of the oxygen carrier generates gaseous O_2 in the FR; (ii) the fuel is oxidized by the oxygen carrier in the FR; and (iii) the reduced oxygen carrier is re-oxidized by air in the AR. The simplified reaction equations of CLOU are shown below.

$$2MeO_x \rightarrow O_2 + 2MeO_{x-1} \quad (3)$$

$$Fuel + O_2 \rightarrow CO_2 + H_2O \quad (4)$$

$$O_2 + 2MeO_{x-1} \rightarrow 2MeO_x \quad (5)$$

The gaseous O_2 generated by the oxygen carrier in the CLOU process significantly speeds up the conversion rate of solid fuels at the desired combustion temperatures (800–1000 °C) [7]. The reactivity and stability of the oxygen carrier are critical for the CLOU process applicability [7]. Many materials appropriate for CLOU have been studied to date, including perovskite-type oxides [8], natural ores [9], synthetic materials [10], and transitional metal oxides [11]. As a consequence of advanced research, transitional metal oxides in a variety of forms, such as nickel, cobalt, iron, manganese, or copper oxides, were confirmed as potential candidates for oxygen carriers [11–16]. Among these metal complexes, copper-based oxygen carriers are appealing due to their high reactivity, high oxygen capacity, and environmental friendliness [17]. On the other hand, CuO has certain drawbacks, including (i) a low mechanical strength, which may lead to defluidization, and (ii) a strong agglomeration trend, which may result in low reactivity due to its low melting point [18]. De Diego et al. [19] found that the reaction rate of CuO dropped rapidly with repeated cycles due to agglomeration and sintering.

Experimental studies revealed that using inert materials as a support for CuO can improve chemical stability and mechanical strength [20,21]. Yet, the addition of inert support inevitably reduces the oxygen transfer capacity and redox ability per unit mass. Therefore, further work has to be conducted to increase the reactivity of the Cu-based oxygen carrier in CLOU. According to the principle described above, the performance of oxygen release for oxygen carriers is a crucial issue that determines the reactivity of CLOU. Numerous experimental and theoretical studies have been conducted on the mechanisms of the reaction in CLOU [22–25]. The results show that oxygen carriers release oxygen at the solid surface, and lattice oxygen diffusion occurs at the same time; consequently, the total reaction rate is controlled by both surface reactions and bulk oxygen diffusion. When exposed to a gaseous atmosphere (reducing gases), oxygenic species on the surface of oxygen carrier react with reducing gases at specific active sites to produce the desired products. As a result of the surface reactions, surface oxygen is consumed, and an oxygen chemical potential gradient develops; oxygen anions must migrate from the bulk phase to the surface [26–28]. To achieve a high reactant conversion and improve the desired product selectivity, the control of the oxygen diffusion and the oxygen release from surface pathways is required [29–31].

Despite the fact that many experimental methods can effectively observe and track changes in bulk phase oxygen carriers, experimental technologies for studying the microscopic mechanisms of CLOU are severely limited because experimental characterization is difficult to capture and it is difficult to observe the transformation of reactants at extreme-speed time scales [32,33]. Computational simulation has demonstrated significant benefits in the research of material surface structure and characteristics, as well as chemical reaction mechanisms. In the areas of solid material structure simulation, property prediction, and an in-depth understanding of reaction mechanisms, first-principle calculations can often provide knowledge and understanding that cannot be achieved by experiments [34–36]. In recent years, first-principles calculations have been widely used for oxygen carriers, such as NiO, Fe_2O_3, Fe_3O_4, and perovskites [31,37–41]. Meanwhile, a number of studies research Cu-based oxygen carriers by density functional theory (DFT) calculations. Zhao et al. [42] investigated the CuO–support interaction in Cu-based oxygen carriers using DFT calculations. The effects of several inert supports on the decomposition reactivity and sintering resistance of Cu-based oxygen carriers were compared. Wu et al. [43] systematically calculated the kinetic mechanisms of CO oxidation on the CuO (111) surface using the DFT calculation to consider the adsorption, reaction, and desorption processes. Zhang et al. [44] studied the decomposition mechanisms of $CuO/CuAl_2O_4$ and the effects of inert supports on the Cu-based oxygen carriers. The results show that the O anion migrates from the lattice to the surface and forms an O_2 complex. Zheng et al. [45] reported the detailed oxidation steps of the Cu-based oxygen carrier by DFT calculations. According to the findings, the bulk Cu_2O phase forms preferentially as a result of copper outward diffusion in the Cu(111) bulk, before being oxidized to produce the bulk CuO phase. Mishra et al. [46] correlated the oxygen vacancy creation energy of the perovskite oxygen carriers with their redox properties by DFT calculations and experiment research. The results indicate that vacancy creation energy is an effective indicator for oxygen release properties of metal oxide oxygen carriers.

Although there has been a lot of research on Cu-based oxygen carriers, the study of oxygen release mechanisms in CLOU is still not comprehensive enough. For instance, the CuO(110) surface has not been examined in CLOU, which is a significant crystal face due to the presence of active sites on it, and X-ray diffraction characterization of CuO also revealed that one of the most observed peaks relates to the (110) facet [46–51]. In addition, the current research exclusively considers O anion diffusion from the lattice to the surface. The diffusion of the O anion between the sites on the surface has not been studied. The objective of this work is to investigate the oxygen release mechanisms of the Cu-based oxygen carrier during CLOU more comprehensively. The oxygen release processes of CuO(111) and CuO(110) were simulated through DFT calculations. More O anion diffusion paths are considered. Additionally, the effects of oxygen vacancy formed during O anion migration on CuO were studied, and the oxygen vacancy creation energy in CuO was calculated. This study aims to understand the main oxygen release mechanisms and real reaction active sites of CLOU on the Cu-based oxygen carrier, which is helpful in controlling and improving the CLOU process.

2. Results and Discussion

2.1. Calculations of Bulk CuO and O_2

The calculated Cu–O bond length (0.196 nm) of the CuO agrees well with the experimental value (0.195 nm) [52]. The calculated lattice parameters for CuO are a = 4.723 Å, b = 3.418 Å, c = 5.161 Å and β = 99.34°, which is comparable to the experimental results (a = 4.683 Å, b = 3.422 Å, c = 5.129 Å and β = 99.54°) [53]. In addition, the bond length of molecule O_2 predicted using our approach is 0.124 nm, which is consistent with experimental result of 0.121 nm [54]. The above results in this investigation support the accuracy of the chosen computation method.

2.2. The Mechanisms of Oxygen Release from the CuO Surface

First, the oxygen release reaction on the CuO(110) surface was studied. In the preliminary work, the oxygen release process on the CuO(110) surface of Cu-terminated was simulated. However, the calculation results show that it is difficult to form stable O_2, probably because the oxygen atoms of the Cu-terminated surface are not in the surface layer, which is very stable, and it is difficult to break the bond with the Cu atoms and migrate to the outermost layer. Therefore, the oxygen release process on the CuO(110) surface of O-terminated is simulated and analyzed here.

The IS column in Figure 1 shows the relaxed structure of the CuO(110) surface of O-terminated. The oxygen atoms on the CuO(110) surface of O-terminated are of the same chemical type as the three-fold coordinated atoms. Yet, oxygen atoms may be in different environments and combine in different paths. O1, O2, O3, and O4 were assigned to the four oxygen atoms on the CuO(110) surface of O-terminated (see Figure 1). In order to determine the possible O_2 formation process, an O1 atom was removed and reintroduced to bond with three neighboring O atoms, including O2, O3, and O4. Subsequently, these configurations were optimized. The optimization results show that the O1 atom forms bonds with the O2, O3 and O4 atoms with bond lengths of 1.466 Å, 1.276 Å, and 1.370 Å, respectively. These bond lengths are close to those of the free O_2 molecule (1.124), implying that an O_2 complex is formed in the three structures. Therefore, the O_2 formation processes of O1-O2, O1-O3, and O1-O4 were considered and denoted by path 1 (P1), path 2 (P2), and path 3 (P3), respectively. For the calculation of O_2 formation processes, the perfect CuO(110) surface of O-terminated was set as the initial state (IS). The final structure was the stable configuration with the O_2 complex, which was denoted by the IM (intermediate state). Then, a transition state (TS) search calculation was performed. The TS structure and the corresponding energy profile for the process are shown in Figures 1 and 2. The energy barriers of the path 1, path 2, and path 3 are 1.89 eV, 4.47 eV, and 4.13 eV, respectively. In addition, all three processes are endothermic by 1.72 eV, 2.38 eV, and 1.91 eV, respectively. These results suggest relatively large differences in the three paths of O_2 formation processes.

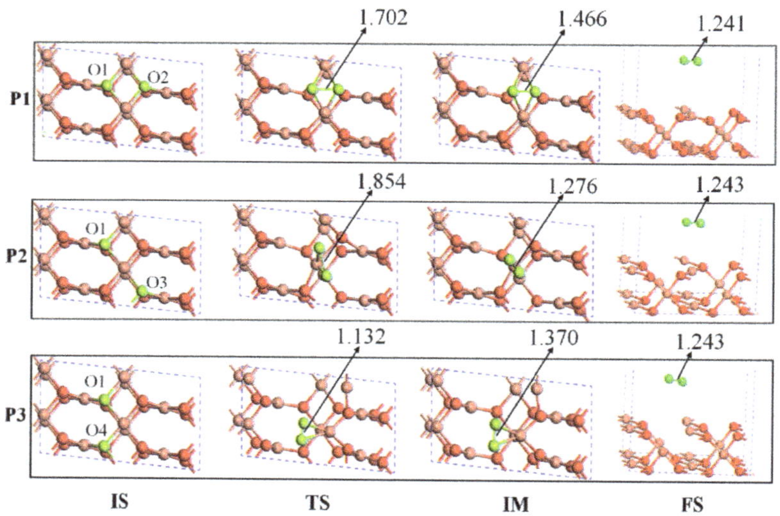

Figure 1. The initial state (IS), transition state (TS), intermediate state (IM), and final state (FS) structures for the O_2 formation and O_2 desorption processes in different pathways on the CuO(110) surface of O-terminated. The bond lengths are presented in Å. The salmon pink and red spheres denote Cu and O atoms, respectively.

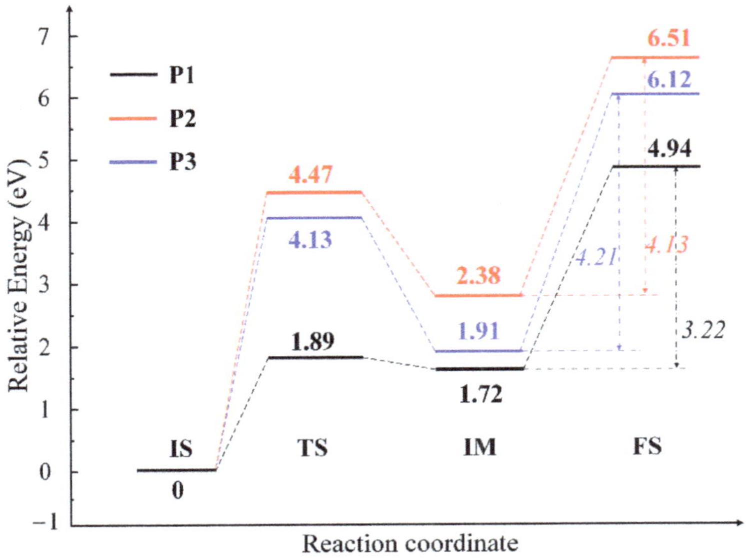

Figure 2. Potential energy diagram for the O_2 formation and desorption processes following different pathways on the CuO(110) surface of O-terminated.

Subsequently, the O_2 desorption processes following the O_2 formation were simulated. The O_2 complex formed in the intermediate state was removed and placed at a distance of 8 Å from the CuO(110) surface of O-terminated. The structures were then optimized and labeled with the FS (final state) (see Figure 1). Thus, a TS search calculation was used to determine the energy barrier between the IM and the FS. The results (see Figure 1) show that all the three O_2 desorption processes are largely endothermic by 3.22 eV, 4.13 eV, and 4.21 eV, respectively, which indicates that the oxygen is only released at high temperatures. This is consistent with the thermodynamic calculation result that the temperature of oxygen release on CuO needs to be greater than 800 °C [7]. Furthermore, due to the low energy barriers for O_2 formation and desorption, path 1 is regarded as the most favorable pathway for oxygen release on the CuO(110) surface of O-terminated. In the path 1 process, an O_2 complex is formed in the IM structure, and the O-O bond length is 1.466 Å. Then, the Cu–O bond is ruptured with a desorption energy of 3.22 eV, and the O_2 is desorbed from the CuO(110) surface. The length of the new O–O bond is 1.241 Å, which is similar to the length of a free O_2 molecule (1.240 Å).

To provide further insight into the oxygen release reaction, the density of states (DOS) was investigated. Figure 3 shows the DOS for O atoms (O1, O2), O_2 complex, and free O_2 molecule in the path 1 process, respectively. The DOS of O atoms in the IS delocalizes over a wide range due to the strong Cu–O bonds that exist in the CuO. In the IM, two peaks are observed in s orbitals, which indicates the formation of an O_2 complex. However, a large delocalization still appears in the p orbitals, which indicates that the O_2 complex still has strong interaction with the CuO(110) surface. After O_2 desorption, the DOS of O_2 in the FS was nearly the same as that of a free O_2 molecule, implying that free O_2 molecules are generated.

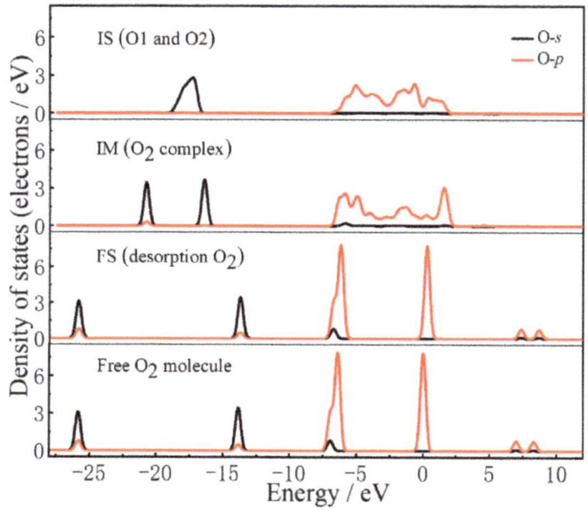

Figure 3. The DOS for O atoms (O1, O2), O_2 complex, and free O_2 molecule in the path 1 process.

Next, the oxygen release reaction on the CuO(111) surface was simulated. Of particular note, the oxygen release reaction on the CuO(111) surface was studied by the work of Zhang et al. [44]. However, the calculation methods were slightly different from those in this paper. In order to make the results in this work as comparable as possible, the O_2 formation and desorption processes were still calculated here, and the results are shown in Figures 4 and 5. There are two pathways for oxygen release on the CuO(111) surface, which are denoted as path 1 (P1) and path 2 (P2). The details of model construction are not repeated here. The energy barriers of O_2 formation in path 1 and path 2 are 3.18 eV and 2.47 eV, respectively. Both O_2 desorption processes are largely endothermic by 3.61 eV and 3.31 eV, respectively. The calculated results are quite similar to those that are reported in the literature [44] (see Table 1). Comparing the calculation results of the CuO(110) surface with that of the CuO(111) surface reveals that there is the lowest energy barrier and total energy in path 1 on the CuO(110) surface. In particular, the activation energy values of oxygen release on CuO studied by experiments (3.26 eV and 3.41 eV) [55,56] were close to the energy barrier of path 1 on the CuO(110) surface, while largely smaller than that of other pathways. The analysis above indicates that the CuO(110) surface may also play an important role in the oxygen release reaction, and the active sites in the oxygen release process on the CuO surface exist in pairs.

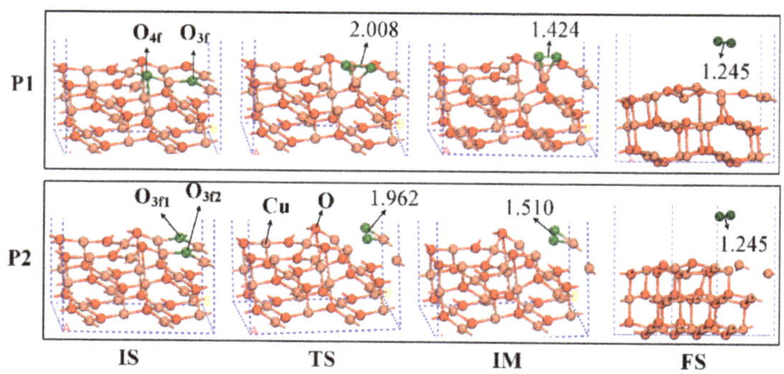

Figure 4. The initial state (IS), transition state (TS), intermediate state (IM), and final state (FS) structures for the O_2 formation and O_2 desorption processes in different pathways on the CuO(111) surface. The bond lengths are presented in Å. The salmon pink and red spheres denote Cu and O atoms, respectively.

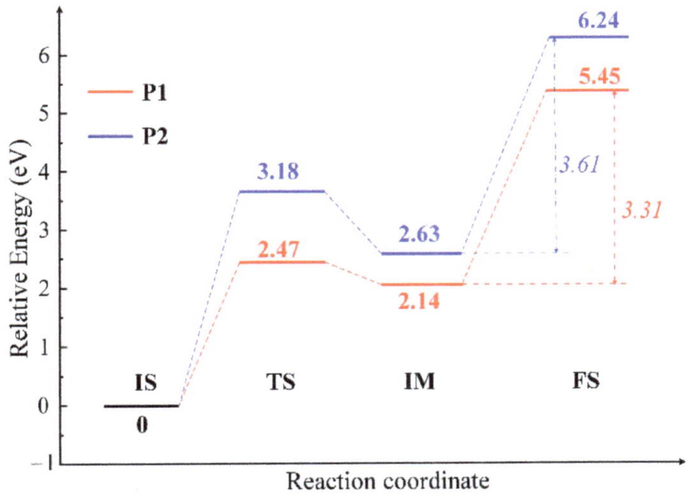

Figure 5. Potential energy diagram for the O_2 formation and desorption processes following different pathways on the CuO(111) surface.

Table 1. The energy barriers of O_2 formation (E_f), O_2 desorption (E_d), total energy (E_t), and activation energy (E_a).

	Pathway	E_f (eV)	E_d (eV)	E_t (eV)	E_a (eV)
CuO(110) surface	P1	1.89	3.22	4.94	–
	P2	4.47	4.13	6.51	–
	P3	4.13	4.21	6.12	–
CuO(111) surface	P1	2.47	3.31	5.45	–
	P2	3.18	3.61	6.24	–
CuO(111) surface in literature	P1	2.3	3.5	5.2	
	P2	3.0	3.1	6.0	
Experiment values	–	–	–	–	3.26
	–	–	–	–	3.41

2.3. The O Anion Diffusion Mechanisms in the CuO

As a result of the O_2 desorption from the CuO surface, surface oxygen is consumed. Therefore, oxygen anions must diffuse to the surface to maintain the oxygen release reaction. First, the O anion diffusion processes in the CuO(110) slab were studied. According to the research results of the oxygen release from CuO(110) surface, there are active sites on the surface during the oxygen release process. Thus, the O anion will diffuse to the active sites. In view of this, one active site for the O atom was labeled as O1, and the three neighboring atoms on the CuO(110) surface were labeled as O2, O3, and O4, respectively (see Figure 6).

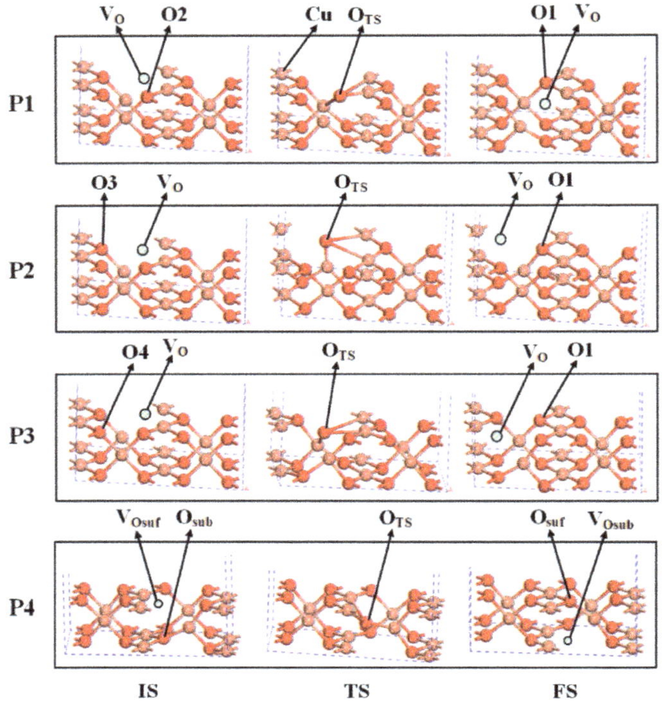

Figure 6. The initial state (IS), transition state (TS), and final state (FS) structures for the O anion diffusion processes in different pathways on the CuO(110) surface of O-terminated. The salmon pink and red spheres denote Cu and O atoms, respectively.

For this calculation, the initial state (IS) was defined as an optimized CuO(110) structure with a surface O1 vacancy. The structure CuO(110) with a surface O2 (O3 or O4) vacancy was optimized and denoted as the final state (FS). Accordingly, three O anion diffusion paths were constructed and denoted as path 1 (P1), path 2 (P2), and path 3 (P3), respectively. In addition, considering that O anion may diffuse from the lattice to the surface, the fourth pathway was constructed and denoted by path 4 (P4), in which IS was defined as an optimized CuO(110) structure with a surface oxygen vacancy (denoted as V_{Osuf}) and the FS was defined as an optimized CuO(110) structure with a subsurface oxygen vacancy (denoted as V_{Osub}). Figures 6 and 7 show the TS search results and the corresponding energy profile for the processes. The energy barriers of path 1, path 2, path 3, and path 4 are 0.51 eV, 1.75 eV, 1.32 eV, and 0.24 eV, respectively, indicating that the energy barriers of O anion diffusion in CuO(110) are extremely low. In particular, the energy barrier of O anion diffusion in path 4 (0.24 eV) is significantly lower than that of the O_2 formation (1.89 eV, 4.13 eV, 4.47 eV) and O_2 desorption processes (3.22 eV, 4.13 eV, 4.21 eV). In addition, all the four processes are exothermic by −0.43 eV, −0.48 eV, −0.47 eV, and −0.02 eV, respectively. These results imply that O anion diffusion in CuO(110) can react readily and is energetically favorable.

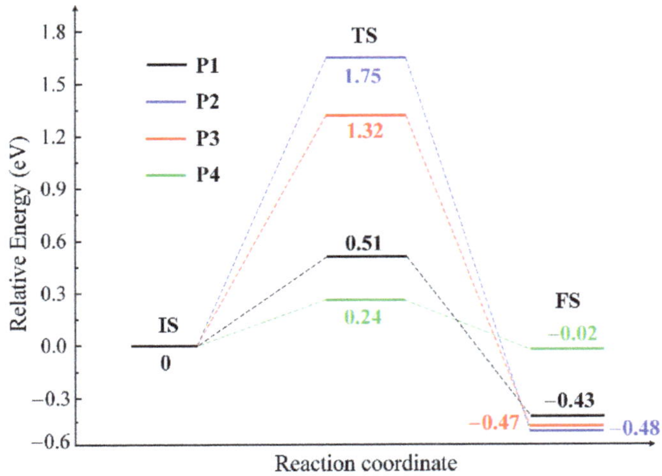

Figure 7. Potential energy diagram for the O anion diffusion processes in different pathways on the CuO(110) surface of O-terminated.

Next, the O anion diffusion processes in the CuO(111) slab were studied. The O_{4f} atom migrating to O_{3f} vacancy, O_{4f} atom migrating to O_{4f} vacancy, O_{3f} atom migrating to O_{3f} vacancy, and subsurface oxygen atom (O_{sub}) migrating to surface oxygen vacancy (V_{Osuf}) were considered as four possible O anion diffusion paths in the CuO(110) slab. Correspondingly, the four paths were denoted by path 5 (P5), path 6 (P6), path 7 (P7), and path 8 (P8), respectively. The optimized structures of IS and FS are shown in Figure 8. The reaction pathway and energy profile for the O anion diffusion in the CuO(110) slab are shown in Figures 8 and 9. The energy barriers of path 5, path 6, path 7, and path 8 are 1.67 eV, 1.58 eV, 1.16 eV, and 0.74 eV, respectively, which are also rather small. The results indicate that the O_2 formation and desorption are the rate limiting steps for oxygen release from CuO. Therefore, the key to improving the reactivity of Cu-based oxygen carriers is to decrease the energy barrier of O_2 formation and desorption processes. In addition, the average energy barrier of O anion diffusion in the CuO(111) slab is higher than that in the CuO(110) slab. Moreover, the path 5 and path 6 are endothermic by 0.40 eV and 0.28 eV, respectively. On the contrary, path 7 and path 8 are exothermic by −0.13 eV and −0.47 eV, which suggests that path 7 and path 8 are energetically favorable. In addition, the average reaction energy of O anion diffusion in the CuO(111) slab is also higher than that in the CuO(110) slab. By comparing the two slabs, path 4 in the CuO(110) slab is regarded as the most favorable pathway for O anion diffusion due to the lowest energy barrier and exothermic process. In the diffusion process of path 4, the O_{sub} atom migrates from the original location to the intermediate position between the surface and subsurface O vacancy. In addition, the subsurface Cu-O_{sub} bond is gradually rupturing. Before populating the surface oxygen vacancy, the O_{sub} atom bonds to the surface Cu atom. Given the above results, it can be found that the O anion migrates more easily in the CuO(110) slab, which further confirms the possibility that CuO(110) is important during oxygen release from the CuO. If the ratio of CuO(110) can be increased, though it is a difficult challenge, the oxygen release performance of the Cu-based oxygen carrier will be effectively improved. Nevertheless, there is a similarity between CuO(110) and CuO(111) slabs, in that paths 4 and 8 have the lowest energy barriers in their respective slabs, indicating that the O anions diffuse more easily from the lattice to the surface.

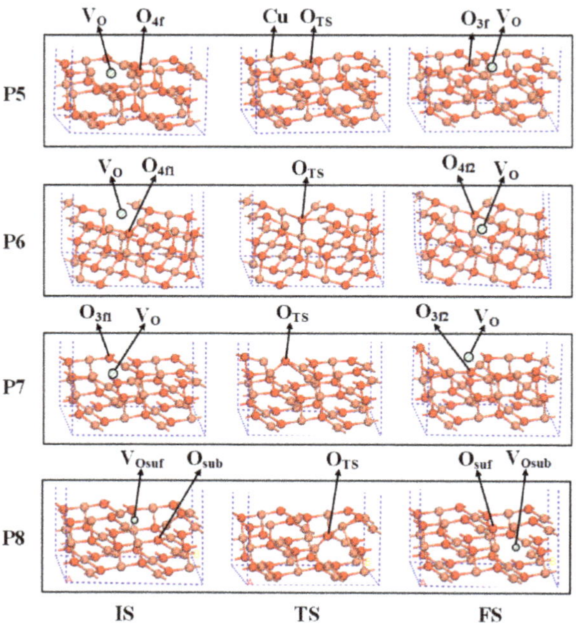

Figure 8. The initial state (IS), transition state (TS), and final state (FS) structures for the O anion diffusion processes in different pathways on the CuO(111) surface. The salmon pink and red spheres denote Cu and O atoms, respectively.

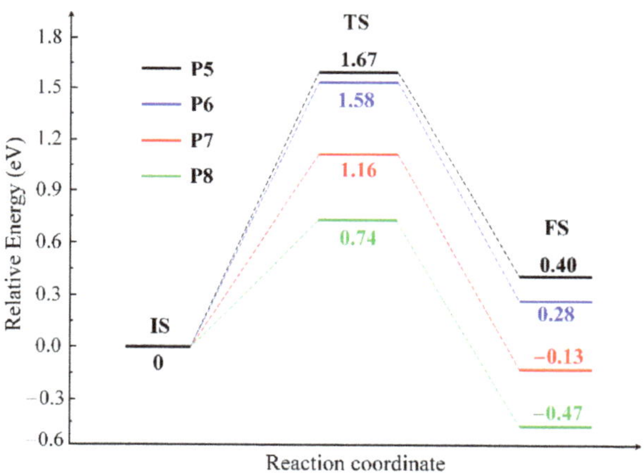

Figure 9. Potential energy diagram for the O anion diffusion processes in different pathways on the CuO(111) surface.

To provide further insight into the O anion diffusion process, the density of states in the IS, TS, and FS was investigated. The partial density of states (PDOS) of the O atom during the diffusion process in path 4 is presented in Figure 10. The electrons of the O atom in IS delocalize in a wide range near the Fermi energy level (E_f = 0 eV), implying that the O atom is more chemically active. In comparison to IS, the DOS of the O atom in TS has a higher peak at approximately −4.98 eV, indicating that an energy barrier is required to transition from IS to TS. In addition, the p-orbital of the O atom in FS is similar to that in IS, while the s-orbital of the O atom in FS is slightly lower than that in IS, which is consistent with the finding that the O anion diffusion process in path 4 has an energy barrier of 0.24 eV and is exothermic by −0.02 eV.

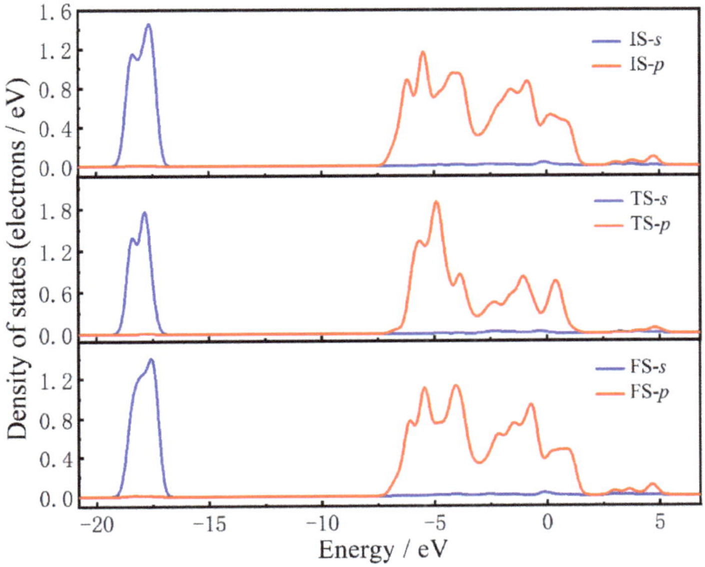

Figure 10. The partial density of states (PDOS) of the O atom during diffusion process in the path 4.

Moreover, the energy barriers of the O anion diffusion process are substantially lower than those of the O_2 formation and desorption processes, which could be caused by the O vacancy in CuO. In order to illustrate this, the density of states of the perfect CuO(110) slab and that with an O vacancy were studied. The total density of states (TDOS) and the PDOS of the O atoms are presented in Figure 11. It is observed that the TDOS for the CuO(110) slab with an O vacancy is more flat than that for the perfect CuO(110) slab, suggesting a larger delocalization in the former. In addition, for the perfect CuO(110) slab, the Cu d-orbital is hybridized with the O p-orbital as shown by the state overlap at approximately −5.84 eV and 0.10 eV, respectively. While the hybridized peaks occur at approximately −5.70 eV and 0.32 eV in the CuO(110) slab with an O vacancy. As seen from Figure 9, both hybridized peaks shift lower in energy after the formation of an O vacancy, indicating a weaker interaction of Cu with the O atom in the CuO(110) slab with an O vacancy, which facilitates O anion migration. This result may provide an idea for increasing the reactivity of oxygen release, that is, increasing oxygen vacancies. However, the oxygen vacancy creation energy in CuO was calculated to be 7.4 eV. The positive value means that energy is needed to create an oxygen vacancy in CuO.

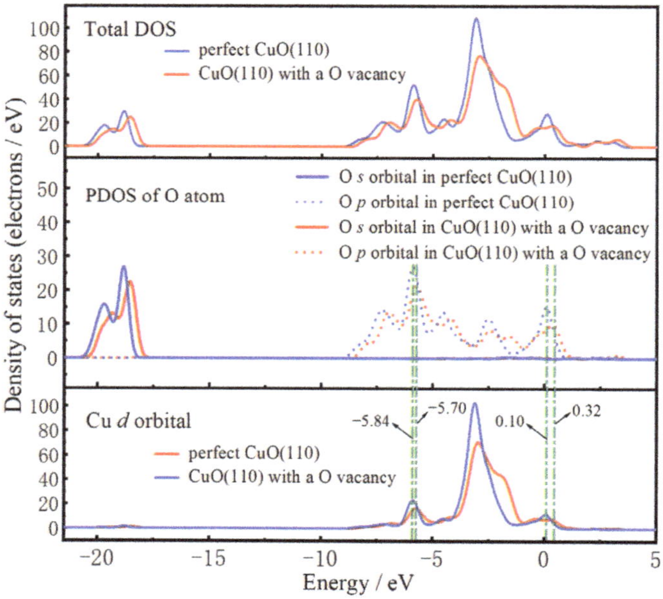

Figure 11. The TDOS of Cu(110) slab and the PDOS of the O atoms.

3. Calculation Method and Model

All DFT calculations were performed using the Cambridge Serial Total Energy Package (CASTEP) program suite [57]. Ultrasoft pseudopotentials were used to represent electron–ion interactions [58]. According to previous work conducted by our research group [47,59], the electron exchange correlation functionals of the Perdew–Wang (PW91) approximation based on the generalized gradient approximation (GGA) [60] offered the most accurate geometrical parameters of bulk CuO. As a result, the GGA-PW91 was employed to perform the calculations. The Hubbard parameter (U) was proposed for the Cu 3d orbitals to correctly reflect the high electron correlations in transition metals. We evaluated the DFT + U with U ranging from 0 to 10 eV and found that for the Cu 3d orbitals, U = 7.5 eV produced appropriate results. Thus, we applied U = 7.5 eV to the Cu 3d orbitals throughout this study. The linear/quadratic synchronous transit (LST/QST) approach [61] was used to study the transition states for each reaction.

It is critical to develop proper models for oxygen release reaction research. The unit cell of CuO that is a monoclinic structure and belongs to the C2/c1 space group is depicted in Figure 12. The surface of CuO(110) was modeled using the supercell method, in which a periodic boundary condition was added to the center supercell, causing it to be replicated periodically throughout space. Taking into consideration that the CuO (110) surface can be terminated by O or Cu atoms, the two models that p(2 × 2) super lattice CuO(110) slab models of the O-terminated and the Cu-terminated four atomic layers were constructed, respectively. Similarly, the p(2 × 2) super lattice CuO(111) slab models of three layers were constructed. There are four chemically distinguishable types of atoms at the CuO(111) surface: three-fold coordinated Cu atoms, four-fold coordinated Cu atoms, three-fold coordinated O atoms, and four-fold coordinated O atoms, which have been labeled as Cu_{3f}, Cu_{4f}, O_{3f}, and O_{4f}, respectively. To avoid interactions between the periodic slabs of atomic layers, a 15 Å vacuum layer was used to separate the slab models. In the calculations, the bottom layer was kept fixed while the remaining layers were allowed to fully relax. The O_2 molecules were allowed to relax. All the surface configurations are shown in Figure 13.

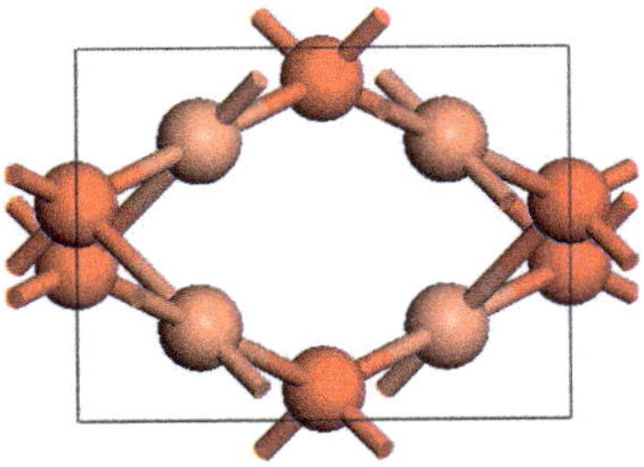

Figure 12. Schematic representation of a CuO unit cell. The salmon pink spheres denote Cu atoms, and the red spheres denote O atoms, respectively.

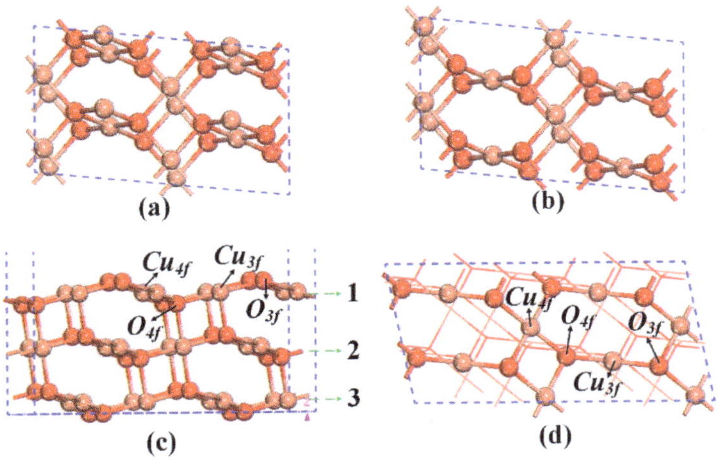

Figure 13. (**a**) O-terminated CuO(110) surface, (**b**) Cu-terminated CuO(110) surface, (**c**) the top view of a CuO(111) slab with different types of Cu and O atoms, and (**d**) the side view of a CuO(111) slab with different types of Cu and O atoms. The salmon pink spheres denote Cu atoms, and the red spheres denote O atoms, respectively.

The cut-off energy (350 eV) and the density of the Monkhorst Pack scheme k-point of $4 \times 4 \times 1$ were chosen to be sufficiently high. The kinetic energy cut-off employed in all calculations was 350 eV, and was used together with the Monkhorst Pack scheme k-point of $4 \times 4 \times 1$. The criteria for the convergence of all calculations in this work were set to (i) an energy tolerance of 2.0×10^{-5} eV/atom; (ii) a maximum force tolerance of 0.05 eV/Å; (iii) a self-consistent field tolerance of 2.0×10^{-6} eV/atom; and (iv) a maximum displacement tolerance of 0.002 Å. The following equation was used to calculate the energy barriers (E_b) of various reactions:

$$E_b = E_{TS} - E_{IS} \quad (6)$$

where E_{IS} and E_{TS} are the energies of the initial state (IS) and the transition state (TS), respectively. The oxygen vacancy creation energy (E_{vac}) is calculated by the following equation:

$$E_{vac} = E_{def} + 0.5E_{O_2} - E_{per} \quad (7)$$

where E_{def} is the energy of a defective CuO supercell containing one oxygen vacancy, and E_{O_2} is the energy of the ground state of an isolated diatomic oxygen molecule in the gas phase. E_{per} is the energy of the CuO supercell. A positive value of E_{vac} means that energy is needed to create an oxygen vacancy.

4. Conclusions

In this study, the detailed oxygen release mechanisms of CuO(110) and CuO(111) were researched at an atomic level using the DFT method. A total of five oxygen release paths on CuO(110) and CuO(111) surfaces were considered. The most favorable pathway for oxygen formation and desorption occurs on the CuO(110) surface of O-terminated. The energy barriers of oxygen formation and desorption processes are 1.89 eV and 3.22 eV, respectively, which are largely lower than other paths. The results indicate that there are active sites on the CuO surface during the oxygen release process, and they exist in pairs. The O anion diffusion processes in the CuO were further investigated. A total of eight O anion diffusion paths in CuO(110) and CuO(111) slabs were considered. The most favorable pathway for O anion diffusion is also found in the CuO(110) slab with the lowest energy barrier of 0.24 eV. By comparing both surfaces, it can be concluded that the CuO(110) surface may play a significantly important role in the oxygen release reaction due to the lowest energy barriers. On the other hand, the energy required for the subsurface oxygen to migrate to the surface oxygen vacancy is the lowest among all the O anion diffusion paths, suggesting that the O anions diffuse more easily from the lattice to the surface. In addition, the energy barriers of the O anion diffusion process are found to be substantially lower than those of the O_2 formation and desorption processes, indicating the O_2 formation and desorption are the rate-limiting steps for oxygen release from CuO, and decreasing the energy barriers of O_2 formation and desorption is the key to improving the reactivity of Cu-based oxygen carriers in CLOU. The density of states of the perfect CuO(110) slab and that with an O vacancy were studied. The results show that the TDOS for the O atoms in the CuO(110) slab shifts lower in energy after an O vacancy formation, which implies that the oxygen vacancy defect is conducive to O anion migration, which may provide an idea for increasing the reactivity of oxygen release. However, the oxygen vacancy creation energy in CuO is considerably high (7.4 eV), which means that energy is needed. Therefore, how to reduce oxygen vacancy creation energy is a topic worth studying in the future. This study would be helpful in rationalizing the development of a high-performance oxygen carrier for CLOU.

Author Contributions: Data curation, M.W. (Minjun Wang) and S.Z.; Investigation, M.W. (Minjun Wang), S.Z., M.X. and M.W. (Mengke Wang); Writing—original draft, M.W. (Minjun Wang) and M.X.; Writing—review and editing, M.W. (Minjun Wang), S.Z., M.X. and M.W. (Mengke Wang). All authors have read and agreed to the published version of the manuscript.

Funding: This research was funded by the National Natural Science Foundation of China (No. 51806089), the General Project of Natural Science Research of Jiangsu Universities (No. 18KJD470001).

Conflicts of Interest: The authors declare no conflict of interest.

References

1. Abuelgasim, S.; Wang, W.; Abdalazeez, A. A brief review for chemical looping combustion as a promising CO_2 capture technology: Fundamentals and progress. *Sci. Total Environ.* **2021**, *764*, 142892. [CrossRef]
2. Wang, M.; Liu, J.; Hu, J.; Liu, F. O_2–CO_2 Mixed Gas Production Using a Zr-Doped Cu-Based Oxygen Carrier. *Ind. Eng. Chem. Res.* **2015**, *54*, 9805–9812. [CrossRef]
3. Li, F.; Luo, S.; Sun, Z.; Bao, X.; Fan, L.-S. Role of metal oxide support in redox reactions of iron oxide for chemical looping applications: Experiments and density functional theory calculations. *Energy Environ. Sci.* **2011**, *4*, 3661–3667. [CrossRef]
4. Hossain, M.M.; Lasa, H. Chemical-looping combustion (CLC) for inherent CO_2 separations—A review. *Chem. Eng. Sci.* **2008**, *63*, 4433–4451. [CrossRef]
5. Sarshar, Z.; Sun, Z.; Zhao, D.; Kaliaguine, S. Development of Sinter-Resistant Core-Shell $LaMn_xFe_{1-x}O_3$@$mSiO_{(2)}$ Oxygen Carriers for Chemical Looping Combustion. *Energy Fuels* **2012**, *26*, 3091–3102. [CrossRef]
6. Ishida, M.; Jin, H. A Novel Chemical-Looping Combustor without NOx Formation. *Ind. Eng. Chem. Res.* **1996**, *35*, 2469–2472. [CrossRef]
7. Mattisson, T.; Lyngfelt, A.; Leion, H. Chemical-looping with oxygen uncoupling for combustion of solid fuels. *Int. J. Greenh. Gas Control* **2009**, *3*, 11–19. [CrossRef]
8. Ahmad, A.; Al Mamun, M.A.; Al-Mamun, M.; Huque, S.; Ismail, M. LFO Perovskites as Oxygen Carriers for Chemical Looping Oxygen Uncoupling (CLOU). *J. Therm. Anal. Calorim.* **2021**, 1–9. [CrossRef]

9. Kuang, C.; Wang, S.; Lv, S.; Cai, J.; Luo, M.; Zhao, J. Comparison of metallic oxide, natural ore and synthetic oxygen carrier in chemical looping combustion process. *Int. J. Hydrogen Energy* **2021**, *46*, 18032–18041. [CrossRef]
10. Ksepko, E. Examining $SrCuO_2$ as an oxygen carrier for chemical looping combustion. *J. Therm. Anal. Calorim.* **2015**, *122*, 621–633. [CrossRef]
11. Ksepko, E.; Lysowski, R. Stable Mixed Fe-Mn Oxides Supported on ZrO_2 Oxygen Carriers for Practical Utilization in CLC Processes. *Catalysts* **2021**, *11*, 1047. [CrossRef]
12. Ksepko, E.; Lysowski, R. Reactivity Study of Bimetallic Fe-Mn Oxides with Addition of TiO_2 for Chemical Looping Combustion Purposes. *Catalysts* **2021**, *11*, 1437. [CrossRef]
13. Siriwardane, R.V.; Ksepko, E.; Tian, H.; Poston, J.; Simonyi, T.; Sciazko, M. Interaction of iron–copper mixed metal oxide oxygen carriers with simulated synthesis gas derived from steam gasification of coal. *Appl. Energy* **2013**, *107*, 111–123. [CrossRef]
14. Adánez, J.; Abad, A. Chemical-looping combustion: Status and research needs. *Proc. Combust. Inst.* **2018**, *37*, 4303–4317. [CrossRef]
15. Alalwan, H.A.; Cwiertny, D.M.; Grassian, V.H. Co_3O_4 nanoparticles as oxygen carriers for chemical looping combustion: A materials characterization approach to understanding oxygen carrier performance. *Chem. Eng. J.* **2017**, *319*, 279–287. [CrossRef]
16. Vos, Y.D.; Jacobs, M.; Voort, P.; Driessche, I.V.; An, V. Development of Stable Oxygen Carrier Materials for Chemical Looping Processes—A Review. *Catalysts* **2020**, *10*, 926. [CrossRef]
17. Xu, Z.; Zhao, H.; Wei, Y.; Zheng, C. Self-assembly template combustion synthesis of a core–shell $CuO@TiO_2$ –Al_2O_3 hierarchical structure as an oxygen carrier for the chemical-looping processes. *Combust. Flame* **2015**, *162*, 3030–3045. [CrossRef]
18. Imtiaz, Q.; Kierzkowska, A.M.; Broda, M.; Mueller, C.R. Synthesis of Cu-Rich, Al_2O_3-Stabilized Oxygen Carriers Using a Coprecipitation Technique: Redox and Carbon Formation Characteristics. *Environ. Sci. Technol.* **2012**, *46*, 3561–3566. [CrossRef] [PubMed]
19. De Diego, L.F.; García-Labiano, F.; Adánez, J.; Gayán, P.; Abad, A.; Corbella, B.M.; Palacios, J.M. Development of Cu-based oxygen carriers for chemical-looping combustion. *Fuel* **2004**, *83*, 1749–1757. [CrossRef]
20. Wang, K.; Yu, Q.; Qin, Q. Reduction Kinetics of Cu-Based Oxygen Carriers for Chemical Looping Air Separation. *Energy Fuels* **2013**, *27*, 5466–5474. [CrossRef]
21. Gayán, P.; Adánez-Rubio, I.; Abad, A.; Luis, F.; García-Labiano, F.; Adánez, J. Development of Cu-based oxygen carriers for Chemical-Looping with Oxygen Uncoupling (CLOU) process. *Fuel* **2011**, *90*, 226–238. [CrossRef]
22. Li, Z. First-principles-based microkinetic rate equation theory for oxygen carrier reduction in chemical looping. *Chem. Eng. Sci.* **2021**, *247*, 117042. [CrossRef]
23. Adánez-Rubio, I.; Abad, A.; Gayán, P.; De Diego, L.F.; García-Labiano, F.; Adánez, J. Biomass combustion with CO_2 capture by chemical looping with oxygen uncoupling (CLOU). *Fuel Process. Technol.* **2014**, *124*, 104–114. [CrossRef]
24. Adánez-Rubio, I.; Abad, A.; Gayán, P.; De Diego, L.F.; García-Labiano, F.; Adánez, J. Performance of CLOU process in the combustion of different types of coal with CO_2 capture. *Int. J. Greenh. Gas Control* **2013**, *12*, 430–440. [CrossRef]
25. Abad, A.; Adánez-Rubio, I.; Gayán, P.; García-Labiano, F.; Luis, F.; Adánez, J. Demonstration of chemical-looping with oxygen uncoupling (CLOU) process in a 1.5 kWth continuously operating unit using a Cu-based oxygen-carrier. *Int. J. Greenh. Gas Control* **2012**, *6*, 189–200. [CrossRef]
26. Qin, L.; Cheng, Z.; Fan, J.A.; Kopechek, D.; Xu, D.; Deshpande, N.; Fan, L.-S. Nanostructure formation mechanism and ion diffusion in iron–titanium composite materials with chemical looping redox reactions. *J. Mater. Chem. A* **2015**, *3*, 11302–11312. [CrossRef]
27. Su, M.A.; Cao, J.A.; Tian, X.A.; Zhang, Y.B.; Zhao, H.A. Mechanism and kinetics of Cu_2O oxidation in chemical looping with oxygen uncoupling. *Proc. Combust. Inst.* **2019**, *37*, 4371–4378. [CrossRef]
28. Li, F.; Sun, Z.; Luo, S.; Fan, L.S. Ionic diffusion in the oxidation of iron—Effect of support and its implications to chemical looping applications. *Energy Environ. Sci.* **2011**, *4*, 876–880. [CrossRef]
29. Shafiefarhood, A.; Galinsky, N.; Huang, A.P.Y.; And, Y.C.; Li, A.P.F. $Fe_2O_3@La_xSr_{1-x}FeO_3$ Core-Shell Redox Catalyst for Methane Partial Oxidation. *ChemCatChem* **2014**, *6*, 790–799. [CrossRef]
30. Neal, L.M.; Shafiefarhood, A.; Li, F. Dynamic Methane Partial Oxidation Using a $Fe_2O_3@La_{0.8}Sr_{0.2}FeO_{3-\delta}$ Core–Shell Redox Catalyst in the Absence of Gaseous Oxygen. *ACS Catal.* **2014**, *4*, 3560–3569. [CrossRef]
31. Cheng, Z.; Qin, L.; Guo, M.; Xu, M.; Fan, L.S. Oxygen vacancy promoted methane partial oxidation over iron oxide oxygen carriers in the chemical looping process. *Phys. Chem. Chem. Phys.* **2016**, *18*, 32418–32428. [CrossRef] [PubMed]
32. Day, M.; Tachibana, S.; Bell, J.; Lijewski, M.; Beckner, V.; Cheng, R.K. A combined computational and experimental characterization of lean premixed turbulent low swirl laboratory flames I. Methane flames. *Combust. Flame* **2012**, *159*, 275–290. [CrossRef]
33. Meng, Y. Water Gas Shift Reaction Activity on Fe (110): A DFT Study. *Catalysts* **2021**, *12*, 27.
34. Li, H.; Shi, L.; Jin, C.; Ye, R.; Zhang, R. Co and Ni Incorporatedγ-Al_2O_3(110) Surface: A Density Functional Theory Study. *Catalysts* **2022**, *12*, 111. [CrossRef]
35. Bruix, A.; Margraf, J.T.; Andersen, M.; Reuter, K. First-principles-based multiscale modelling of heterogeneous catalysis. *Nat. Catal.* **2019**, *2*, 659–670. [CrossRef]
36. Meng, Y.; Liu, X.W.; Bai, M.; Guo, W.P.; Cao, D.B.; Yang, Y.; Li, Y.W.; Wen, X.D. Prediction on morphologies and phase equilibrium diagram of iron oxides nanoparticles. *Appl. Surf. Sci.* **2019**, *480*, 478–486. [CrossRef]

37. Bennett, J.W.; Huang, X.; Fang, Y.; Cwiertny, D.M.; Grassian, V.H.; Mason, S.E. Methane Dissociation on α-Fe_2O_3(0001) and Fe3O4(111) Surfaces: First-Principles Insights into Chemical Looping Combustion. *J. Phys. Chem. C* **2019**, *123*, 6450–6463. [CrossRef]
38. Dong, C.; Sheng, S.; Qin, W.; Lu, Q.; Zhao, Y.; Wang, X.; Zhang, J. Density functional theory study on activity of α-Fe_2O_3 in chemical-looping combustion system. *Appl. Surf. Sci.* **2011**, *257*, 8647–8652. [CrossRef]
39. Yuan, Y.; Dong, X.; Ricardez-Sandoval, L. Insights into Syngas Combustion on a Defective NiO Surface for Chemical Looping Combustion: Oxygen Migration and Vacancy Effects. *J. Phys. Chem. C* **2020**, *124*, 28359–28370. [CrossRef]
40. Feng, Y.; Wang, N.; Guo, X. Influence mechanism of supports on the reactivity of Ni-based oxygen carriers for chemical looping reforming: A DFT study. *Fuel* **2018**, *229*, 88–94. [CrossRef]
41. Feng, Y.; Guo, X. Study of reaction mechanism of methane conversion over Ni-based oxygen carrier in chemical looping reforming. *Fuel* **2017**, *210*, 866–872. [CrossRef]
42. Zhao, H.; Zhang, Y.; Wei, Y.; Gui, J. Understanding CuO-support interaction in Cu-based oxygen carriers at a microcosmic level. *Proc. Combust. Inst.* **2017**, *36*, 4069–4077. [CrossRef]
43. Wu, L.N.; Tian, Z.Y.; Kasmi, A.E.; Arshad, M.F.; Qin, W. Mechanistic study of the CO oxidation reaction on the CuO(111) surface during chemical looping combustion. *Proc. Combust. Inst.* **2021**, *38*, 5289–5297. [CrossRef]
44. Zhang, Y.; Zhao, H.; Guo, L.; Zheng, C. Decomposition mechanisms of Cu-based oxygen carriers for chemical looping with oxygen uncoupling based on density functional theory calculations. *Combust. Flame* **2015**, *162*, 1265–1274. [CrossRef]
45. Zheng, C.; Cao, J.; Zhang, Y.; Zhao, H. Insight into the Oxidation Mechanism of a Cu-Based Oxygen Carrier (Cu$\rightarrow Cu_2O \rightarrow$CuO) in Chemical Looping Combustion. *Energy Fuels* **2020**, *34*, 8718–8725. [CrossRef]
46. Mishra, A.; Li, T.; Li, F.; Santiso, E.E. Oxygen Vacancy Creation Energy in Mn-Containing Perovskites: An Effective Indicator for Chemical Looping with Oxygen Uncoupling. *Chem. Mater.* **2018**, *31*, 689–698. [CrossRef]
47. Xiang, W.; Liu, J.; Chang, M.; Zheng, C. The adsorption mechanism of elemental mercury on CuO(110) surface. *Chem. Eng. J.* **2012**, *200-202*, 91–96. [CrossRef]
48. Moreno, J.L.V.; Padama, A.A.B.; Kasai, H. A density functional theory-based study on the dissociation of NO on a CuO(110) surface. *CrystEngComm* **2014**, *16*, 2260–2265. [CrossRef]
49. Chang, Y.; Zeng, H.C. Controlled Synthesis and Self-Assembly of Single-Crystalline CuO Nanorods and Nanoribbons. *Cryst. Growth Des.* **2004**, *4*, 397–402. [CrossRef]
50. Bari, R.H.; Patil, S.B.; Bari, A.R. Spray-pyrolized nanostructured CuO thin films for H_2S gas sensor. *Int. Nano Lett.* **2013**, *3*, 12. [CrossRef]
51. Li, L.; Zhao, J.; Li, W.; Zhu, C.; Mao, L.-F.; Huang, Q. First-Principles investigation on the behavior of Pt single and triple atoms supported on monolayer CuO (110) in CO oxidation. *Appl. Surf. Sci.* **2021**, *564*, 150435. [CrossRef]
52. Massarotti, V.; Capsoni, D.; Bini, M.; Altomare, A.; Moliterni, A.G.G. X-ray powder diffraction ab initio structure solution of materials from solid state synthesis: The copper oxide case. *Z. Kristallogr.* **1998**, *213*, 259–265. [CrossRef]
53. Åsbrink, S.; Norrby, L.-J. A refinement of the crystal structure of copper(II) oxide with a discussion of some exceptional e.s.d.'s. *Acta Crystallogr. B* **1970**, *26*, 8–15. [CrossRef]
54. Zhang, R.; Liu, H.; Zheng, H.; Ling, L.; Li, Z.; Wang, B. Adsorption and dissociation of O_2 on the Cu_2O(1 1 1) surface: Thermochemistry, reaction barrier. *Appl. Surf. Sci.* **2011**, *257*, 4787–4794. [CrossRef]
55. Chadda, D.; Ford, J.D.; Fahim, M.A. Chemical energy storage by the reaction cycle cupric oxide/cuprous oxide. *Int. J. Energy Res.* **1989**, *13*, 63–73. [CrossRef]
56. Eyring, E.M.; Konya, G.; Lighty, J.S.; Sahir, A.H.; Sarofim, A.F.; Whitty, K. Chemical Looping with Copper Oxide as Carrier and Coal as Fuel. *Oil Gas Sci. Technol.* **2011**, *66*, 209–221. [CrossRef]
57. Segall, M.D.; Lindan, P.J.D.; Probert, M.J.; Pickard, C.J.; Hasnip, P.J.; Clark, S.J.; Payne, M.C. First-principles simulation: Ideas, illustrations and the CASTEP code. *J. Phys. Condens. Matter* **2002**, *14*, 2717–2744. [CrossRef]
58. Vanderbilt, D. Soft self-consistent pseudopotentials in a generalized eigenvalue formalism. *Phys. Rev. B Condens. Matter* **1990**, *41*, 7892–7895. [CrossRef]
59. Wang, M.; Liu, J.; Shen, F.; Cheng, H.; Dai, J.; Long, Y. Theoretical study of stability and reaction mechanism of CuO supported on ZrO_2 during chemical looping combustion. *Appl. Surf. Sci.* **2016**, *367*, 485–492. [CrossRef]
60. Perdew, J.P.; Burke, K.; Ernzerhof, M. Generalized Gradient Approximation Made Simple. *Phys. Rev. Lett.* **1996**, *77*, 3865–3868. [CrossRef]
61. Govind, N.; Petersen, M.; Fitzgerald, G.; King-Smith, D.; Andzelm, J. A generalized synchronous transit method for transition state location. *Comput. Mater. Sci.* **2003**, *28*, 250–258. [CrossRef]

catalysts

MDPI

Article

Reactivity Study of Bimetallic Fe-Mn Oxides with Addition of TiO_2 for Chemical Looping Combustion Purposes

Ewelina Ksepko * and Rafal Lysowski

Department of Chemistry and Technology of Fuels, Wroclaw University of Science and Technology, 7/9 Gdanska, 50-373 Wroclaw, Poland; rafal.lysowski@pwr.edu.pl
* Correspondence: ewelina.ksepko@pwr.edu.pl; Tel.: +48-71-320-65-321

Abstract: The objective of the research was to prepare Mn-based materials for use as oxygen carriers and investigate their reactivity in terms of their applicability to energy systems. The family of Fe_2O_3-MnO_2 with the addition of TiO_2 was prepared by mechanical mixing method and calcination. Five samples with addition of Fe_2O_3 (20, 30, 35, and 50 wt.%) to MnO_2 (65, 55, 50, 35, and 85 wt.%) with constant amount of inert TiO_2 (15 wt.%) were prepared. The performance of TiO_2 supported Fe-Mn oxides oxygen carriers with hydrogen/air in an innovative combustion technology known as chemical looping combustion (CLC) was evaluated. Thermogravimetric analysis was used for reactivity studies within a wide temperature range (800–1000 °C). Comprehensive characterization contained multipurpose techniques for newly synthesized materials. Moreover, post-reaction experiments considered morphology analysis by SEM, mechanical strength testing by dynamometry, and crystal phase study by XRD. Based on wide-ranging testing, the F50M35 sample was indicated as the most promising for gaseous fuel combustion via CLC at 850–900 °C temperature.

Keywords: manganese-iron-titanium oxides; mixed metal oxygen carriers; CLC

1. Introduction

Rising concern about global warming in combination with the necessity of limitation emission of greenhouse gasses (GHG) combined with the policy of reducing environmental pollution is leading to the exploration of new energy generation methods. These also include new and more environmentally friendly fuels usage [1].

Nowadays, the technology of generating energy from gaseous fuels, such methane, hydrogen, or hydrogen-containing fuels obtained from biomass or coal gasification, is receiving widespread attention from researchers [2]. In principle, fumes after combusting those fuels should only contain carbon dioxide and water vapors, and in combination with carbon dioxide sequestration, it enables the creation of near zero emission energy generation technology. However, in practice, the major concern during the combustion of the mentioned fuels is their high adiabatic flame temperature which is leading to thermal NO_x emissions [2].

Chemical Looping Combustion (CLC) is a new, promising technology that allows one to burn fuels without direct contact with atmospheric air. During the CLC process, oxygen for the reaction is provided by a material called the oxygen carrier (OC). Oxygen could be released during direct reaction between the OC particle and the fuel, or it could be released spontaneously, due to change of partial oxygen pressure in the reactor, during a process called chemical looping with oxygen uncoupling (CLOU) [3–5]. The oxygen carrier is usually a metal oxide that could be either synthetic or natural, when considering the raw material used for OC production. As a result, a variety of materials may be applied, for example, perovskite type OCs [6] or $CaSO_4$, or natural ores [5]. During the CLC process, usually a system of two reactors is used. In the fuel reactor, combustion of fuel takes place, by evolving oxygen from the OC structure. While in the oxidation reactor (air reactor), the oxygen carrier is oxidized back with the application of atmospheric air, and then is transported into the fuel reactor. Lack of direct contact between fuel and atmospheric nitrogen limits the creation of thermal NO_x; therefore, the fumes coming from this process contain mostly H_2O and CO_2 [5]. CLC technology could be applied for the combustion of a large variety of fuels such as solids (hard coal [7], lignite, biomass [8–10], liquid [11] or gaseous (syngas, natural gas, methane or hydrogen etc.) [12–14].

Suitable materials for the oxygen carrier should be characterized with certain properties such as high oxygen transport capacity and low production cost. The main issue that affects the cost of applying OCs in the fuel combustion process is their chemical and physical stability during muliple redox cycles. According to the National Energy Technology Laboratory (NETL), the make-up cost of OC for practical application in the CLC unit used for power generation should not exceed USD 5 per $1MW_{th}$-hr of energy generated in such an installation [15]. To remedy this different approach, different solutions are used like applying natural ores, using industrial wastes as OCs or designing new synthetic OCs.

Naturally occurring ores usually exhibit poor mechanical durability and also show agglomeration tendency [16]. Although it is possible to use single metal oxides such as Fe_2O_3 or Mn_2O_3 as an oxygen carrier, a widely used solution is to apply mixed oxide systems due to their improved characteristics. While single active phase OCs are relatively simple in application, they have numerous disadvantages that could be eliminated using two or more active phases [17]. For example, iron oxide, despite being relatively cheap, exhibits a decent reactivity towards methane and a tendency to agglomerate, which shortens its lifetime in CLC reactors. Moreover, Fe_2O_3 does not exhibit CLOU properties that could be particularly valuable during the combustion of solid fuels [18]. In comparison, single phase manganese OCs have CLOU properties (evolving of O_2 from oxide structure due to high temperature) but their reoxidation is undergone with relatively slow rates. They also exhibit poor reactivity toward numerous gaseous fuels [4,19]. Forming combined systems with two or more active phases may be a solution to those problems. Potentially, it may eliminate such disadvantages and make emerge new properties due to the synergy effect [20,21].

For example, in the case of the Fe-Mn OC system, the formation of bixbyite $(Fe_xMn_{1-x})_2O_3$ has been reported. Bixbyite phase tends to transition into $(Fe_xMn_{1-x})_3O_4$ spinel with the releasing of gaseous oxygen. Moreover, spinel could be further reduced to so-called manganowüstite $(Fe_xMn_{1-x})O$. Those reactions are occurring in specific conditions, depending on temperature and partial oxygen pressure [9,19]. The bixbyite and spinel phases are characterized by the reactivity greater than the reactivity observed for pure manganese oxide. They also exhibit CLOU properties in contrast to pure Fe_2O_3 [10,21,22].

However, mixed oxide carriers could undergo deactivation due to the formation of less reactive phases such as manganese-rich spinels of Fe-Mn systems [23]. Thus, a vital part of designing new OCs is upgrading their stability. It could be achieved, for example, by mixing the active phases of OC with an inert material (support). Supporting phases are responsible for the task of improving mechanical and chemical stability. Furthermore, the support phase could be inert or can react with the active phases of OCs, occasionally leading to improving or aggravating their properties [21].

Mn-Fe systems have previously attracted considerable attention from scientists, because of promising CLC and CLOU properties [19,24–26]. Both manganese and iron oxides are characterized by low cost, and they are also relatively safe for the environment. However, some studies showed that they exhibit fairly low stability while being used as OC carriers during continuous operation in fluidized bed CLC reactors [19]. Other studies showed a common practice to improve those OCs characteristics by, for example, an application of additional inert phases that could positively affect their mechanical and chemical stability [21,26].

Titanium oxide, due to its relatively low price, seems to be a promising support material. Although it tends to react with both manganese and iron, some reaction products, such as an ilmenite ($FeTiO_3$), could profitably exhibit promising CLC capability with both gaseous and solid fuels. It could be also obtained by oxidation of iron (III) titanate (Fe_2TiO_5) [27,28]. Moreover, the appearance of ilmenite phase could strengthen the mechanical durability of OC. Some papers discussed the presence of $MnTiO_3$, i.e., a product of manganese and titanium oxide reaction showing a poor reactivity [29]. On the contrary, it has been also reported that titanium oxide can improve releasing oxygen from, for example, manganese-rich bixbyite and spinel phases by forming $MnTiO_3$ and low content manganese spinel phases [30]. That effect is particularly interesting since it showed that manganese-rich spinels are characterized by poor reactivity and CLOU properties [7,20]; thus, the addition of titanium oxide into manganese-rich oxygen carries could further improve their characteristics.

Due to the promising properties such as stability during multiple working cycles and resistance to the influence of ash coming from combustion products, mixed Fe-Mn-Ti systems were recently studied as promising materials for CLC and as efficient materials for biomass and solid fuel combustion. They also have been reported to have ferromagnetic properties, which could be useful for future industrial applications as an option for recovery reacted OC particles from ash [5,29,31].

In this work, Fe-Mn OCs (containing Fe_2O_3 20, 30, 35, and 50 wt.% and MnO_2 of 65, 55, 50, 35, and 85 wt.%) with addition of 15% wt. TiO_2 were synthesized by mechanical mixing and calcination. Their CLC properties were tested with the application of model gaseous fuel 4% H_2 and thermogravimetric analyzer period, The aim of this work is to broaden our knowledge of bimetallic Fe-Mn OCs and natural continuation of our research on supported low cost oxygen carriers with CLC and CLOU properties [14]. Hydrogen is an environmentally friendly energy carrier derived from biomass or coal [32]. Hydrogen is also a part of the synthesis gas, which can be used as a gaseous fuel for CLC for power generation [13]. Thus, it was applied in this work to examine the reactivity of synthesized Fe-Mn based OCs. Furthermore, H_2 was used to explore the oxygen transfer capacity of the materials under severe reducing conditions [6,13]. This study aims to evaluate the redox reactivity in the function of temperature of the Fe-Mn-Ti oxide family.

The results of this study will greatly contribute to chemical looping combustion bimetallic oxygen carriers knowledge.

2. Results

2.1. Phase Composition of the Ceramics

Analysis of the obtained diffractograms, shown in Figure 1, indicates that the ceramics were well crystalized materials. The samples F20M65–F50M35 synthetized with both iron (III) oxide and manganese (IV) oxide showed the presence of bixbyite and mixed iron-manganese titanate phases.

Manganese (IV) oxide has been reported to undergo reduction into Mn_2O_3 in temperatures above 500 °C, and Mn_3O_4 in temperatures over 900 °C; while Mn_2O_3 is considered the most thermodynamically stable at lower temperatures, although reoxidation of Mn_3O_4 in temperatures above 800 °C occurs very slowly [33].

Mixed iron manganese oxides were present in all bimetallic OCs; however, they differ in terms of manganese content. For example, for the F20M65 sample, it was specifically $(Fe_{0.2}Mn_{0.8})_2O_3$ detected with ICSD 264710, and for samples less rich in Mn samples, from F30M55 to F50M35, the iron manganese oxide of $(Fe_{0.42}Mn_{0.58})_2O_3$ type was detected. Furthermore, in the F50M35 sample containing 35 wt.% of MnO_2, the iron manganese trioxide of $(Fe_{0.42}Mn_{0.58})_2O_3$ type, iron titanate Fe_2TiO_5, mixed iron (II) manganese (II) titanate $Fe_{0.6}Mn_{0.4}TiO_3$, and iron oxide Fe_2O_3 were detected. Although the mixed iron–manganese titanate $(Fe_{0.6}Mn_{0.4})TiO_3$ in XRD diffractogram peaks partially cover iron (III) titanate and bixbyite peaks, the presence of that phase in the fresh F50M35 sample and reacted samples was confirmed by using the EDS method.

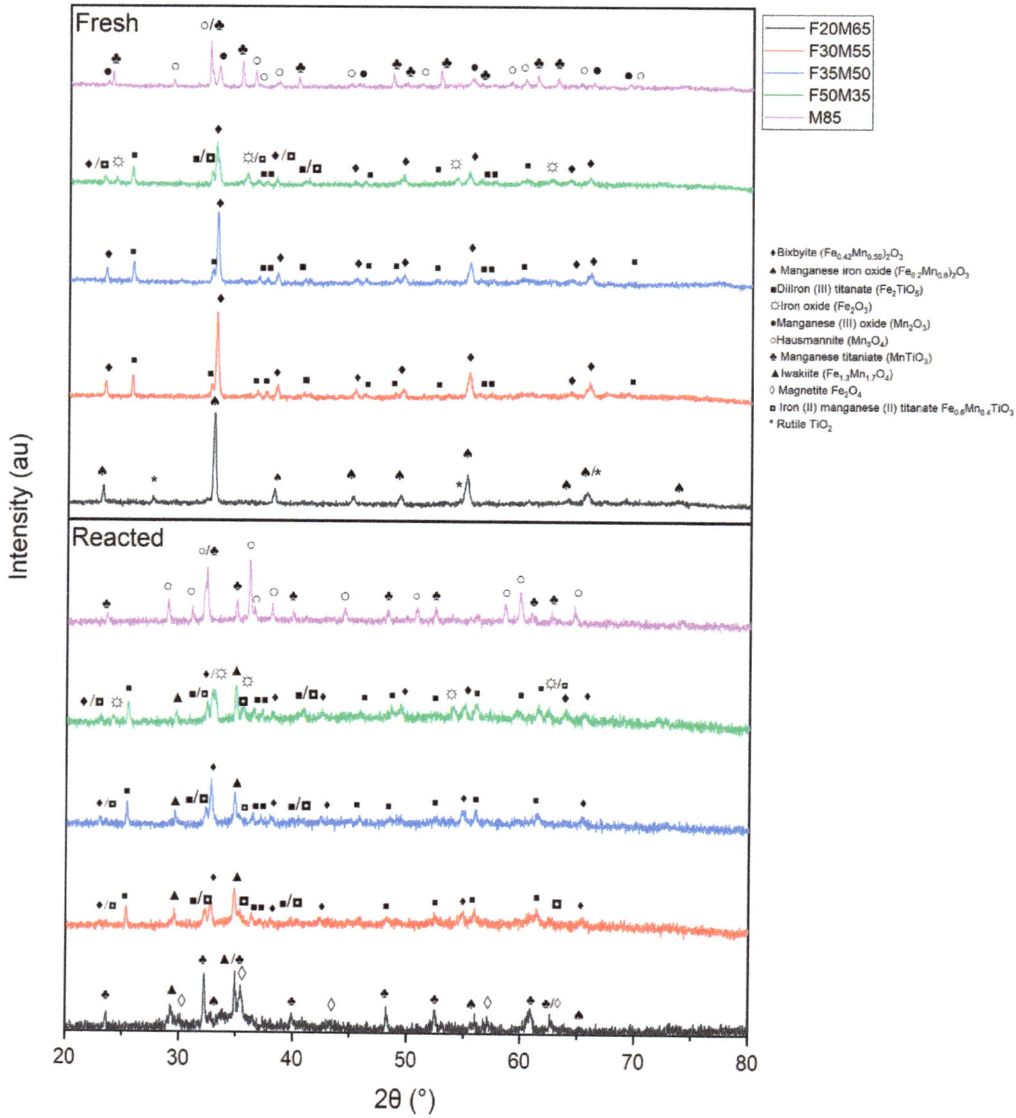

Figure 1. Comparison of X-ray diffraction pattern between fresh and reacted samples.

The diffraction peaks clearly corresponded with those oxides, with the ICSD card number of 154347, 51225, 247551 and 33643, respectively. In the case of fresh F20M65 sample, unreacted rutile was detected; meanwhile, in the reacted samples, instead of iron titanate detected in the more abundant in iron samples, manganese titanate was formed (Table 1). Formation of manganese titanate instead of iron titanate in rich manganese Fe-Mn-Ti systems has been previously reported [30]. The reference ceramics M85 showed the presence of two forms of manganese oxide, i.e., manganese (III) oxide Mn_2O_3, and hausmannite Mn_3O_4, with some perovskite type manganese titanate $MnTiO_3$ (Table 1).

Table 1. Phase composition of as-synthesized and spent OCs.

	Fresh		Spent	
Sample	**Phase Composition/Standard Number**			
F20M65	Manganese iron oxide $(Fe_{0.2}Mn_{0.8})_2O_3$ Rutile TiO_2	ICSD 264710 ICSD 167953	Iwakiite $Fe_{1.3}Mn_{1.7}O_4$ Manganese titanate $MnTiO_3$ Magnetite Fe_3O_4 Manganese iron oxide $(Fe_{0.2}Mn_{0.8})_2O_3$	ICSD 76612 ICSD 44407 ICSD 158504 ICSD 264710
F30M55	Bixbyite $(Fe_{0.42}Mn_{0.58})_2O_3$ Iron (III) titanate Fe_2TiO_5	ICSD 154347 ICSD 51225	Iwakiite $Fe_{1.3}Mn_{1.7}O_4$ Bixbyite $(Fe_{0.42}Mn_{0.58})_2O_3$ Iron (III) titanate Fe_2TiO_5 Iron(II) manganese (II) titanate $Fe_{0.6}Mn_{0.4}TiO_3$	ICSD 76612 ICSD 154347 ICSD 51225 ICSD 247551
F35M50	Bixbyite $(Fe_{0.42}Mn_{0.58})_2O_3$ Iron (III) titanate Fe_2TiO_5	ICSD 154347 ICSD 51225	Iwakiite $(Fe_{1.3}Mn_{1.7})O_4$ Bixbyite $(Fe_{0.42}Mn_{0.58})_2O_3$ Iron (III) titanate Fe_2TiO_5 Iron (II) manganese (II) titanate $Fe_{0.6}Mn_{0.4}TiO_3$	ICSD 76612 ICSD 154347 ICSD 51225 ICSD 247551
F50M35	Bixbyite $(Fe_{0.42}Mn_{0.58})_2O_3$ Iron (III) titanate Fe_2TiO_5 Iron oxide Fe_2O_3 Iron (II) manganese (II) titanate $Fe_{0.6}Mn_{0.4}TiO_3$	ICSD 154347 ICSD 51225 ICSD 33643 ICSD 247551	Iwakiite $(Fe_{1.3}Mn_{1.7})O_4$ Bixbyite $(Fe_{0.42}Mn_{0.58})_2O_3$ Iron oxide Fe_2O_3 Iron (III) titanate Fe_2TiO_5 Iron (II) manganese (II) titanate $Fe_{0.6}Mn_{0.4}TiO_3$	ICSD 76612 ICSD 154347 ICSD 33643 ICSD 51225 ICSD 247551
M85	Manganese (III) oxide Mn_2O_3 Hausmannite Mn_3O_4 Manganese titanate $MnTiO_3$	ICSD 76087 ICSD 77478 ICSD 44407	Hausmannite Mn_3O_4 Manganese titanate $MnTiO_3$	ICSD 77478 ICSD 44407

Figure 1 shows the XRD patterns obtained at RT for the five oxygen carriers of the Fe-Mn-Ti-O family. Furthermore, phase identification was also carried out for the used materials. Those carriers after reaction with fuel were processed with a stream of air. X-ray powder diffraction patterns recorded for both unreacted and reacted samples over teens of redox cycles are shown below.

As the mechanical strength is one of the key parameters shaping the functional properties of the obtained mixed compounds, for this purpose 30 grains of each composition were subjected to crushing strength tests in this work. Figure 2 shows the crushing strengths obtained for fresh and reacted OCs samples. The average value obtained for 30 grains of each composition could be arranged in the following row: 0.98 N (F20M65) > 0.94 N (F30M55) > 0.91 N (F35M50) and 0.89 N (F50M35). Those values are still acceptable since they are close to the required 1 N. In the paper [29], for some Fe-Mn-Ti materials, the mechanical strength values were in the range of 0.7 to 0.9 N; for example, for F17MT51 and F33MT9 it was 0.7 N, while for F46MT9 and F61MT9 it was 0.8, and 0.9 N, respectively. For those samples, both higher (50 and 51 wt.%) and lower (9 wt.%) titania content was used in contrast to this paper. Contrary to what has been reported by [29], these results indicate different patterns. Although this study also showed Fe-Mn-Ti OCs, there is a significant gap between mechanical strengths trends. For example, a slight increase in crushing strength was observed with the increase of Fe content, even if the amount of Fe was doubled as for F17MT51 and M34MT50 (increase from 0.7 to 0.8 N) as reported; which is opposite to our findings, showing the reverse trend. This can be explained by the utilization of different amounts of inert material 15 wt.%, and by applying higher manganese oxide (IV) addition, which is 65–50 wt.%, leading to substantially different final phase compositions.

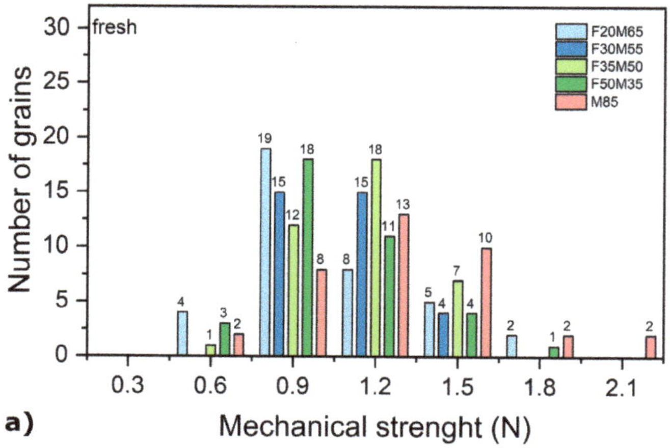

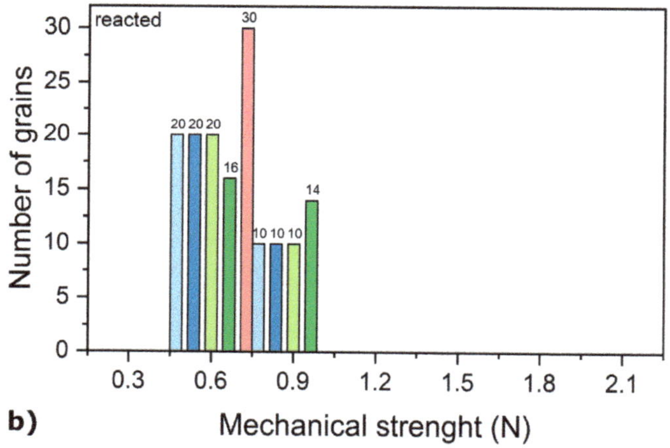

Figure 2. Comparison of crushing strengths obtained for fresh (**a**) and reacted (**b**) OC samples.

Based on the analysis of Figure 2, where the crushing strengths were obtained for fresh materials, it is clear to see the presence of three extracted groups of mechanical strength, with the maximum at 0.9, 1.2 and 1.5 N. This observation correlates well with the phase composition data summarized in Table 1. The obtained perovskite structure of iron manganese oxides and iron titanate may be responsible for improved strength. Below those ranges, there are very few grains with low strength of about 0.5 N, and they may be attributed to the presence of iron oxide or rutile rich phases. On the other hand, for the reference sample, there is a group of grains with a strength of 1.3 N and 1.6 N, which correlates well with hausmannite and manganese (III) oxide occurrence. There are also 8 grains showing 1 N strength, which are attributed to the evidence of manganese titanate $MnTiO_3$.

In general, the highest possible value of the compressive and abrasive forces acting on the OC grains during the CLC process is desirable. Following the experiments for fresh samples, additional measurements were carried out for reused samples. After multicycle CLC tests, the samples were subjected to the compressive forces again to evaluate the complete OCs performance. Figure 2 shows that the measured strength values decreased slightly for all OCs, both for the mixed Fe-Mn-Ti and for the reference monometallic Mn-Ti sample.

2.2. Examination of the Ability to Transport Oxygen

Reactivity of Fe-Mn-Ti oxides was studied by TG method. This is shown in Figure 3. A small amount of a sample of ca. 50 mg of each composition was placed in alumina crucible and heated with a 20 K/min rate to 800 °C temperature. As the required process temperature was reached, alternatively, fuel and air were used for the fuel combustion stage and oxygen carrier material regeneration for 3 cycles at each temperature. As the results of the test are presented in Figure 3, it can be seen that all the obtained oxide materials take part in fuel combustion, following by reduced OCs reoxidation within a wide temperature range, which is 800–1000 °C.

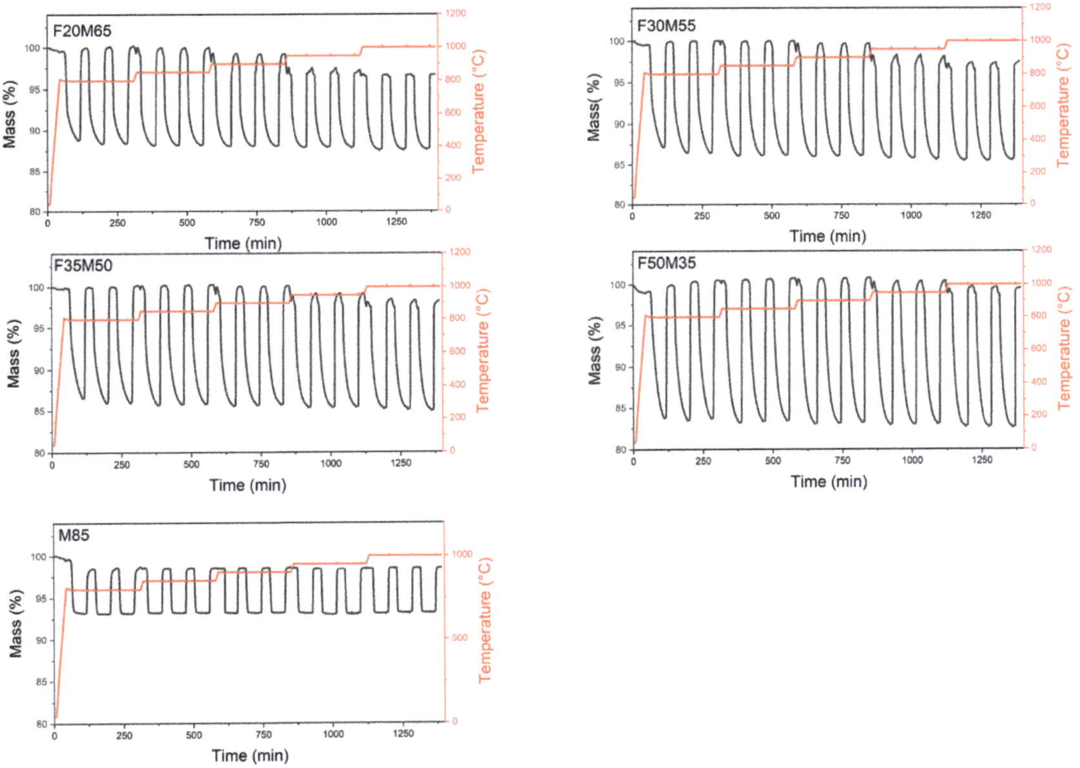

Figure 3. Thermogravimetric measurement data for cycling redox tests: 4% H_2/Ar was used for the reduction, while 20% O_2/N_2 was used for the oxidation reactions.

The data were used further for calculation of the reaction rates and for evaluation of the oxygen transport capacity.

A principal factor for OC choice for industrial applications is the oxygen carrier's ability to transport oxygen needed for fuel combustion. Therefore, the oxygen transport capacity was assessed based on the prediction and experienced from thermogravimetric examinations. Theoretical oxygen transport capacity of as-synthesized materials was calculated based on the amount of active compounds such as manganese and iron oxide constructing the OCs. Those theoretical reflections are considered, depending on the process conditions, different final reduction states, and are shown in Table 2. For bimetallic MnO_2-Fe_2O_3 oxides, both CLC and CLOU properties may be expected. This is because the existence of manganese oxide may contribute to both CLOU and CLC capacity; while the presence of iron oxide in OC contributes to CLC capacity.

Table 2. Predicted oxygen transfer capacity of the oxygen carriers under different reduction states.

Oxidized Form/Reduced Form	Theoretical Oxygen Transfer Capacity (wt.%)				
	F20M65	F30M55	F35M50	F50M15	M85
$Fe_2O_3 + MnO_2/Fe + Mn$	29.9	29.3	28.9	27.9	-
$Fe_2O_3 + MnO_2/Fe + MnO$	18.0	19.1	19.7	21.5	-
$Fe_2O_+ MnO_2/Fe + Mn_3O_4$	14.0	15.8	16.7	19.3	-
$Fe_2O_3 + MnO_2/Fe + Mn_2O_3$	12.0	14.1	15.1	18.3	-
					-
$Fe_2O_3 + MnO_2/FeO + Mn$	25.9	23.3	21.9	17.9	-
$Fe_2O_3 + MnO_2/FeO + MnO$	14.0	13.1	12.7	11.5	-
$Fe_2O_3 + MnO_2/FeO + Mn_3O_4$	10.0	9.8	9.6	9.3	-
$Fe_2O_3 + MnO_2/FeO + Mn_2O_3$	8.0	8.1	8.1	8.2	-
					-
$Fe_2O_3 + MnO_2/Fe_3O_4 + Mn$	24.6	21.3	19.6	14.6	-
$Fe_2O_3 + MnO_2/Fe_3O_4 + MnO$	12.6	11.1	10.4	8.1	-
$Fe_2O_3 + MnO_2/Fe_3O_4 + Mn_3O_4$	8.6	7.7	7.3	5.9	-
$Fe_2O_3 + MnO_2/Fe_3O_4 + Mn_2O_3$	6.7	6.1	5.8	4.9	-
					-
MnO_2/Mn	-	-	-	-	31.3
MnO_2/MnO	-	-	-	-	15.6
MnO_2/Mn_3O_4	-	-	-	-	10.4
MnO_2/Mn_2O_3					7.8

The produced samples contain different rations of Fe:Mn oxide, which is 20–50 wt.% of Fe_2O_3, and a parallel 65–15 wt.% of MnO_2, for the F20M65–F50M15 family. Hypothetically, if reduced to the metallic form, the ceramics should evolve from 27.9 up to 29.9 wt.% of free oxygen. Due to the wide range of concentration, both metal oxides can significantly contribute to the capacity. In a case where both oxides are partially reduced, in this scenario, 4.9–25.9 mass % of oxygen can be transferred. The reference sample, containing only one active metal oxide, i.e., MnO_2, can release from 7.8 to 31.3 wt.%.

These theoretical maximum extents of mass reduction data will be used and compared further with the experimentally observed capacities.

Figure 4 shows the experimental oxygen capacities obtained for Fe-Mn-Ti oxide family members. They were observed in the 800–1000 °C temperature range. The obtained data for iron–manganese materials containing 20–50 wt.% of Fe_2O_3 showed that the experimental oxygen transport capacities determined via TGA were lower than the predicted maximal values (Table 2). At a temperature of 800 °C, the F20M65 sample can deliver 10.5 wt.% of O_2. With an increased amount of iron oxide, this resulted in an increase of capacity to 12.8, 13.2, 15.9 for F30M55, F35M50, F50M15 samples, respectively. Additionally, the positive maintaining capacities through cycling were detected. On the other hand, the monometallic sample shows the lowest capacity which is estimated at ca. 5.0 wt.%, which is far from the predicted values (Table 2).

When increasing temperature, the stable oxygen evolving from the oxides structure to the fuel was observed and maintained as stable with the cycling number. This behavior supports the beneficial use of bimetals instead of one. However, further increase of temperature leads, first, to a small increase of capacities for manganese-rich samples with formal composition F20M65, F30M55 and F35M50, reaching maximal values at 850 °C, and at that point, slightly decreasing. Additionally, the F50M15 sample shows different behavior. As a sample is capable to transport more oxygen than the other bimetallic carriers, here, the capacity reaches a maximum at 850 and 900 °C temperatures, which was experimentally proven to be ca. 16.7 wt.%.

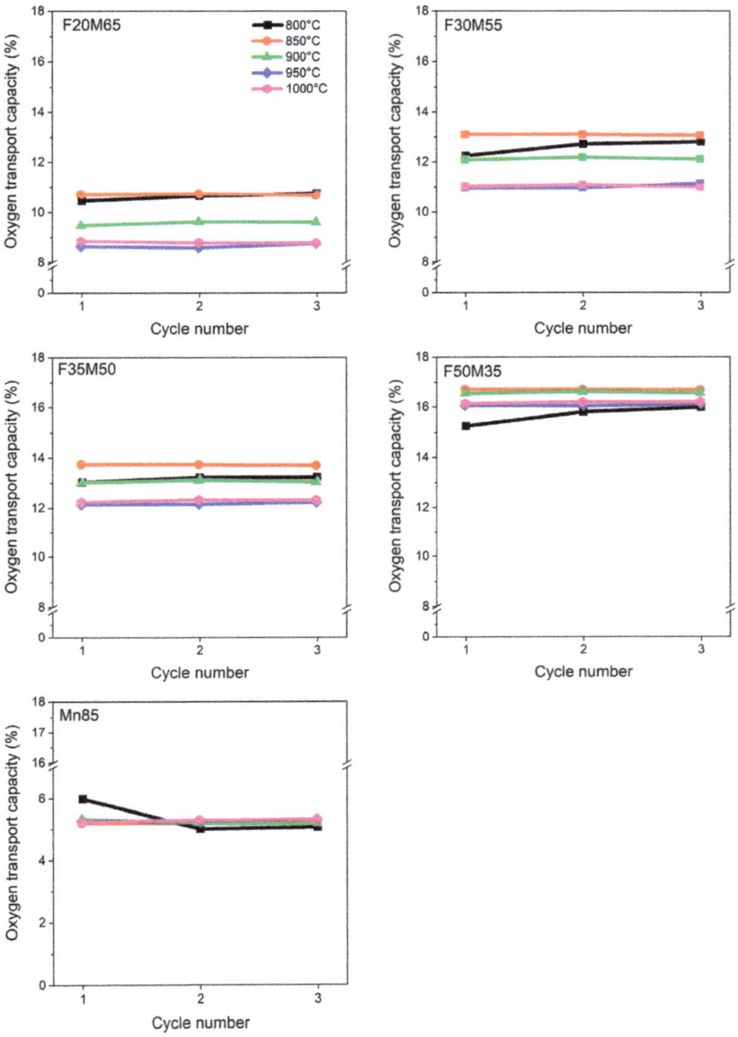

Figure 4. Experimental oxygen capacity versus cycle number for F20M65, F30M55, F35M50, F50M35, and M85 oxygen carriers.

Interestingly, for all mixed iron manganese oxides, stable performance is observed within increasing cycling number (1–3 cycles). Moreover, some activation between the 1st and 2nd cycle is observed for ternary compounds at the very beginning of cycling at 800 °C. This behavior may be due to the variety of phase compositions. For F20M65, F30M55 and F35M50, OC active form of oxides are both $(Fe_{0.42}Mn_{0.58})_2O_3$ (bixbyite) and Fe_2TiO_5 (iron (III) titanate); while the different redox behavior enabling more oxygen to release detected for F50M15 OC may be explained by the presence of active iron manganese trioxide ($FeMnO_3$) aside from iron (III) titanate (Fe_2TiO_5). Additionally, the sample contained also mixed iron (II) manganese (II) titanate ($Fe_{0.6}Mn_{0.4}TiO_3$) and finally, some iron oxide (Fe_2O_3). Experimental capacities increased with increase of temperature from 800 °C to 850 °C for all oxygen carriers as predicted. However, the most striking observation to emerge from the analysis was the decrease of capacities for some compositions above 850 °C, while for others, over 900 °C. This is a rather surprising result. In the first group, compositions F20M65, F30M55 and F35M50 show such behavior; while the F50M15 sample shows a decrease of oxygen capacities above a 900 °C temperature. These findings also confirm those of earlier studies for other group of FeMn oxides [7,20]. The decreases observed in this work are certainly due to the multiplex phase compositions of obtained samples (Table 1). One possible explanation for such an observation is that both bixbyite, iron (II) manganese (II) titanate reactivity increases with an increase of temperature. The different behavior of the F50M15 sample is due to the additional presence of Fe_2O_3, enabling higher capacities to be reached; while a decrease of capacities may be due to the unfavorable presence of high-manganese spinels owning some restrictions. Similar trends have been reported by Ksepko et al. in their work related to the FeMn/Zr carriers [7]. It was proven in the work that high-manganese content spinel may initially show partial and finally, poor reactivity during CLOU tests.

For pure MnO_2/TiO_2 OC, some decrease in cycling number (1–2) is observed at 800 °C. For M85 at 800–1000 °C, the most potential reduction stage according to the experimentally proven 5.0 wt.% and compared to the predicted capacity 7.8 wt.% is the partial reduction of MnO_2 to Mn_2O_3. For mixed metal OCs, observed capacities were well suited for $Fe_2O_3 + MnO_2/FeO + Mn_3O_4$ reduction for F20M65 material; while observed maximal extents of reduction fit well with $Fe_2O_3 + MnO_2/FeO + MnO$ and $Fe_2O_3 + MnO_2/Fe_3O_4 + MnO$ reactions for F30M55, and $Fe_2O_3 + MnO_2/FeO + MnO$ for F35M50. Samples containing the highest amount of Fe, i.e., F50M15 well match with $Fe_2O_3 + MnO_2/FeO + Mn$ reduction.

The relatively high oxygen capacities obtained and proven via TGA in this work are advantages. Primarily, they enable a lower bed inventory to be used. Subsequently, they allow one to decrease OC circulation rates in industrial CLC applications.

2.3. Reactivity with Fuel and Air and Thermal Stability

Evaluation of the reactivity of the synthesized OC ceramics towards fuel and air was carried out through TGA experiments at 800–1000° C temperature suitable for the chemical looping combustion processes. The characterization of the oxygen carriers using thermogravimetric method was performed to learn about:

i. the effect of oxygen carrier chemical composition on its reactivity,
ii. the effective temperature for efficient gaseous fuel combustion when Fe-Mn-Ti oxides are applied,
iii. thermal stability and resistivity of the ceramics.

Reduction reaction rates as a function of cycling number for specific OC compositions are shown in Figure 5, at each temperature. At 800 °C, all doped samples show similar behavior. At the first cycle, the highest reduction rate was observed for F20M65 material, then it slightly increased as the amount of manganese content increased in the sample. Between the first and second cycle, some clear increase in the rate by 68% was observed for all samples. It means some activation occurs at this stage. For the M85 sample, the lowest rate 1.2 %/min was observed and slightly decreased with a cycle number. When the chemical looping hydrogen combustion process temperature was increased by 50 °C, a further increase of rate was observed. For example, for the sample with the lowest Fe content, i.e., F20M65, ca. 2.5 times faster combustion processed at 850 °C, while for the sample with the highest Fe content, i.e., F50M35, the rates were 2.6 times faster than that observed at 800 °C. However, further sample heating led slightly to the rates decreasing, and finally, both at 950 °C and 1000 °C, there were no significant differences observed for the doped samples. For the reference material, i.e., M85, in general, the reduction reaction rates were of similar value at 850–1000 °C, and simultaneously, significantly lower values were achieved compared to doped materials. Additionally, it can be concluded that the introduction of the second metal to OC helped and positively influenced the rates. Based on the behavior of the samples, the best temperature range for Fe-Mn-Ti working duty is 850–900 °C.

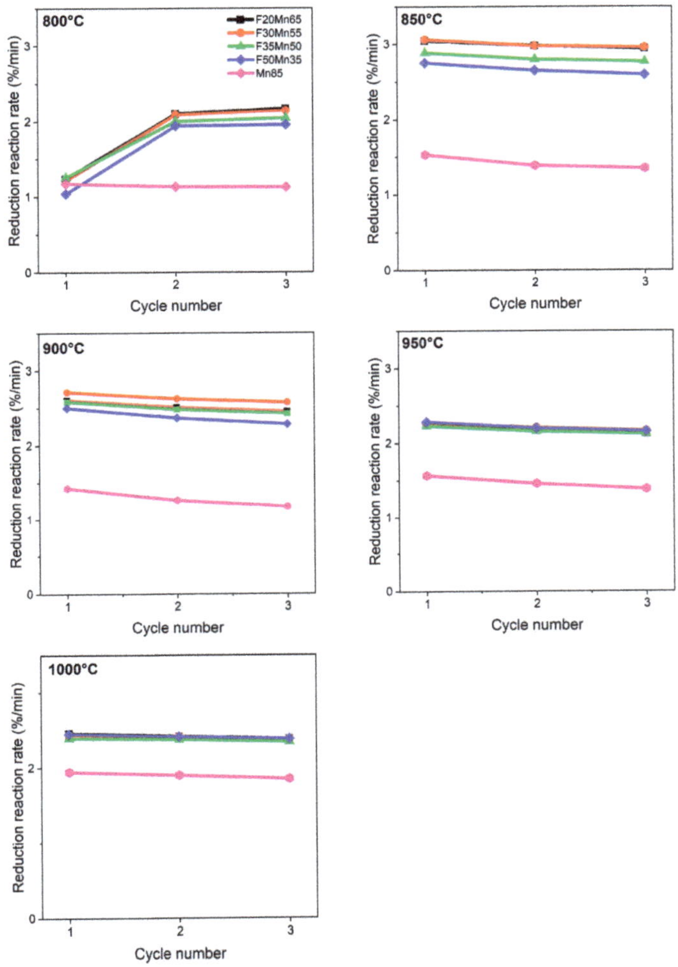

Figure 5. Reduction reaction rates for OCs versus cycle number at given temperatures (800, 850, 900, 950, 1000 °C).

Regeneration (Figure 6) of previously reduced in looping process OCs processed significantly faster than the reduction reaction (Figure 5). At 800 °C, the fastest reactions 5.48%/min were for F50M35 composition and slightly decreased when the amount of Fe decreased in the sample. The slowest oxidation was observed for undoped M85 material. When the regeneration temperature increased to 850 °C, reaction sped up evidently. As a positive result of temperature, 10.21%/min rate was observed for the F50M35 composition, which is almost doubled compared to 850 °C, and those rates maintained unchanged through a wide temperature range. Regeneration reactions were stable at all temperatures tested. The reference M85 showed both the lowest reduction rates and the lowest regeneration rate, which corresponds well with other works [9].

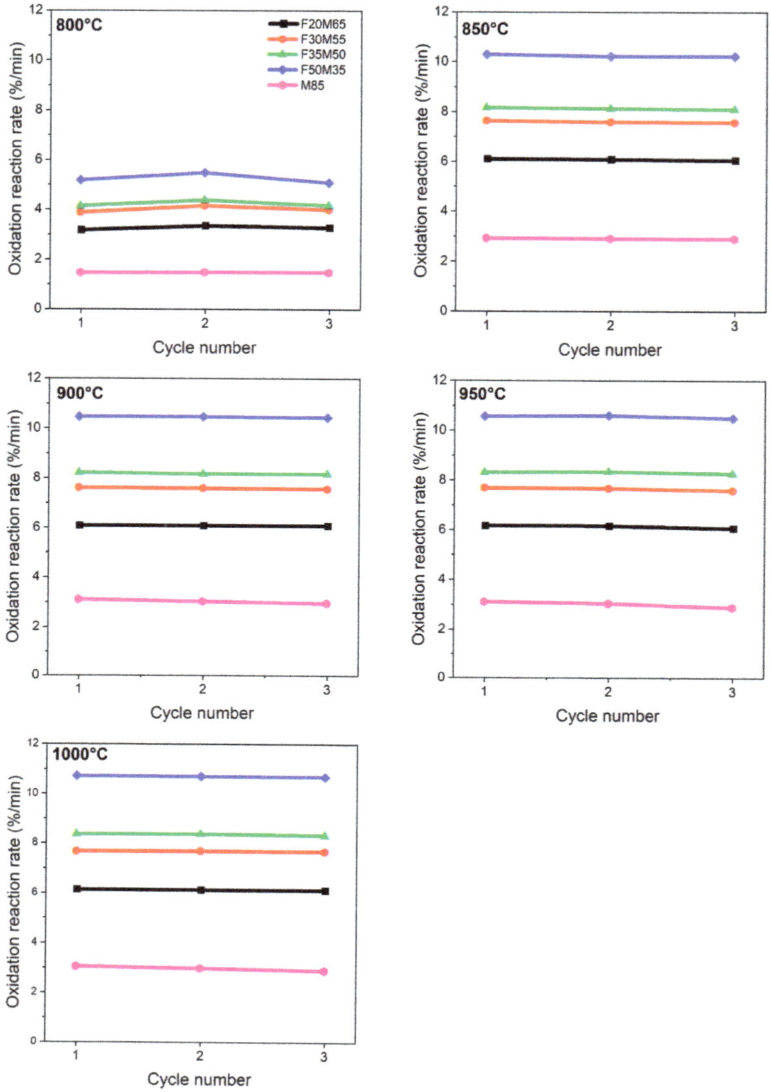

Figure 6. Oxidation reaction rates for OCs versus cycle number at given temperatures (800, 850, 900, 950, 1000 °C).

2.4. Characterization after Redox Tests

The obtained Fe-Mn-Ti-O samples were subjected to mechanical strength test after cycling testing, as mentioned earlier, which is also shown in Figure 2. Comparing these results with those for as-synthesized samples, it can be concluded with no doubt as a decrease in mechanical strength. Now, for 30 grains of each composition, the strength is moved to lower values for all of the carrier compositions. Their values oscillated between 0.5 and 0.95 N within the two subgroups. The deeper insight may clarify the reasons of such behavior. Some new phases were formed during cycling redox reactions in OC samples (see Figures 1 and 7). EDS analysis exhibited the existence of mixed Fe-Mn titanate fuzzy structures on the surface of OC grains (Figure 7). Meanwhile, bixbyite and spinel-like phases were apparent as compact crystalline forms. In the case of fresh samples, they were partially covered by titanate. or had the form of compact blocks, clearly visible on the surface of the reacted samples (Figures 7 and 8). Spinel and perovskite oxides, specifically iwakiite ($Fe_{1.3}Mn_{1.7}O_4$) and manganese titanate ($MnTiO_3$), are the new phases that occurred in the samples, as shown in X-ray diffractograms of the reacted samples. Moreover, some pure iron oxide forms have appeared in F20M65, F30M55 and F50M35 samples. In F20M65 manganese-rich samples, magnetite (Fe_3O_4) was confirmed, while for manganese-poor F50M35 ceramics, it was precisely the iron oxide (Fe_2O_3) detected. For the reference sample, the X-ray diffraction peaks can be attributed to the presence of hausmannite (Mn_3O_4) and manganese titanate ($MnTiO_3$), since the manganese (III) oxide, observed in the fresh samples, completely transformed into hausmannite. The presence of iwakiite in manganese-rich samples and bixbyite to spinel transition in reacted Fe-Mn OC systems were also reported in our previous works [7,14].

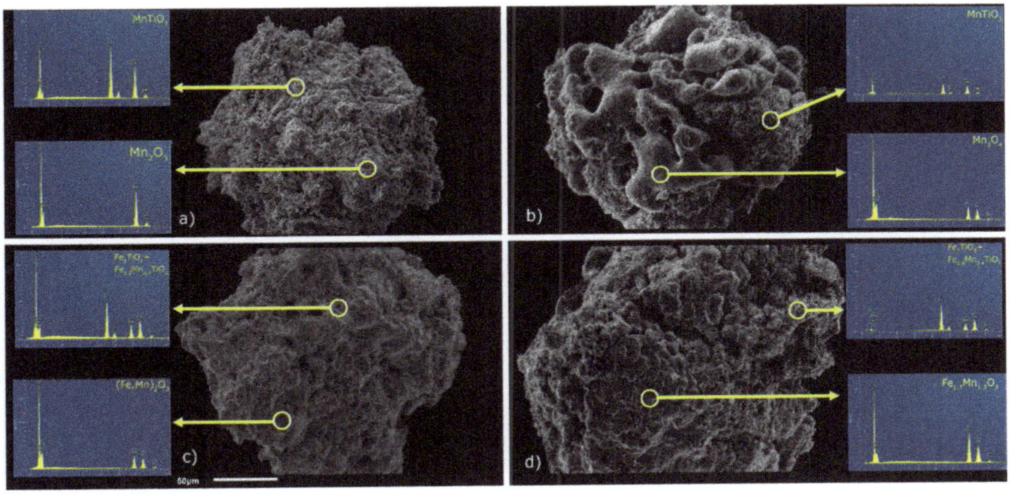

Figure 7. EDS analysis for selected OC grains: fresh M85 (**a**), reacted M85 (**b**), fresh F50M35 (**c**) and reacted F50M35 (**d**).

In the case of the M85 sample, a complete conversion of Mn_2O_3 to hausmannite (Mn_3O_4) was observed. Although Mn_2O_3 is considered to be the most thermodynamically stable, manganese oxide reoxidation of hausmannite occurs very slowly at high temperatures. It has been experimentally proven that after 2 h of continuous oxidation in 800 °C, the obtaining of Mn_2O_3 was not possible [33]. Moreover, continuous reduction oxidation cycling could lead to increased crystallinity of the sample [33], which was also reported during the current research (Figures 7 and 8 j). As such, the poorer mechanical strength of the used samples is a result of phase transformation mainly to iwakiite ($Fe_{1.3}Mn_{1.7}O_4$) and iron oxides which are now the dominant phases in iron–manganese doped samples (Figure 1). Furthermore, the reacted samples exhibited a partially molten surface with visible big pores (Figure 8). The appearance of those pores could be partially responsible for the lowering of OCs mechanical durability presented in Figure 2.

As specified in previous work [14], the high content of iron is positively affecting the OC reactivity. However, when comparing the redox behavior of Fe-Mn OCs doped with two different inerts such as ZrO_2 (published works) and TiO_2 (this work), those carriers with titanium dioxide addition show better higher temperature stability. Correspondingly, in this study, the lesser amount of spinel-type phase and substantially more of bixbyite was detected in the reacted samples (Figure 1). An addition of titanium oxide, that is able to react with both of the active phases such as MnO_2 and Fe_2O_3 of the examined OCs [5,21,34], could positively activate manganese-rich spinels. Similar trends have been reported by [30] in their work. As a consequence of TiO_2 addition to Fe-Mn oxides, fruitful regeneration of Fe-rich spinels to bixbyite structure may be gained [7,23,35].

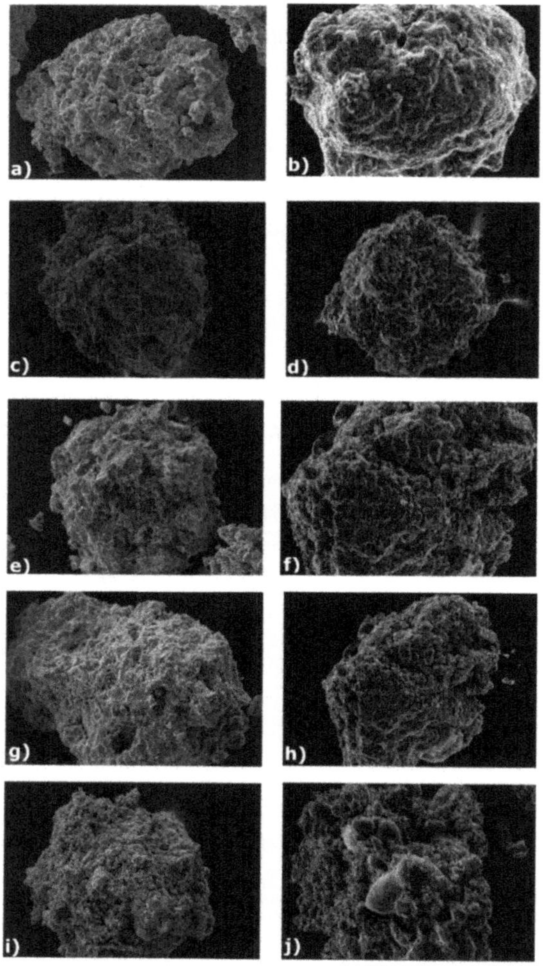

Figure 8. Morphology comparison of fresh (left column) and reacted (right column) oxygen carrier samples. SEM photomicrograph taken at x500 magnification for F20M65 (**a**,**b**), F30M55 (**c**,**d**), F35M50 (**e**,**f**) F50M35 (**g**,**h**) and M85 (**i**,**j**).

3. Materials and Methods

3.1. Synthesis and Materials Quality Examination

Fe_2O_3-MnO_2 supported on 15 wt.% TiO_2 samples with different chemical compositions OCs were prepared by using mechanical mixing method (Table 3). High quality powders of Fe_2O_3 and MnO_2 (<99%, Sigma Aldrich, Steinheim, Gemany) were mixed roughly with inert TiO_2, and then deionized water was added to obtain a paste. The paste was dried at 115 °C overnight, crushed and calcined at 1050 °C in air for 10 h. Ten wt.% of graphite was added to the mixture to improve the porosity/surface area of the samples. The graphite is oxidized to CO_2 during high temperature calcination in air and is well known to be a porosity increasing agent. After cooling, the procedure was repeated. In other words, the sample was crushed and thoroughly mixed with water and graphite addition again. It was then calcined at 1050 °C in air for 10 h. Calcined samples were subjected to sieving to isolate the desired fraction of 125–180 µm. As a result, five different oxygen carriers were obtained (Table 3), The symbol F20M65 refers to 20 wt.% Fe_2O_3–65wt.% MnO_2 and 15wt.% TiO_2 samples, etc. To examine the effect of manganese addition, different concentrations of Mn oxides were prepared and finally, the M85 sample containing only MnO_2 and TiO_2 was obtained.

Table 3. The obtained OC samples.

Sample	Fe_2O_3 (wt.%)	>MnO_2 (wt.%)	TiO_2 (wt.%)
F20M65	20	65	15
F30M55	30	55	15
F35M50	35	50	15
F50M35	50	35	15
M85	x	85	15

3.2. Phase Composition

XRD patterns of "fresh" Fe_2O_3-MnO_2/TiO_2 oxygen carriers and "used" after reaction with hydrogen and then regenerated with air were collected at room temperature to the X-ray powder diffractometer MiniFlex600 by RIGAKU (Tokyo, Japan). The experiments were carried out at the following conditions: 40 kV, 15 mA, and with filtered CuKα radiation of λ = 1.54056 Å. The diffraction patterns were obtained with 20–80° 2Θ with a step of 0.02°. Phase analysis of the obtained XRD diffractograms was performed using SmartLab Studio II software (ver. 4.3.147.0 (2014) by Rigaku Corporation, Tokyo, Japan) and crystallographic data provided by ICSD (Inorganic Crystal Structure Database) (Karlsruhe, Germany).

3.3. Microstructure

The microstructure of the material surface was examined with the application of a scanning electron microscope instrument, JEOL JSM–6610 LV (Tokyo, Japan). The surface morphology was studied by gluing carbon tape onto the samples; carbon tape was not used in the chemical analysis of the samples. The study was conducted using a low vacuum detector at an accelerating voltage of 15 kV and different magnification of image 50-1000x.

An energy dispersion X-ray spectrometer (EDS) was applied in this study for chemical microanalysis evaluation. Therefore, Oxford Aztec Energy (Abingdon, Great Britain), with an Si (Li) X-ray detector was used. Both the composition of the entire grain surface together with a point analysis of selected grain parts to examine the composition and possible presence of the expected structures was analyzed. EDS analysis was performed using Aztec software (ver 2.1 (2013), by Oxford Instruments, Abingdon, Great Britain).

3.4. Thermogravimetric Analysis (TG)

Thermogravimetric experiments were conducted using a thermal analyzer (STA 409 F5 Jupiter by Netzsch Selb, Germany) which was coupled to a quadrupole mass spectrometer QMS 403 Aëolos Quadro (Netzsch Selb, Germany). The OCs' mass changes were measured isothermally. First, the sample was heated in Ar atmosphere with a heating rate of 20 K/min to 800 °C. When the desired temperature was reached, the sample was held for about 10 min. and flushed with synthetic air, then the experiment was started with repeated introduction of reducing and oxidizing gases. As soon as the sample was stabilized, the OCs' particles were exposed to three consecutive redox cycles, conducted at atmospheric pressure to determine their reactivity.

In the CLC measurements, for each test, a sample weighing ca. 50 mg was placed in an alumina made crucible. A mixture of 4% H_2/Ar was used for the reduction, while 20% O_2/N_2 was used for the oxidation reactions. The reduction time was 45 min. and the oxidation time 20 min. The reaction gas flow rates were 100 mL/min. Between the fuel combustion stage and OC regeneration stage, the TGA chamber was flushed with Ar for 10 min. This was proceeded to ensure avoiding the reduction gases mixing with air. Prior to the main testing, additional examinations were carried out to ensure the quality of the experiments were achieved. The OCs' mass and gas flow rates, used in the reactivity experiments, were chosen to avoid limitations in the external film mass transfer and inter-particle diffusion. To determine the stability of the carriers, the redox cycles were conducted at atmospheric pressure. To understand the effect of temperature, a reactivity test was carried out at 800–1000 °C. The reaction rates were calculated by differentiating the mass data with respect to time.

3.5. Mechanical Strength

Mechanical strength test was carried out to assess both the strength and durability of the synthesized OC. The Digital Force Gauge Shimpo FG-5000A, by Shimpo Instruments (Kyoto, Japan), model was used. The instrument provides solid force testing within a capacity of 5.0 kg and a resolution of 0.001 kg. For the testing, 30 particles of each composition of the Fe-Mn-Ti-O materials have been analyzed. Both as-synthesized and used materials were subjected to mechanical strength testing.

4. Conclusions

In this work, we showed that high contents of iron undeniably positively affect OC reactivity. However, when comparing redox reactivity of Fe-Mn OCs doped with two different supports such as zirconium oxide (other published works) and titanium oxide (this work), oxygen carriers with the addition of titanium oxide show better stability, specifically in higher process temperatures. Addition of titanium oxide, which is known as capable to react with both active phases (MnO_2 and Fe_2O_3), may certainly partially inhibit the negative manganese-rich spinel phase formation. As the most capable OC from the Fe-Mn/TiO_2 family tested in this work, the F50M35 sample is indicated as the most perspective. This material shows the highest ca. 16–17 wt.% oxygen transport capacity, outstanding oxidation and fine reduction rates. This phenomenon was attributed to the particular phase composition, where in the fresh sample, mostly iron manganese trioxide ($FeMnO_3$), iron (III) titanate (Fe_2TiO_5) and mixed iron (II) manganese titanate (II) with some iron oxide (Fe_2O_3) were detected. The deep reduction of OC when combusting the gaseous fuel may lead to unmodifiable phase and morphology changes. Those changes led to the formation of iwakiite ($(Fe_{1.3}Mn_{1.7})O_4$), followed by bixbyite ($(Fe_{0.42}Mn_{0.58})_2O_3$), and finally followed by iron (III) titanate (Fe_2TiO_5) and mixed iron (II)-manganese (II) titanate phases, which are reactive on the one hand, but on another, show lower mechanical strength. To take full advantage of the OCs' potential, milder reaction conditions are recommended such as 850–900 °C temperature. Our results are encouraging and should be validated by a larger sample size and longer testing. Future research should also consider the potential effects of contaminants in solid fuels such as biomass and coal.

Author Contributions: Conceptualization, methodology, supervision E.K.; sample synthesis and SEM images acquiring R.L. and E.K.; writing—original draft preparation, writing—review and editing E.K.; visualization R.L. All authors have read and agreed to the published version of the manuscript.

Funding: The work was financed in part by the National Science Centre, Poland, Project No. 2020/37/B/ST5/01259.

Data Availability Statement: Data is contained within the article.

Conflicts of Interest: The authors declare no conflict of interest.

References

1. Liu, D.; Guo, X.; Xiao, B. What causes growth of global greenhouse gas emissions? Evidence from 40 countries. *Sci. Total Environ.* **2019**, *661*, 750–766. [CrossRef] [PubMed]
2. Yu, Y.; Gaofeng, W.; Qizhao, L.; Chengbiao, M.; Xianjun, X. Flameless combustion for hydrogen containing fuels. *Int. J. Hydrog. Energy* **2010**, *35*, 2694–2697. [CrossRef]

3. Alalwan, H.A.; Alminshid, A.H. CO_2 capturing methods: Chemical Looping Combustion (CLC) as a promising technique. *Sci. Total Environ.* **2021**, *788*, 147850. [CrossRef] [PubMed]
4. Adanez, J.; Abad, A.; Garcia-Labiano, F.; Gayan, P.; De Diego, L.F. Progress in chemical-looping combustion and reforming technologies. *Prog. Energy Combust. Sci.* **2012**, *38*, 215–282. [CrossRef]
5. Adánez, J.; Abad, A.; Mendiara, T.; Gayán, P.; de Diego, L.F.; García-Labiano, F. Chemical looping combustion of solid fuels. *Prog. Energy Combust. Sci.* **2018**, *65*, 6–66. [CrossRef]
6. Ksepko, E. Perovskite Sr(Fe1-xCux)O3-δ materials for chemical looping combustion applications. *Int. J. Hydrog. Energy* **2018**, *43*, 9622–9634. [CrossRef]
7. Ksepko, E.; Lysowski, R. Extremely stable and durable mixed Fe–Mn oxides supported on ZrO_2 for practical utilization in CLOU and CLC processes. *Catalysts* **2021**, *11*, 1285. [CrossRef]
8. Ohlemüller, P.; Ströhle, J.; Epple, B. Chemical looping combustion of hard coal and torrefied biomass in a 1 MWth pilot plant. *Int. J. Greenh. Gas. Control.* **2017**, *65*, 149–159. [CrossRef]
9. Pérez-Astray, A.; Mendiara, T.; de Diego, L.F.; Abad, A.; García-Labiano, F.; Izquierdo, M.T.; Adánez, J. Behavior of a manganese-iron mixed oxide doped with titanium in reducing the oxygen demand for CLC of biomass. *Fuel* **2021**, *292*, 120381. [CrossRef]
10. Mendiara, T.; Abad, A.; de Diego, L.F.; García-Labiano, F.; Gayán, P.; Adánez, J. Biomass combustion in a CLC system using an iron ore as an oxygen carrier. *Int. J. Greenh. Gas. Control.* **2013**, *19*, 322–330. [CrossRef]
11. Moldenhauer, P.; Serrano, A.; García-Labiano, F.; de Diego, L.F.; Biermann, M.; Mattisson, T.; Lyngfelt, A. Chemical-looping combustion of kerosene and gaseous fuels with a natural and a manufactured Mn–Fe-based oxygen carrier. *Energy Fuels* **2018**, *32*, 8803–8816. [CrossRef]
12. Zerobin, F.; Bertsch, O.; Penthor, S.; Pröll, T. Concept study for competitive power generation from chemical looping combustion of natural gas. *Energy Technol.* **2016**, *4*, 1299–1304. [CrossRef]
13. Ahmed, I.; de Lasa, H. Syngas chemical looping combustion using a highly performing fluidizable oxygen carrier. *Catal. Today* **2020**, *343*, 63–71. [CrossRef]
14. Ksepko, E.; Lysowski, R. Stable mixed Fe-Mn oxides supported on ZrO_2 oxygen carriers for practical utilization in CLC processes. *Catalysts* **2021**, *11*, 1047. [CrossRef]
15. Straub, D.; Bayham, S.; Weber, J. *Operation of the NETL Chemical Looping Reactor with Natural Gas and a Novel Copper-Iron Material*; National Energy Technology Laboratory: Pittsburgh, PA, USA, 2017. [CrossRef]
16. Tian, H.; Siriwardane, R.; Simonyi, T.; Poston, J. Natural ores as oxygen carriers in chemical looping combustion. *Energy Fuels* **2013**, *27*, 4108–4118. [CrossRef]
17. Fan, Y.; Siriwardane, R.; Tian, H. Trimetallic oxygen carriers $CuFeMnO_4$, $CuFeMn_2O_4$, and CuFe0.5Mn1.5O4 for chemical looping combustion. *Energy Fuels* **2015**, *29*, 6616–6624. [CrossRef]
18. Leion, H.; Mattisson, T.; Lyngfelt, A. Using Chemical-Looping with Oxygen Uncoupling (CLOU) for combustion of six different solid fuels. *Energy Procedia* **2009**, *1*, 447–453. [CrossRef]
19. Larring, Y.; Braley, C.; Pishahang, M.; Andreassen, K.A.; Bredesen, R. Evaluation of a mixed Fe-Mn oxide system for chemical looping combustion. *Energy Fuels* **2015**, *29*, 3438–3445. [CrossRef]
20. Rydén, M.; Lyngfelt, A.; Mattisson, T. Combined manganese/iron oxides as oxygen carrier for chemical Looping Combustion with Oxygen Uncoupling (CLOU) in a circulating fluidized bed reactor system. *Energy Procedia* **2011**, *4*, 341–348. [CrossRef]
21. De Vos, Y.; Jacobs, M.; Van Der Voort, P.; Van Driessche, I.; Snijkers, F.; Verberckmoes, A. Development of Stable oxygen carrier materials for chemical looping processes—A review. *Catalysts* **2020**, *10*, 926. [CrossRef]
22. Tian, H.; Simonyi, T.; Poston, J.; Siriwardane, R. Effect of hydrogen sulfide on chemical looping combustion of coal-derived synthesis gas over bentonite-supported metal−oxide oxygen carriers. *Ind. Eng. Chem. Res.* **2009**, *48*, 8418–8430. [CrossRef]
23. Azimi, G.; Rydén, M.; Leion, H.; Mattisson, T.; Lyngfelt, A. (MnzFe1-z)yOxcombined oxides as oxygen carrier for chemical-looping with oxygen uncoupling. *AIChE J.* **2013**, *59*, 582–588. [CrossRef]
24. Ksepko, E.; Siriwardane, R.V.; Tian, H.; Simonyi, T.; Sciazko, M. Effect of H_2S on chemical looping combustion of coal-derived synthesis gas over Fe–Mn oxides supported on sepiolite, ZrO_2, and Al_2O_3. *Energy Fuels* **2012**, *26*, 2461–2472. [CrossRef]
25. Larring, Y.; Pishahang, M.; Sunding, M.F.; Tsakalakis, K. Fe–Mn based minerals with remarkable redox characteristics for chemical looping combustion. *Fuel* **2015**, *159*, 169–178. [CrossRef]
26. Azimi, G.; Leion, H.; Mattisson, T.; Rydén, M.; Snijkers, F.; Lyngfelt, A. Mn–Fe oxides with support of $MgAl_2O_4$, CeO_2, ZrO_2 and Y2O_3–ZrO_2 for chemical-looping combustion and chemical-looping with oxygen uncoupling. *Ind. Eng. Chem. Res.* **2014**, *53*, 10358–10365. [CrossRef]
27. Bidwe, A.R.; Mayer, F.; Hawthorne, C.; Charitos, A.; Schuster, A.; Scheffknecht, G. Use of ilmenite as an oxygen carrier in chemical looping combustion-batch and continuous dual fluidized bed investigation. *Energy Procedia* **2011**, *4*, 433–440. [CrossRef]
28. Cuadrat, A.; Abad, A.; Adánez, J.; De Diego, L.F.; García-Labiano, F.; Gayán, P. Behavior of ilmenite as oxygen carrier in chemical-looping combustion. *Fuel Process. Technol.* **2012**, *94*, 101–112. [CrossRef]
29. Ryden, M.; Kallen, M.; Jing, D.; Hedayati, A.; Mattisson, T.; Lyngfelt, A. (Fe1-xMnx)TiyO_3 based oxygen carriers for chemical-looping combustion and chemical-looping with oxygen uncoupling. *Energy Procedia* **2014**, *51*, 85–98. [CrossRef]
30. Abián, M.; Abad, A.; Izquierdo, M.T.; Gayán, P.; de Diego, L.F.; García-Labiano, F.; Adánez, J. Titanium substituted manganese-ferrite as an oxygen carrier with permanent magnetic properties for chemical looping combustion of solid fuels. *Fuel* **2017**, *195*, 38–48. [CrossRef]

31. Kim, H.R.; Wang, D.; Zeng, L.; Bayham, S.; Tong, A.; Chung, E.; Kathe, M.V.; Luo, S.; McGiveron, O.; Wang, A.; et al. Coal direct chemical looping combustion process: Design and operation of a 25-kWth sub-pilot unit. *Fuel* **2013**, *108*, 370–384. [CrossRef]
32. Liao, B.; Guo, L.; Lu, Y.; Zhang, X. Solar receiver/reactor for hydrogen production with biomass gasification in supercritical water. *Int. J. Hydrog. Energy* **2013**, *38*, 13038–13044. [CrossRef]
33. Stobbe, E.R.; De Boer, B.A.; Geus, J.W. The reduction and oxidation behaviour of manganese oxides. *Catal. Today* **1999**, *47*, 161–167. [CrossRef]
34. Abián, M.; Abad, A.; Izquierdo, M.T.; Gayán, P.; De Diego, L.F.; García-Labiano, F.; Adánez, J. Evaluation of (MnxFe1-x)2TiyOz particles as oxygen carrier for chemical looping combustion. *Energy Procedia* **2017**, *114*, 302–308. [CrossRef]
35. Shafiefarhood, A.; Stewart, A.; Li, F. Iron-containing mixed-oxide composites as oxygen carriers for Chemical Looping with Oxygen Uncoupling (CLOU). *Fuel* **2015**, *139*, 1–10. [CrossRef]

MDPI AG
Grosspeteranlage 5
4052 Basel
Switzerland
Tel.: +41 61 683 77 34

Catalysts Editorial Office
E-mail: catalysts@mdpi.com
www.mdpi.com/journal/catalysts

www.ingramcontent.com/pod-product-compliance
Lightning Source LLC
Chambersburg PA
CBHW061623210926
51850CB00009B/2411
* 9 7 8 3 7 2 5 8 6 7 1 8 9 *